CREATIVITY

ALSO BY MIHALY CSIKSZENTMIHALYI

Being Adolescent

Flow

The Evolving Self

CREATIVITY

The Psychology of
Discovery and Invention

Mihaly Csikszentmihalyi

HARPER**PERENNIAL** MODERN**CLASSICS**

NEW YORK • LONDON • TORONTO • SYDNEY • NEW DELHI • AUCKLAND

HARPER**PERENNIAL** 🔵 MODERN**CLASSICS**

A hardcover edition of this book was published in 1996 by HarperCollins Publishers.

P.S.™ is a trademark of HarperCollins Publishers.

HarperCollins books may be purchased for educational, business, or sales promotional use. For information please e-mail the Special Markets Department at SPsales@harpercollins.com.

First Harper Perennial edition published 1997.
First Harper Perennial Modern Classics edition published 2013.

Designed by Jessica Shatan

The Library of Congress has cataloged the hardcover edition as follows:

Csikszentmihalyi, Mihaly.
      Creativity : flow and the psychology of discover and invention / Mihaly
Csikszentmihalyi — 1st ed.
        p. cm.
      Includes bibliographical references and index.
      ISBN 0-06-017133-2
Creative ability. 2. Creative thinking. I. Title
        BF408.C77   1996
        153.3'—dc20   96–4116

ISBN 978-0-06-228325-2 (pbk.)

23 24 25 26 27 LBC 24 23 22 21 20

CONTENTS

PART III
DOMAINS OF CREATIVITY 235

ACKNOWLEDGMENTS

The idea for this book emerged in a conversation with Larry Cremin, then president of the Spencer Foundation. We agreed that it would be important to study creativity as a process that unfolds over a lifetime, and that no systematic studies of living creative individuals existed. With its customary vision, the Spencer Foundation then financed a research project, which was to last four years, to remedy this gap in our understanding. Without this grant the laborious task of collecting, transcribing, and analyzing the lengthy interviews would have been impossible.

The other contribution without which this book could not have been written is the assistance of the ninety-one respondents whose interviews form the bulk of the book. All of them are extremely busy individuals, whose time is literally invaluable—thus I deeply appreciate their availability for the lengthy interviews. It is indeed difficult to express my gratitude for their help, and I can only hope that they will find the results were worth their time.

A number of graduate students helped with this project and often contributed creatively to it. Several have written or coauthored articles about the project in professional journals. Especially important were four of my students who have been involved in the project since its inception and who have since earned their doctorates: Kevin Rathunde, Keith Sawyer, Jeanne Nakamura, and Carol Mockros. The others who took an active part are listed among the interviewers in appendix A, which describes the sample.

While we collected and analyzed the data, I had many opportunities to consult with fellow scholars whose specialty is creativity. I should mention at the very least Howard Gardner, David Feldman, Howard Gruber, Istvan Magyari-Beck, Vera John-Steiner, Dean Simonton, Robert Sternberg, and Mark Runco—all of whom contributed, knowingly or not, to the development of ideas in this book.

Several colleagues helped with earlier drafts of the manuscript. I am particularly glad to acknowledge the inspiration and critique of my old friend Howard Gardner, of Harvard University. As usual, his comments have been exactly on target. William Damon, of Brown University, made several excellent suggestions that helped reorganize the contents of the volume. Benö Csapó, from the University of Szeged, Hungary, brought a different cultural perspective to the work.

Three chapters of the book were drafted while I was a guest of the Rockefeller Foundation in its Italian Center at Bellagio. The rest were written while I was a fellow at the Center for Advanced Studies in the Behavioral Sciences in Palo Alto, with support from the John D. and Catherine T. MacArthur Foundation grant #8900078, and the National Science Foundation grant #SBR–9022192. I am grateful to them for the opportunity to concentrate on the manuscript without the usual interruptions—and in such glorious surroundings.

In the later stages of the work, Isabella Selega, who had the good grace to consent to marry me some thirty years ago, oversaw the editing of the manuscript and many other important details. She did the same when I wrote my doctoral dissertation in 1965 on the same topic. It is difficult for me to admit how much of whatever I have accomplished in the years in between I owe to her loving, if critical, help.

None of the shortcomings of this book should be attributed to any of those mentioned here, except myself. For whatever is good in it, however, I thank them deeply.

SETTING THE STAGE

T his book is about creativity, based on histories of contemporary people who know about it firsthand. It starts with a description of what creativity is, it reviews the way creative people work and live, and it ends with ideas about how to make your life more like that of the creative exemplars I studied. There are no simple solutions in these pages and a few unfamiliar ideas. The real story of creativity is more difficult and strange than many overly optimistic accounts have claimed. For one thing, as I will try to show, an idea or product that deserves the label "creative" arises from the synergy of many sources and not only from the mind of a single person. It is easier to enhance creativity by changing conditions in the environment than by trying to make people think more creatively. And a genuinely creative accomplishment is almost never the result of a sudden insight, a lightbulb flashing on in the dark, but comes after years of hard work.

Creativity is a central source of meaning in our lives for several reasons. Here I want to mention only the two main ones. First, most of the things that are interesting, important, and *human* are the results of creativity. We share 98 percent of our genetic makeup with chimpanzees. What makes us different—our language, values, artistic expression, scientific understanding, and technology—is the result of individual ingenuity that was recognized, rewarded, and transmitted through learning. Without creativity, it would be difficult indeed to distinguish humans from apes.

The second reason creativity is so fascinating is that when we are involved in it, we feel that we are living more fully than during the rest of life. The excitement of the artist at the easel or the scientist in the lab comes close to the ideal fulfillment we all hope to get from life, and so rarely do. Perhaps only sex, sports, music, and religious ecstasy—even when these experiences remain fleeting and leave no trace—provide as profound a sense of being part of an entity greater than ourselves. But creativity also leaves an outcome that adds to the richness and complexity of the future.

An excerpt from one of the interviews on which this book is based may give a concrete idea of the joy involved in the creative endeavor, as well as the risks and hardships involved. The speaker is Vera Rubin, an astronomer who has contributed greatly to our knowledge about the dynamics of galaxies. She describes her recent discovery that stars belonging to a galaxy do not all rotate in the same direction; the orbits can circle either clockwise or counterclockwise on the same galactic plane. As is the case with many discoveries, this one was not planned. It was the result of an accidental observation of two pictures of the spectral analysis of the same galaxy obtained a year apart. By comparing the faint spectral lines indicating the positions of stars in the two pictures, Rubin noted that some had moved in one direction during the interval of time, and others had moved in the opposite direction. Rubin was lucky to be among the first cohort of astronomers to have access to such clear spectral analyses of nearby galaxies—a few years earlier, the details would not have been visible. But she could use this luck only because she had been, for years, deeply involved with the small details of the movements of stars. The finding was possible because the astronomer was interested in galaxies for their own sake, not because she wanted to prove a theory or make a name for herself. Here is her story:

It takes a lot of courage to be a research scientist. It really does. I mean, you invest an enormous amount of yourself, your life, your time, and nothing may come of it. You could spend five years working on a problem and it could be wrong before you are done. Or someone might make a discovery just as you are finishing that could make it all wrong. That's a very real possibility. I guess I have been lucky. Initially I went into this [career] feeling very much that my role as an astronomer, as an observer, was just to gather very good data. I just looked upon my role as that of gathering valuable data for the astronomical community, and in most cases it turned out to be more than that. I wouldn't be disappointed if it were only that. But discoveries are always nice. I just discovered something this spring that's enchanting, and I remember how fun it was.

With one of the postdocs, a young fellow, I was making a study of galaxies in the Virgo cluster. This is the biggest large cluster near us. Well, what I've learned in looking at these nearby clusters is that, in fact, I have enjoyed very much learning the details of each galaxy.

I mean, I have almost gotten more interested in just their [individual traits], because these galaxies are close to us—well, close to us on a universal scale. This is the first time that I have ever had a large sample of galaxies all of which were close enough so that I could see lots of little details, and I have found that very strange things are happening near the centers of many of these galaxies—very rapid rotations, little discs, all kinds of interesting things—I have sort of gotten hung up on these little interesting things. So, having studied and measured them all and trying to decide what to do because it was such a vast quantity of interesting data, I realized that some of them were more interesting than others for all kinds of reasons, which I won't go into. So I decided that I would write up first those that had the most interesting central properties (which really had nothing to do with why I started the program), and I realized that there were twenty or thirty that were just very interesting, and I picked fourteen. I decided to write a paper on these fourteen interesting galaxies. They all have very rapidly rotating cores and lots of gas and other things.

Well, one of them was unusually interesting. I first took a spectrum of it in 1989 and then another in 1990. So I had two spectra of these objects and I had probably not measured them until 1990 or 1991. At first I didn't quite understand why it was so interesting, but it was unlike anything that I had ever seen. You know, in a galaxy, or in a spiral or disc galaxy, almost all of the stars are orbiting in a plane around the center. Well, I finally decided that in this galaxy some of the stars were going one way and some of the stars were going the other way; some were going clockwise and some were going counterclockwise. But I only had two spectra and one wasn't so good, so I would alternately believe it and not believe it. I mean, I would think about writing this one up alone and then I would think that the spectra were not good enough, and then I would show it to my colleagues and they would believe it and they could see two lines, or they couldn't, and I would worry about whether the sky was doing something funny. So I decided, because the 1991 applications for using the main telescopes had already passed, that in the spring of '92 I would go and get another spectrum. But then I had an idea. Because there were some very peculiar things on the spectrum and I suddenly . . . I don't know . . . months were taken up in trying to understand what I was looking at. I do the thinking in the other room. I sit in front of this very exotic TV screen next to a computer, but it gives me the images of these spectra very carefully and I can play with them. And I don't know, one day I just decided that I had to understand what this complexity was that I was looking at and I made sketches on a piece of paper and suddenly I understood it all. I have no other way of describing it. It was exquisitely clear. I don't know why I hadn't done this two years earlier.

And then in the spring I went observing, so I asked one of my colleagues here to come observing with me. He and I occasionally do things together. We had three nights. On two of them we never opened the telescope, and the third night was a terrible night but we got a little. We got enough on this galaxy that it sort of confirmed it. But on the other hand it really didn't matter because by then I already knew that everything was right.

So that's the story. And it's fun, great fun, to come upon something new. This spring I had to give a talk at Harvard and of course I stuck this in, and in fact it was confirmed two days later by astronomers who had spectra of this galaxy but had not [analyzed them].

This account telescopes years of hard work, doubt, and confusion. When all goes well, the drudgery is redeemed by success. What is remembered are the high points: the burning curiosity, the wonder at a mystery about to reveal itself, the delight at stumbling on a solution that makes an unsuspected order visible. The many years of tedious calculations are vindicated by the burst of new knowledge. But even without success, creative persons find joy in a job well done. Learning for its own sake is rewarding even if it fails to result in a public discovery. How and why this happens is one of the central questions this book explores.

EVOLUTION IN BIOLOGY AND IN CULTURE

For most of human history, creativity was held to be a prerogative of supreme beings. Religions the world over are based on origin myths in which one or more gods shaped the heavens, the earth, and the waters. Somewhere along the line they also created men and women—puny, helpless things subject to the wrath of the gods. It was only very recently in the history of the human race that the tables were reversed: It was now men and women who were the creators and gods the figments of their imagination. Whether this started in Greece or China two and a half millennia ago, or in Florence two thousand years later, does not matter much. The fact is that it happened quite recently in the multimillion-year history of the race.

So we switched our views of the relationship between gods and humans. It is not so difficult to see why this happened. When the first myths of creation arose, humans were indeed helpless, at the mercy of cold, hunger, wild beasts, and one another. They had no idea how to explain the great forces they saw around them—the rising and setting of the sun, the wheeling stars, the alternating seasons. Awe suffused their groping for a foothold in this mysterious world. Then, slowly at first, and with increasing speed in the last thousand years or so, we began to understand how things work—from microbes to planets, from the circulation of the blood to ocean tides—and humans no longer seemed so helpless after all. Great machines were built, energies harnessed, the entire face of the earth transformed by human craft and appetite. It is not surprising that as we ride the crest of evolution we have taken over the title of creator.

Whether this transformation will help the human race or cause its downfall is not yet clear. It would help if we realized the awesome responsibility of this new role. The gods of the ancients, like Shiva, like Yehova, were both builders and destroyers. The universe endured in a precarious balance between their mercy and their wrath. The world we inhabit today also teeters between becoming either the lovely garden or the barren desert that our contrary impulses strive to bring about. The desert is likely to prevail if we ignore the potential for destruction our stewardship implies and go on abusing blindly our new-won powers.

While we cannot foresee the eventual results of creativity—of the attempt to impose our desires on reality, to become the main power that decides the destiny of every form of life on the planet—at least we can try to understand better what this force is and how it works. Because for better or for worse, our future is now closely tied to human creativity. The result will be determined in large part by our dreams and by the struggle to make them real.

This book, which attempts to bring together thirty years of research on how creative people live and work, is an effort to make more understandable the mysterious process by which men and women come up with new ideas and new things. My work in this area has convinced me that creativity cannot be understood by looking only at the people who appear to make it happen. Just as the sound of a tree crashing in the forest is unheard if nobody is there to hear it, so creative ideas vanish unless there is a receptive audience to record and implement them. And without the assessment of competent outsiders, there is no reliable way to decide whether the claims of a self-styled creative person are valid.

According to this view, creativity results from the interaction of a system composed of three elements: a culture that contains symbolic rules, a person who brings novelty into the symbolic domain, and a field of experts who recognize and validate the innovation. All three are necessary for a creative idea, product, or discovery to take place. For instance, in Vera Rubin's account of her astronomical discovery, it is impossible to imagine it without access to the huge amount of information about celestial motions that has been collecting for centuries, without access to the institutions that control modern large telescopes, without the critical skepticism and eventual support of other astronomers. In my view these are not incidental contributors to individual originality but essential components of the creative process, on a par with the individual's own contributions. For this reason, in this book I devote almost as much attention to the domain and to the field as to the individual creative persons.

Creativity is the cultural equivalent of the process of genetic changes that result in biological evolution, where random variations take place in the chemistry of our chromosomes, below the threshold of consciousness. These changes result in the sudden appearance of a new physical characteristic in a child, and if the trait is an improvement over what existed before, it will have a greater chance to be transmitted to the child's descendants. Most new traits do not improve survival chances and may disappear after a few generations. But a few do, and it is these that account for biological evolution.

In cultural evolution there are no mechanisms equivalent to genes and chromosomes. Therefore, a new idea or invention is not automatically passed on to the next generation. Instructions for how to use fire, or the wheel, or atomic energy are not built into the nervous system of the children born after such discoveries. Each child has to learn them again from the start. The analogy to genes in the evolution of culture are *memes,* or units of information that we must learn if culture is to continue. Languages, numbers, theories, songs, recipes, laws, and values are all memes that we pass on to our children so that they will be remembered. It is these memes that a creative person changes, and if enough of the right people see the change as an improvement, it will become part of the culture.

Therefore, to understand creativity it is not enough to study the individuals who seem most responsible for a novel idea or a new thing. Their contribution, while necessary and important, is only a link in a chain, a phase in a process. To say that Thomas Edison invented electricity or that Albert Einstein discovered relativity is a convenient simplification. It satisfies our ancient predilection for stories that are easy to comprehend and involve superhuman heroes. But Edison's or Einstein's discoveries would be inconceivable without the prior knowledge, without the intellectual and social network that stimulated their thinking, and without the social mechanisms that recognized and spread their innovations. To say that the theory of relativity was created by Einstein is like saying that it is the spark that is responsible for the fire. The spark is necessary, but without air and tinder there would be no flame.

This book is not about the neat things children often say, or the creativity all of us share just because we have a mind and we can think. It does not deal with great ideas for clinching business deals, new ways for baking stuffed artichokes, or original ways of decorating the living room for a party. These are examples of creativity with a small *c,* which is an important ingredient of everyday life, one that we definitely should try to enhance. But to do so well it is necessary first to understand Creativity—and that is what this book tries to accomplish.

ATTENTION AND CREATIVITY

Creativity, at least as I deal with it in this book, is a process by which a symbolic domain in the culture is changed. New songs, new ideas, new machines are what creativity is about. But because these changes do not happen automatically as in biological evolution, it is necessary to consider the price we must pay for creativity to occur. It takes effort to change traditions. For example, memes must be learned before they can be changed: A musician must learn the musical tradition, the notation system, the way instruments are played before she can think of writing a new song; before an inventor can improve on airplane design he has to learn physics, aerodynamics, and why birds don't fall out of the sky.

If we want to learn anything, we must pay attention to the information to be learned. And attention is a limited resource: There is just so much information we can process at any given time. Exactly how much we don't know, but it is clear that, for instance, we cannot learn physics and music at the same time. Nor can we learn well while we do the other things that need to be done and require attention, like taking a shower, dressing, cooking breakfast, driving a car, talking to our spouse, and so forth. The point is, a great deal of our limited supply of attention is committed to the tasks of surviving from one day to the next. Over an entire lifetime, the amount of attention left over for learning a symbolic domain—such as music or physics—is a fraction of this already small amount.

Some important consequences follow logically from these simple premises. To achieve creativity in an existing domain, there must be surplus attention available. This is why such centers of creativity as Greece in the fifth century B.C., Florence in the fifteenth century, and Paris in the nineteenth century tended to be places where wealth allowed individuals to learn and to experiment above and beyond what was necessary for survival. It also seems true that centers of creativity tend to be at the intersection of different cultures, where beliefs, lifestyles, and knowledge mingle and allow individuals to see new combinations of ideas with greater ease. In cultures that are uniform and rigid, it takes a greater investment of attention to achieve new ways of thinking. In other words, creativity is more likely in places where new ideas require less effort to be perceived.

As cultures evolve, it becomes increasingly difficult to master more than one domain of knowledge. Nobody knows who the last Renaissance man really was, but sometime after Leonardo da Vinci it became impossible to learn enough about all of the arts and the sciences to be an expert in more than a small fraction of them. Domains have split into subdomains, and a mathematician who has mastered algebra may not know much about number theory, combinatorix, topology—and vice versa. Whereas in the past an artist typically painted, sculpted, cast gold, and designed buildings, now all of these special skills tend to be acquired by different people.

Therefore, it follows that as culture evolves, specialized knowledge will be favored over generalized knowledge. To see why this must be so, let us assume that there are three persons, one who studies physics, one who studies music, and one who studies both. Other things being equal, the person who studies both music and physics will have to split his or her attention between two symbolic domains, while the other two can focus theirs exclusively on a single domain. Consequently, the two specialized individuals can learn their domains in greater depth, and their expertise will be preferred over that of the generalist. With time, specialists are bound to take over leadership and control of the various institutions of culture.

Of course, this trend toward specialization is not necessarily a good thing. It can easily lead to a cultural fragmentation such as described in the biblical story of the building of the Tower of Babel. Also, as the rest of this book amply demonstrates, creativity generally involves crossing the boundaries of domains, so that, for instance, a chemist who adopts quantum mechanics from physics and applies it to molecular bonds can make a more substantive contribution to chemistry than one who stays exclusively within the bounds of chemistry. Yet at the same time it is important to recognize that given how little attention we have to work with, and given the increasing amounts of information that are constantly being added to domains, specialization seems inevitable. This trend might be reversible, but only if we make a conscious effort to find an alternative; left to itself, it is bound to continue.

Another consequence of limited attention is that creative individuals are often considered odd—or even arrogant, selfish, and ruthless. It is important to keep in mind that these are not traits *of* creative people, but traits that the rest of us attribute to them on the basis of our perceptions. When we meet a person who focuses all of his attention on physics or music and ignores us and forgets our names, we call that person "arrogant" even though he may be extremely humble and friendly if he could only spare attention from his pursuit. If that person is so taken with his domain that he fails to take our wishes into account we call him "insensitive" or "selfish" even though such attitudes are far from his mind. Similarly, if he pursues his work regardless of other people's plans, we call him "ruthless." Yet it is practically impossible to learn a domain deeply enough to make a change in it without dedicating all of one's attention to it and thereby appearing to be arrogant, selfish, and ruthless to those who believe they have a right to the creative person's attention.

In fact, creative people are neither single-minded, specialized, nor selfish. Indeed, they seem to be the opposite: They love to make connections with adjacent areas of knowledge. They tend to be—in principle—caring and sensitive. Yet the demands of their role inevitably push them toward specialization and selfishness. Of the many paradoxes of creativity, this is perhaps the most difficult to avoid.

WHAT'S THE GOOD OF STUDYING CREATIVITY?

There are two main reasons why looking closely at the lives of creative individuals and the contexts of their accomplishments is useful. The first is the most obvious one: The results of creativity enrich the culture and so they indirectly improve the quality of all our lives. But we may also learn from this knowledge how to make our own lives directly more interesting and productive. In the last chapter of this volume I summarize what this study suggests for enriching anyone's everyday existence.

Some people argue that studying creativity is an elite distraction from the more pressing problems confronting us. We should focus all our energies on combating overpopulation, poverty, or mental retardation instead. A concern for creativity is an unnecessary luxury, according to this argument. But this position is somewhat short-sighted. First of all, workable new solutions to poverty or overpopulation will not appear magically by themselves. Problems are solved only when we devote a great deal of attention to them and in a creative way. Second, to have a good life, it is not enough to remove what is wrong from it. We also need a positive goal, otherwise why keep going? Creativity is one answer to that question: It provides one of the most exciting models for living. Psychologists have learned much about how healthy human beings think and feel from studying pathological cases. Brain-damaged patients, neurotics, and delinquents have provided contrasts against which normal functioning may better be understood. But we have learned little from the other end of the continuum, from people who are extraordinary in some positive sense. Yet if we wish to find out what might be missing from our lives, it makes sense to study lives that are rich and fulfilling. This is one of the main reasons for writing the book: to understand better a way of being that is more satisfying than most lives typically are.

Each of us is born with two contradictory sets of instructions: a conservative tendency, made up of instincts for self-preservation, self-aggrandizement, and saving energy, and an expansive tendency made up of instincts for exploring, for enjoying novelty and risk—the curiosity that leads to creativity belongs to this set. We need both of these programs. But whereas the first tendency requires little encouragement or support from outside to motivate behavior, the second can wilt if it is not cultivated. If too few opportunities for curiosity are available, if too many obstacles are placed in the way of risk and exploration, the motivation to engage in creative behavior is easily extinguished.

You would think that given its importance, creativity would have a high priority among our concerns. And in fact there is a lot of lip service paid to it. But if we look at the reality, we see a different picture. Basic scientific research is minimized in favor of immediate practical applications. The arts are increasingly seen as dispensable luxuries that must prove their worth in the impersonal mass market. In one company after another, as downsizing continues, one hears CEOs report that this is not an age for innovators but for bookkeepers, not a climate for building and risking but for cutting expenses. Yet as economic competition heats up around the globe, exactly the opposite strategy is needed.

And what holds true for the sciences, the arts, and for the economy also applies to education. When school budgets tighten and test scores wobble, more and more schools opt for dispensing with frills—usually with the arts and extracurricular activities—so as to focus instead on the so-called basics. This would not be bad if the "three Rs" were taught in ways that encouraged originality and creative thinking; unfortunately, they rarely are. Students generally find the basic academic subjects threatening or dull; their chance of using their minds in creative ways comes from working on the student paper, the drama club, or the orchestra. So if the next generation is to face the future with zest and self-confidence, we must educate them to be original as well as competent.

HOW THE STUDY WAS CONDUCTED

Between 1990 and 1995 I and my students at the University of Chicago videotaped interviews with a group of ninety-one exceptional individuals. The in-depth analysis of these interviews helps illustrate what creative people are like, how the creative process works, and what conditions encourage or hinder the generation of original ideas.

There were three main conditions for selecting respondents: The person had to have made a difference to a major domain of culture—one of the sciences, the arts, business, government, or human well-being in general; he or she had to be still actively involved in that domain (or a different one); and he or she had to be at least sixty years old (in a very few cases, when circumstances warranted, we interviewed respondents who were a bit younger). A list of the respondents interviewed thus far is in appendix A.

The selection process was slow and lengthy. I set out to interview equal numbers of men and women who met our criteria. A further desideratum was to get as wide a representation of cultural backgrounds as possible. With these conditions in mind, I began generating lists of people who met these attributes. In this task I availed myself of the best advice of colleagues and experts in different disciplines. After a while the graduate students involved in the project also suggested names, and other leads were provided by the respondents after each interview, producing what is sometimes called a "snowball sample."

When the research team agreed that the achievements of a person nominated for the sample warranted inclusion, he or she was sent a letter that explained the study and requested participation. If there was no response within three weeks or so, we repeated the request, and then tried to contact the person by phone. Of the 275 persons initially contacted, a little over a third declined, the same number accepted, and a quarter did not respond or could not be traced. Those who accepted included many individuals whose creativity had been widely recognized; there were fourteen Nobel prizes shared among the respondents (four in physics, four in chemistry, two in literature, two in physiology or medicine, and one each in peace and in economics). Most of the others' accomplishments were of the same order, even if they were not as widely recognized.

A few declined for health reasons, many more because they could not spare the time. The secretary to novelist Saul Bellow wrote: "Mr. Bellow informed me that he remains creative in the second half of life, at least in part, because he does not allow himself to be the object of other people's 'studies.' In any event, he's gone for the summer." The photographer Richard Avedon just scrawled the answer "Sorry—too little time left!" The secretary of composer George Ligeti had this to say:

> He is creative and, because of this, totally overworked. Therefore, the very reason you wish to study his creative process is also the reason why he (unfortunately) does not have the time to help you in this study. He would also like to add that he cannot answer your letter personally because he is trying desperately to finish a Violin Concerto which will be premiered in the Fall. He hopes very much you will understand.
>
> Mr. Ligeti would like to add that he finds your project extremely interesting and would be very curious to read the results.

Occasionally the refusal was due to the belief that studying creativity is a waste of time. Poet and novelist Czeslaw Milosz wrote back: "I am skeptical as to the investigation of creativity and I do not feel inclined to submit myself to interviews on that subject. I guess I suspect some methodological errors at the basis of all discussions about 'creativity.'" The novelist Norman Mailer replied: "I'm sorry but I never agree to be interviewed on the process of work. Heisenberg's principle of uncertainty applies." Peter Drucker, the management expert and professor of Oriental art, excused himself in these terms:

> I am greatly honored and flattered by your kind letter of February 14th—for I have admired you and your work for many years, and I have learned much from it. But, my dear Professor Csikszentmihalyi, I am afraid I have to disappoint you. I could not possibly answer your questions. I am told I am creative—I don't know what that means. . . . I just keep on plodding. . . .
>
> . . . I hope you will not think me presumptuous or rude if I say that one of the secrets of productivity (in which I believe whereas I do not believe in creativity) is to have a VERY BIG waste paper basket to take care of ALL invitations such as yours—productivity in my experience consists of NOT doing anything that helps the work of other people but to spend all one's time on the work the Good Lord has fitted one to do, and to do well.

The rate of acceptance varied among disciplines. More than half of the natural scientists, no matter how old or busy they were, agreed to participate. Artists, writers, and musicians, on the other hand, tended to ignore our letters or declined—less than a third of those approached accepted. It would be interesting to find out the causes of this differential attrition.

The same percentage of women and men accepted, but since in certain domains well-known creative women are underrepresented, we were unable to achieve the fifty-fifty gender ratio we were hoping for. Instead, the split is about seventy-thirty in favor of men.

Usually in psychological research, you must make sure that the individuals studied are "representative" of the "population" in question—in this case, the population of creative persons. If the sample is not representative, what you find cannot be generalized to the population. But here I don't even attempt to come up with generalizations that are supposed to hold for all creative persons. What I try to do occasionally is to *disprove* certain widespread assumptions. The advantage of disproof over proof in science is that whereas a single case can disprove a generalization, even all the cases in the world are not enough for a conclusive positive proof. If I could find just one white raven, that would be enough to disprove the statement: "All ravens are black." But I can point at millions of black ravens without confirming the statement that all ravens are black. Somewhere there may be a white raven hiding. The same lack of symmetry between what is called falsification and proof holds even for the most sacred laws of physics.

For the purposes of this book, the strategy of disproof is amply sufficient. The information we collected could not prove, for instance, that all creative individuals had a happy childhood, even if all the respondents had said that their childhood had been happy. But even one unhappy child can disprove that hypothesis—just as one happy child could disprove the opposite hypothesis, that creative individuals must have unhappy childhoods. So the relatively small size of the sample, or its lack of representativeness, is no real impediment to deriving solid conclusions from the data.

It is true that in the social sciences statements are usually neither true nor false but only claim the statistical superiority of one hypothesis over another. We would say that there are so many more black ravens than white ravens that chance alone cannot account for it. Therefore, we conclude that "most ravens are black," and we are glad that we can say this much. In this book I do not avail myself of statistics to test the comparisons that will be reported, for a variety of reasons. First of all, the ability to disprove some deeply held assumptions about creativity seems to me sufficient, and here we are on solid ground. Second, the characteristics of this unique sample violate most assumptions on which statistical tests can be safely conducted. Third, there is no meaningful "comparison group" against which to test the patterns found in this sample.

With a very few exceptions, the interviews were conducted in the offices or homes of the respondents. The interviews were videotaped and then transcribed verbatim. They generally lasted about two hours, although a few were shorter and some lasted quite a bit longer. But the interviews are only the tip of the iceberg as far as information about this sample is concerned. Most of the respondents have written books and articles; some have written autobiographies or other works that could be inspected. In fact, each of them left such an extensive paper trail that to follow it all the way would take several lifetimes; however, the material is extremely useful to round out our understanding of each person and his or her life.

Our interview schedule had a number of common questions that we tried to ask each respondent (a copy of it is in appendix B). However, we did not necessarily ask the questions in the same order, nor did we always use exactly the same wording; my priority was to keep the interview as close to a natural conversation as possible. Of course, there are advantages and disadvantages to both methods. I felt, however, that it would be insulting, and therefore counterproductive, to force these respondents to answer a mechanically structured set of questions. Because I hoped to get genuine and reflective answers, I let the exchanges develop around the themes I was interested in, instead of forcing them into a mold. The interviews are rich as well as being comprehensive—thanks in large measure also to the excellent cadre of graduate students who helped collect them.

When I started to write the book I was confronted with an embarrassment of riches. Thousands of pages clamored for attention, yet I could not do justice to more than a tiny fraction of the material. The choices were often painful—so many beautiful accounts had to be dropped or greatly compressed. The interviews I quote extensively are not necessarily those from the most famous or even the most creative people but the ones that most clearly address what I thought were important theoretical issues. So the choice is personal. Yet I am confident that I have not distorted the meaning of any of the respondents or the consensus of the group as a whole.

Even though the voice of some respondents is not represented by even a single quotation, the content of their statements is included in the generalizations that occasionally are presented, in verbal or numerical form. And I hope that either I, my students, or other scholars will eventually tap those parts of this rich material that I was forced to shortchange.

TOO GOOD TO BE TRUE?

Contrary to the popular image of creative persons, the interviews present a picture of creativity and creative individuals that is upbeat and positive. Instead of suspecting these stories of being self-serving fabrications, I accept them at face value—provided they are not contradicted by other facts known about the person or by internal evidence.

Yet many social scientists in the last hundred years have made it their task to expose the hypocrisy, self-delusion, and self-interest underlying human behavior traits that were never questioned scientifically before the end of the nineteenth century. Poets like Dante or Chaucer were of course intimately acquainted with the foibles of human nature. But it was not until Freud explained the possibility of repression, Marx argued the power of false consciousness, and sociobiologists showed how our actions are the outcome of selective pressures that we had systematic insights into why our reports about ourselves may be so deceptive.

Unfortunately, the understanding for which we owe Freud and the rest of those great thinkers an immense intellectual debt has been marred to a certain extent by the indiscriminate application of their ideas to every aspect of behavior. As a result, in the words of the philosopher Hannah Arendt, our discipline runs the risk of degenerating into a "de-bunking enterprise," based more on ideology than evidence. Even the novice student of human nature learns to distrust appearances—not as a sensible methodological precaution that any good scientist would endorse but as a certainty in the dogma that nothing can be trusted at face value. I can imagine what some sophisticated colleagues would do with the following claim made by one of our respondents: "I have been married for forty-four some years to someone I adore. He is a physicist. We have four children, each of whom has a Ph.D. in science; each of whom has a happy life."

They would probably smile with refined irony and see in these sentences an attempt on the speaker's part to deny an unhappy family life. Others would see it as an attempt to impress the audience. Still others may think that this person's optimistic outburst is simply a narrative device that arose in the context of the interview, not because it is literally true, but because conversations have their own logic and their own truth. Or they would see it as the expression of a bourgeois ideology where academic degrees and comfortable middle-class status are equated with happiness.

But what if there is actual evidence that this woman has been married for forty-four years, that despite her busy schedule as a leading scientist she brought up four children who worked themselves into demanding professional careers, and that she spends most of her free time with her husband at home or traveling? And that her children appear contented with their lives, visit her often, and are in frequent contact with the parents? Should we not relent and admit, however grudgingly, that the meaning of the passage is closer to what the speaker intended than to the alternative meanings I attributed to the imaginary critic?

Let me present a passage from another interview that also illustrates the optimism that is typical of these accounts. This is from the sculptor Nina Holton, married to a well-known (and also creative) scholar.

I like the expression "It makes the spirit sing," and I use it quite often. Because outside my house on the Cape we have this tall grass and I watch it and I say "It is singing grass, I hear it singing." I have a need inside me, of a certain joy, you see? An expression of joy. I feel it. I suppose that I am glad to be alive, glad that I have a man whom I love and a life that I enjoy and the things which I work on which sometimes make my spirit sing. And I hope everybody has that feeling inside. I am grateful that I have a spirit inside me which often sings.

I feel that I do things that make a difference to me and give me great satisfaction. And I can always discuss things with my husband, and we find great parallels, you see, of when he has an idea when he works on something and when we come together and discuss our days and what we have been doing. Not always but often. It is a great bond between us. And also he has been very interested in what I am doing and so in a way he is very much involved in my world. He photographs the things which I do and he is very, very much interested. I can discuss everything with him. It is not like I am working in the dark. I can always come to him and he will give me some advice. I may not always take it, but still there it is. Life feels rich with it. It does.

Again, a cynical reading might lead one to conclude that, well, it must be nice for a two-career couple to have a good time while being creative, but isn't it common knowledge that to achieve anything new and important, especially in the arts, a person must be poor and suffering and tired of the world? So lives like these either represent only a small minority of the creative population, or they must not be accepted at face value, even if all the evidence suggests their truth.

I am not saying that all creative persons are well-off and happy.

Family strain, professional jealousies, and thwarted ambitions were occasionally evident in the interviews. Moreover, it is probable that a selection bias has affected the sample I have collected. Focusing on people beyond sixty years of age eliminated those who may have led a more high-risk lifestyle and thus died early. Some of the individuals we asked to participate and who did not respond or refused may have been less happy and less adjusted than those who accepted. Two or three of those who initially agreed to be interviewed became so infirm and despondent that after the appointment was made they asked to be excused. Thus the individuals who ended up as part of the sample are skewed in the direction of positive health, physical and psychological.

But after several years of intensive listening and reading, I have come to the conclusion that the reigning stereotype of the tortured genius is to a large extent a myth created by Romantic ideology and supported by evidence from isolated and—one hopes—atypical historical periods. In other words, if Dostoyevsky and Tolstoy showed more than their share of pathology it was due less to the requirements of their creative work than to the personal sufferings caused by the unhealthful conditions of a Russian society nearing collapse. If so many American poets and playwrights committed suicide or ended up addicted to drugs and alcohol, it was not their creativity that did it but an artistic scene that promised much, gave few rewards, and left nine out of ten artists neglected if not ignored.

Because of these considerations, I find it more realistic, if more difficult, to approach these interviews with an open skepticism, keeping in mind the bias in favor of happiness these people display and what we have learned about the human tendency to disguise and embellish reality. Yet at the same time, I am ready to accept a positive scenario when it appears to be warranted. It seems to me a risk worth running because I agree with these sentiments of the Canadian novelist Robertson Davies:

Pessimism is a very easy way out when you're considering what life really is, because pessimism is a short view of life. If you look at what is happening around us today and what has happened just since you were born, you can't help but feel that life is a terrible complexity of problems and illnesses of one sort or another. But if you look back a few thousand years, you realize that we have advanced fantastically from the day when the first amoeba crawled out of the slime and made its adventure on land. If you take a long view, I do not see how you can be pessimistic about the future of man or the future of the world. You can take a short view and think that everything is a mess, that life is a cheat and a deceit, and of course you feel miserable. And I become very much amused by some of my colleagues, particularly in the study of literature, who say the pessimistic, the tragic view, is the only true key to life— which I think is just self-indulgent nonsense. It's very much easier to be tragic than it is to be comic. I have known people to embrace the tragic view of life, and it is a cop-out. They simply feel rotten about everything, and that is terribly easy. And if you try to see things a little more evenly, it's surprising what complexities of comedy and ambiguity and irony appear in it. And that, I think, is what is vital to a novelist. Just writing tragic novels is rather easy.

Davies's critique applies more broadly, and not just to the literary field. It is equally easy to explain creativity in a way that only exposes, debunks, reduces, deconstructs, and rationalizes what creative persons do, while ignoring the genuine joy and fulfillment their life contains. But to do so blinds us to the most important message we can learn from creative people: how to find purpose and enjoyment in the chaos of existence.

I did not, however, write this book to prove a point. The findings I discuss emerged from the data. They are not my recycled preconceptions, nor those of anyone else. It is the extraordinary people whose voices fill these pages who tell the story of the unfolding of creativity. Its plot cannot be reduced to glib definitions or superficial techniques. But in its richness and complexity, it is a story that reveals the deep potentials of the human spirit. Having introduced some of the themes that the following chapters will develop, it is now time to get on with the show.

PART I

THE CREATIVE
PROCESS

TWO

WHERE IS CREATIVITY?

The answer is obvious: Creativity is some sort of mental activity, an insight that occurs inside the heads of some special people. But this short assumption is misleading. If by creativity we mean an idea or action that is new and valuable, then we cannot simply accept a person's own account as the criterion for its existence. There is no way to know whether a thought is new except with reference to some standards, and there is no way to tell whether it is valuable until it passes social evaluation. Therefore, creativity does not happen inside people's heads, but in the interaction between a person's thoughts and a sociocultural context. It is a systemic rather than an individual phenomenon. Some examples will illustrate what I mean.

When I was a graduate student I worked part-time for a few years as an editor for a Chicago publishing house. At least once a week we would get in the mail a manuscript from an unknown author who claimed to have made a great discovery of one sort or another. Perhaps it was an eight-hundred-page tome that described in minute detail how a textual analysis of the *Odyssey* showed that, contrary to received opinion, Ulysses did not sail around the Mediterranean.

Instead, according to the author's calculations, if one paid attention to the landmarks, the distances traveled, and the pattern of the stars mentioned by Homer, it was obvious that Ulysses actually traveled around the coast of Florida.

Or it might be a textbook for building flying saucers, with extremely precise blueprints—which on closer inspection turned out to be copied from a service manual for a household appliance. What made reading these manuscripts depressing was the fact that their authors actually believed they had found something new and important and that their creative efforts went unrecognized only because of a conspiracy on the part of philistines like myself and the editors of all the other publishing houses.

Some years ago the scientific world was abuzz with the news that two chemists had achieved cold fusion in the laboratory. If true, this meant that something very similar to the perpetual motion machine—one of the oldest dreams of mankind—was about to be realized. After a few frenetic months during which laboratories around the world attempted to replicate the initial claims—some with apparent success, but most without—it became increasingly clear that the experiments on which the claims were based had been flawed. So the researchers who at first were hailed as the greatest creative scientists of the century became somewhat of an embarrassment to the scholarly establishment. Yet, as far as we know, they firmly believed that they were right and that their reputations had been ruined by jealous colleagues.

Jacob Rabinow, himself an inventor but also an evaluator of inventions for the National Bureau of Standards in Washington, has many similar stories to tell about people who think they have invented perpetual motion machines:

> I've met many of these inventors who invent something that cannot work, that is theoretically impossible. But they spent three years developing it, running a motor without electricity, with magnets. You explain to them it won't work. It violates the second law of thermodynamics. And they say, "Don't give me your goddamn Washington laws."

Who is right: the individual who believes in his or her own creativity, or the social milieu that denies it? If we take sides with the individual, then creativity becomes a subjective phenomenon. All it takes to be creative, then, is an inner assurance that what I think or do is new and valuable. There is nothing wrong with defining creativity this way, as long as we realize that this is not at all what the term originally was supposed to mean—namely, to bring into existence something genuinely new that is valued enough to be added to the culture. On the other hand, if we decide that social confirmation is necessary for something to be called creative, the definition must encompass more than the individual. What counts then is whether the inner certitude is validated by the appropriate experts—such as the editors of the publishing house in the case of far-out manuscripts, or other scientists in the case of cold fusion. And it isn't possible to take a middle ground and say that sometimes the inner conviction is enough, while in other cases we need external confirmation. Such a compromise leaves a huge loophole, and trying to agree on whether something is creative or not becomes impossible.

The problem is that the term "creativity" as commonly used covers too much ground. It refers to very different entities, thus causing a great deal of confusion. To clarify the issues, I distinguish at least three different phenomena that can legitimately be called by that name.

The first usage, widespread in ordinary conversation, refers to persons who express unusual thoughts, who are interesting and stimulating—in short, to people who appear to be unusually bright. A brilliant conversationalist, a person with varied interests and a quick mind, may be called creative in this sense. Unless they also contribute something of permanent significance, I refer to people of this sort as *brilliant* rather than creative—and by and large I don't say much about them in this book.

The second way the term can be used is to refer to people who experience the world in novel and original ways. These are individuals whose perceptions are fresh, whose judgments are insightful, who may make important discoveries that only they know about. I refer to such people as *personally creative*, and try to deal with them as much as possible (especially in chapter 14, which is devoted to this topic). But given the subjective nature of this form of creativity, it is difficult to deal with it no matter how important it is for those who experience it.

The final use of the term designates individuals who, like

Leonardo, Edison, Picasso, or Einstein, have changed our culture in some important respect. They are the *creative* ones without qualifications. Because their achievements are by definition public, it is easier to write about them, and the persons included in my study belong to this group.

The difference among these three meanings is not just a matter of degree. The last kind of creativity is not simply a more developed form of the first two. These are actually different ways of being creative, each to a large measure unrelated to the others. It happens very often, for example, that some persons brimming with brilliance, whom everyone thinks of as being exceptionally creative, never leave any accomplishment, any trace of their existence—except, perhaps, in the memories of those who have known them. Whereas some of the people who have had the greatest impact on history did not show any originality or brilliance in their behavior, except for the accomplishments they left behind.

For example, Leonardo da Vinci, certainly one of the most creative persons in the third sense of the term, was apparently reclusive, and almost compulsive in his behavior. If you had met him at a cocktail party, you would have thought that he was a tiresome bore and would have left him standing in a corner as soon as possible. Neither Isaac Newton nor Thomas Edison would have been considered assets at a party either, and outside of their scientific concerns they appeared colorless and driven. The biographers of outstanding creators struggle valiantly to make their subjects interesting and brilliant, yet more often than not their efforts are in vain. The accomplishments of a Michelangelo, a Beethoven, a Picasso, or an Einstein are awesome in their respective fields—but their private lives, their everyday ideas and actions, would seldom warrant another thought were it not that their specialized accomplishments made everything they said or did of interest.

By the definition I am using here, one of the most creative persons in this study is John Bardeen. He is the first person to have been awarded the Nobel prize in physics twice. The first time it was for developing the transistor; the second for his work on superconductivity. Few persons have ranged as widely and deeply in the realm of solid state physics, or come out with such important insights. But talking with Bardeen on any issue besides his work was not easy; his mind followed abstract paths while he spoke slowly, haltingly, and without much depth or interest about "real life" topics.

It is perfectly possible to make a creative contribution without being brilliant or personally creative, just as it is possible—even likely—that someone personally creative will never contribute a thing to the culture. All three kinds of creativity enrich life by making it more interesting and fulfilling. But in this context I focus primarily on the third use of the term, and explore what is involved in the kind of creativity that leaves a trace in the cultural matrix.

To make things more complicated, consider two more terms that are sometimes used interchangeably with creativity. The first is *talent*. Talent differs from creativity in that it focuses on an innate ability to do something very well. We might say that Michael Jordan is a talented athlete, or that Mozart was a talented pianist, without implying that either was creative for that reason. In our sample, some individuals were talented in mathematics or in music, but the majority achieved creative results without any exceptional talent being evident. Of course, talent is a relative term, so it might be argued that in comparison to "average" individuals the creative ones are talented.

The other term that is often used as a synonym for "creative" is *genius*. Again, there is an overlap. Perhaps we should think of a genius as a person who is both brilliant and creative at the same time. But certainly a person can change the culture in significant ways without being a genius. Although several of the people in our sample have been called a genius by the media, they—and the majority of creative individuals we interviewed—reject this designation.

THE SYSTEMS MODEL

We have seen that creativity with a capital *C*, the kind that changes some aspect of the culture, is never only in the mind of a person. That would by definition *not* be a case of cultural creativity. To have any effect, the idea must be couched in terms that are understandable to others, it must pass muster with the experts in the field, and finally it must be included in the cultural domain to which it belongs. So the first question I ask of creativity is not *what* is it but *where* is it?

The answer that makes most sense is that creativity can be observed only in the interrelations of a system made up of three main parts. The first of these is the *domain*, which consists of a set of symbolic rules and procedures. Mathematics is a domain, or at a finer resolution algebra and number theory can be seen as domains. Domains are in turn nested in what we usually call culture, or the symbolic knowledge shared by a particular society, or by humanity as a whole.

The second component of creativity is the *field,* which includes all the individuals who act as gatekeepers to the domain. It is their job to decide whether a new idea or product should be included in the domain. In the visual arts the field consists of art teachers, curators of museums, collectors of art, critics, and administrators of foundations and government agencies that deal with culture. It is this field that selects what new works of art deserve to be recognized, preserved, and remembered.

Finally, the third component of the creative system is the individual *person.* Creativity occurs when a person, using the symbols of a given domain such as music, engineering, business, or mathematics, has a new idea or sees a new pattern, and when this novelty is selected by the appropriate field for inclusion into the relevant domain. The next generation will encounter that novelty as part of the domain they are exposed to, and if they are creative, they in turn will change it further. Occasionally creativity involves the establishment of a new domain: It could be argued that Galileo started experimental physics and that Freud carved psychoanalysis out of the existing domain of neuropathology. But if Galileo and Freud had not been able to enlist followers who came together in distinct fields to further their respective domains, their ideas would have had much less of an impact, or none at all.

So the definition that follows from this perspective is: Creativity is any act, idea, or product that changes an existing domain, or that transforms an existing domain into a new one. And the definition of a creative person is: someone whose thoughts or actions change a domain, or establish a new domain. It is important to remember, however, that a domain cannot be changed without the explicit or implicit consent of a field responsible for it.

Several consequences follow from this way of looking at things. For instance, we don't need to assume that the creative person is necessarily different from anyone else. In other words, a personal trait of "creativity" is not what determines whether a person will be creative. What counts is whether the novelty he or she produces is accepted for inclusion in the domain. This may be the result of chance, perseverance, or being at the right place at the right time. Because creativity is jointly constituted by the interaction among domain, field, and person, the trait of personal creativity may help generate the novelty that will change a domain, but it is neither a sufficient nor a necessary condition for it.

A person cannot be creative in a domain to which he or she is not exposed. No matter how enormous mathematical gifts a child may have, he or she will not be able to contribute to mathematics without learning its rules. But even if the rules are learned, creativity cannot be manifested in the absence of a field that recognizes and legitimizes the novel contributions. A child might possibly learn mathematics on his or her own by finding the right books and the right mentors, but cannot make a difference in the domain unless recognized by teachers and journal editors who will witness to the appropriateness of the contribution.

It also follows that creativity can be manifested only in existing domains and fields. For instance, it is very difficult to say "This woman is very creative at nurturing" or "This woman is very creative in her wisdom," because nurturance and wisdom, although extremely important for human survival, are loosely organized domains with few generally accepted rules and priorities, and they lack a field of experts who can determine the legitimacy of claims. So we are in the paradoxical situation that novelty is more obvious in domains that are often relatively trivial but easy to measure; whereas in domains that are more essential novelty is very difficult to determine. There can be agreement on whether a new computer game, rock song, or economic formula is actually novel, and therefore creative, less easy to agree on the novelty of an act of compassion or of an insight into human nature.

The model also allows for the often mysterious fluctuations in the attribution of creativity over time. For example, the reputation of Raphael as a painter has waxed and waned several times since his heyday at the court of Pope Julius II. Gregor Mendel did not become famous as the creator of experimental genetics until half a century after his death. Johann Sebastian Bach's music was dismissed as old-fashioned for several generations. The conventional explanation is that Raphael, Mendel, and Bach were always creative, only their reputation changed with the vagaries of social recognition. But the systems model recognizes the fact that creativity cannot be separated from its recognition. Mendel was not creative during his years of relative obscurity because his experimental findings were not that important until a group of British geneticists, at the end of the nineteenth century, recognized their implications for evolution.

The creativity of Raphael fluctuates as art historical knowledge, art critical theories, and the aesthetic sensitivity of the age change. According to the systems model, it makes perfect sense to say that Raphael was creative in the sixteenth and in the nineteenth centuries but not in between or afterward. Raphael is creative when the community is moved by his work, and discovers new possibilities in his paintings. But when his paintings seem mannered and routine to those who know art, Raphael can only be called a great draftsman, a subtle colorist—perhaps even a personally creative individual—but not creative with a capital C. If creativity is more than personal insight and is cocreated by domains, fields, and persons, then creativity can be constructed, deconstructed, and reconstructed several times over the course of history. Here is one of our respondents, the poet Anthony Hecht, commenting on this issue:

> Literary reputations are constantly shifting. Sometimes in trifling, frivolous ways. There was a former colleague of mine who, at a recent meeting of the English Department, said that she thought it was now no longer important to teach Shakespeare because among other things he had a very feeble grasp of women. Now that seems to me as trifling an observation as can be made, but it does mean that, if you take this seriously, nobody's place in the whole canon is very secure, that it's constantly changing. And this is both good and bad. John Donne's position was in the nineteenth century of no consequence at all. The *Oxford Book of English Verse* had only one poem of his. And now, of course, he was resurrected by Herbert Grierson and T. S. Eliot and he's one of the great figures of seventeenth-century poetry. But he wasn't always. This is true of music, too. Bach was eclipsed for two hundred years and rediscovered by Mendelssohn. This means that we are constantly reassessing the past. And that's a good, valuable, and indeed necessary thing to do.

This way of looking at things might seem insane to some. The usual way to think about this issue is that someone like van Gogh was a great creative genius, but his contemporaries did not recognize this. Fortunately, now we have discovered what a great painter he was after all, so his creativity has been vindicated. Few flinch at the presumption implicit in such a view. What we are saying is that we know what great art is so much better than van Gogh's contemporaries did—those bourgeois philistines. What—besides unconscious conceit—warrants this belief? A more objective description of van Gogh's contribution is that his creativity came into being when a sufficient number of art experts felt that his paintings had something important to contribute to the domain of art. Without such a response, van Gogh would have remained what he was, a disturbed man who painted strange canvases.

Perhaps the most important implication of the systems model is that the level of creativity in a given place at a given time does not depend only on the amount of individual creativity. It depends just as much on how well suited the respective domains and fields are to the recognition and diffusion of novel ideas. This can make a great deal of practical difference to efforts for enhancing creativity. Today many American corporations spend a great deal of money and time trying to increase the originality of their employees, hoping thereby to get a competitive edge in the marketplace. But such programs make no difference unless management also learns to recognize the valuable ideas among the many novel ones, and then finds ways of implementing them.

For instance, Robert Galvin at Motorola is justly concerned about the fact that in order to survive among the hungry Pacific Rim electronic manufacturers, his company must make creativity an intentional part of its productive process. He is also right in perceiving that to do so he first has to encourage the thousands of engineers working for the company to generate as many novel ideas as possible. So various forms of brainstorming are instituted, where employees free-associate without fear of being ridiculously impractical. But the next steps are less clear. How does the field (in this case, management) choose among the multitude of new ideas the ones worth pursuing? And how can the chosen ideas be included in the domain (in this case, the production schedule of Motorola)? Because we are used to thinking that creativity begins and ends with the person, it is easy to miss the fact that the greatest spur to it may come from changes outside the individual.

CREATIVITY IN THE RENAISSANCE

A good example is the sudden spurt in artistic creativity that took place in Florence between 1400 and 1425. These were the golden years of the Renaissance, and it is generally agreed that some of the most influential new works of art in Europe were created during that quarter century. Any list of the masterpieces would include the dome of the cathedral built by Brunelleschi, the "Gates of Paradise" crafted for the baptistery by Ghiberti, Donatello's sculptures for the chapel of Orsanmichele, the fresco cycle by Masaccio in the Brancacci Chapel, and Gentile da Fabriano's painting of the Adoration of the Magi in the Church of the Trinity.

How can this flowering of great art be explained? If creativity is something entirely within a person, we would have to argue that for some reason an unusually large number of creative artists were born in Florence in the last decades of the fourteenth century. Perhaps some freak genetic mutation occurred, or a drastic change in the education of Florentine children suddenly caused them to become more creative. But an explanation involving the domain and the field is much more sensible.

As far as the domain is concerned, the Renaissance was made possible in part by the rediscovery of ancient Roman methods of building and sculpting that had been lost for centuries during the so-called Dark Ages. In Rome and elsewhere, by the end of the thirteen hundreds, eager scholars were excavating classical ruins, copying down and analyzing the styles and techniques of the ancients. This slow preparatory work bore fruit at the turn of the fifteenth century, opening up long-forgotten knowledge to the artisans and craftsmen of the time.

The cathedral of Florence, Santa Maria Novella, had been left open to the skies for eighty years because no one could find a way to build a dome over its huge apse. There was no known method for preventing the walls from collapsing inward once the curvature of the dome had advanced beyond a certain height. Every year eager young artists and established builders submitted plans to the Opera del Duomo, the board that supervised the building of the cathedral, but their plans were found unpersuasive. The Opera was made up of the political and business leaders of the city, and their personal reputations were at stake in this choice. For eighty years they did not feel that any proposed solution for the completion of the dome was worthy of the city, and of themselves.

But eventually humanist scholars became interested in the Pantheon of Rome, measured its enormous dome, and analyzed how it had been constructed. The Pantheon had been rebuilt by the emperor Hadrian in the second century. The diameter of its 71-foot-high dome was 142 feet. Nothing on that scale had been built for well over a thousand years, and the methods that allowed the Romans to build such a structure that would stand up and not collapse had been long forgotten in the dark centuries of barbarian invasions. But now that peace and commerce were reviving the Italian cities, the knowledge was slowly being pieced back together.

Brunelleschi, who in 1401 appears to have visited Rome to study its antiquities, understood the importance of the studies of the Pantheon. His idea for how to complete the dome in Florence was based on the framework of internal stone arches that would help contain the thrust, and the herringbone brickwork between them. But his design was not just a restatement of the Roman model—it was influenced also by all the architecture of the intervening centuries, especially the Gothic models. When he presented his plan to the Opera, they recognized it as a feasible and beautiful solution. And after the dome was built, it became a liberating new form that inspired hundreds of builders who came after him, including Michelangelo, who based on it his design for the cupola of St. Peter's in Rome.

But no matter how influential the rediscovery of classical art forms, the Florentine Renaissance cannot be explained only in terms of the sudden availability of information. Otherwise, the same flowering of new artistic forms would have taken place in all the other cities exposed to the ancient models. And though this actually did happen to a certain extent, no other place matched Florence in the intensity and depth of artistic achievement. Why was this so?

The explanation is that the field of art became particularly favorable to the creation of new works at just about the same time as the rediscovery of the ancient domains of art. Florence had become one of the richest cities in Europe first through trading, then through the manufacture of wool and other textiles, and finally through the financial expertise of its rich merchants. By the end of the fourteenth century there were a dozen major bankers in the city—the Medici being only one of the minor ones—who were getting substantial interest every year from the various foreign kings and potentates to whom they had lent money.

But while the coffers of the bankers were getting fuller, the city itself was troubled. Men without property were ruthlessly exploited, and political tensions fueled by economic inequality threatened at any moment to explode into open conflict. The struggle between pope and emperor, which divided the entire continent, was reproduced inside the city in the struggle between the Guelf and Ghibelline factions. To make matters worse, Florence was surrounded by Siena, Pisa, and Arezzo, cities jealous of its wealth and ambitions and always ready to snatch away whatever they could of Florentine trade and territory.

It was in this atmosphere of wealth and uncertainty that the urban leaders decided to invest in making Florence the most beautiful city in Christendom—in their words, "a new Athens." By building awesome churches, impressive bridges, and splendid palaces, and by commissioning great frescoes and majestic statues, they must have felt that they were weaving a protective spell around their homes and businesses. And in a way, they were not wrong: When more than five hundred years later Hitler ordered the retreating German troops to blow up the bridges on the Arno and level the city around them, the field commander refused to obey on the grounds that too much beauty would be erased from the world—and the city was saved.

The important thing to realize is that when the Florentine bankers, churchmen, and heads of great guilds decided to make their city intimidatingly beautiful, they did not just throw money at artists and wait to see what happened. They became intensely involved in the process of encouraging, evaluating, and selecting the works they wanted to see completed. It was because the leading citizens, as well as the common people, were so seriously concerned with the outcome of their work that the artists were pushed to perform beyond their previous limits. Without the constant encouragement and scrutiny of the members of the Opera, the dome over the cathedral would probably not have been as beautiful as it eventually turned out to be.

Another illustration of how the field of art operated in Florence at this time concerns the building of the north and especially the east door of the baptistery, one of the uncontested masterpieces of the period, which Michelangelo declared was worthy of being the "Gate of Paradise" when he saw its heart-wrenching beauty. In this case also a special commission had been formed to supervise the building of the doors for this public edifice. The board was composed of eminent individuals, mostly the leaders of the guild of wool weavers that was financing the project. The board decided that each door should be of bronze and have ten panels illustrating Old Testament themes. Then they wrote to some of the most eminent philosophers, writers, and churchmen in Europe to request their opinion of which scenes from the Bible should be included in the panels, and how they should be represented. After the answers came in, they drew up a list of specifications for the doors and in 1401 announced a competition for their design.

From the dozens of drawings submitted the board chose five finalists—Brunelleschi and Ghiberti among them. The finalists on the short list were given a year to finish a bronze mock-up of one of the door panels. The subject was to be "The Sacrifice of Isaac" and had to include at least one angel and one sheep in addition to Abraham and his son. During that year all five finalists were paid handsomely by the board for time and materials. In 1402 the jury reconvened to consider the new entries and selected Ghiberti's panel, which showed technical excellence as well as a wonderfully natural yet classical composition.

Lorenzo Ghiberti was twenty-one years old at the time. He spent the next twenty years finishing the north door and then another twenty-seven finishing the famed east door. He was involved with perfecting the baptistery doors from 1402 to 1452, a span of a half century. Of course, in the meantime he finished many more commissions and sculpted statues for the Medicis, the Pazzis, the guild of merchant bankers, and other notables, but his reputation rests on the Gates of Paradise, which changed the Western world's conception of decorative art.

If Brunelleschi had been influenced by Roman architecture, Ghiberti studied and tried to emulate Roman sculpture. He had to relearn the technique for casting large bronze shapes, and he studied the classic profiles carved on Roman tombs on which he modeled the expressions of the characters he made emerge from the door panels. And again, he combined the rediscovered classics with the more recent Gothic sculpture produced in Siena. However, one could claim without too much risk of exaggeration that what made the

Gates of Paradise so beautiful was the care, concern, and support of the entire community, represented by the field of judges who supervised their construction. If Ghiberti and his fellows were driven to surpass themselves, it was by the intense competition and focused attention their work attracted. Thus the sociologist of art Arnold Hauser rightly assesses this period: "In the art of the early Renaissance . . . the starting point of production is to be found mostly not in the creative urge, the subjective self-expression and spontaneous inspiration of the artist, but in the task set by the customer."

Of course, the great works of Florentine art would never have been made just because the domain of classical art had been rediscovered, or because the rulers of the city had decided to make it beautiful. Without individual artists the Renaissance could not have taken place. After all, it was Brunelleschi who built the dome over Santa Maria Novella, and it was Ghiberti who spent his life casting the Gates of Paradise. At the same time, it must be recognized that without previous models and the support of the city, Brunelleschi and Ghiberti could not have done what they did. And that with the favorable conjunction of field and domain, if these two artists had not been born, some others would have stepped in their place and built the dome and the doors. It is because of this inseparable connection that creativity must, in the last analysis, be seen not as something happening within a person but in the relationships within a system.

DOMAINS OF KNOWLEDGE AND ACTION

It seems that every species of living organism, except for us humans, understands the world in terms of more or less built-in responses to certain types of sensations. Plants turn toward the sun. There are amoebas sensitive to magnetic attraction that orient their bodies toward the North pole. Baby indigo buntings learn the patterns of the stars as they look out of their nests and then are able to fly great distances at night without losing their way. Bats respond to sounds, sharks to smell, and birds of prey have incredibly developed vision. Each species experiences and understands its environment in terms of the information its sensory equipment is programmed to process.

The same is true for humans. But in addition to the narrow windows on the world our genes have provided, we have managed to open up new perspectives on reality based on information mediated by symbols. Perfect parallel lines do not exist in nature, but by postulating their existence Euclid and his followers could build a system for representing spatial relations that is much more precise than what the unaided eye and brain can achieve. Different as they are from each other, lyric poetry and magnetic resonance spectroscopy are both ways to make accessible information that otherwise we would never have an inkling about.

Knowledge mediated by symbols is extrasomatic; it is not transmitted through the chemical codes inscribed in our chromosomes but must be intentionally passed on and learned. It is this extrasomatic information that makes up what we call a culture. And the knowledge conveyed by symbols is bundled up in discrete domains—geometry, music, religion, legal systems, and so on. Each domain is made up of its own symbolic elements, its own rules, and generally has its own system of notation. In many ways, each domain describes an isolated little world in which a person can think and act with clarity and concentration.

The existence of domains is perhaps the best evidence of human creativity. The fact that calculus and Gregorian chants exist means that we can experience patterns of order that were not programmed into our genes by biological evolution. By learning the rules of a domain, we immediately step beyond the boundaries of biology and enter the realm of cultural evolution. Each domain expands the limitations of individuality and enlarges our sensitivity and ability to relate to the world. Each person is surrounded by an almost infinite number of domains that are potentially able to open up new worlds and give new powers to those who learn their rules. Therefore, it is astounding how few of us bother to invest enough mental energy to learn the rules of even one of these domains, and live instead exclusively within the constraints of biological existence.

For most people, domains are primarily ways to make a living. We choose nursing or plumbing, medicine or business administration because of our ability and the chances of getting a well-paying job. But then there are individuals—and the creative ones are usually in this group—who choose certain domains because of a powerful calling to do so. For them the match is so perfect that acting within the rules of the domain is rewarding in itself; they would keep doing what they do even if they were not paid for it, just for the sake of doing the activity.

Despite the multiplicity of domains, there are some common reasons for pursuing them for their own sake. Nuclear physics, microbiology, poetry, and musical composition share few symbols and rules, yet the calling for these different domains is often astonishingly similar. To bring order to experience, to make something that will endure after one's death, to do something that allows humankind to go beyond its present powers are very common themes.

When asked why he decided to become a poet at the age of seven, György Faludy answered, "Because I was afraid to die." He explained that creating patterns with words, patterns that because of their truth and beauty had a chance to survive longer than the body of the poet, was an act of defiance and hope that gave meaning and direction to his life for the next seventy-three years. This urge is not so very different from physicist John Bardeen's description of his work on superconductivity that might lead to a world without friction, the physicist Heinz Maier-Leibnitz's hope that nuclear energy will provide unlimited power, or the biochemical physicist Manfred Eigen's attempt to understand how life evolved. Domains are wonderfully different, but the human quest they represent converges on a few themes. In many ways, Max Planck's obsession with understanding the Absolute underlies most human attempts to transcend the limitations of a body doomed to die after a short span of years.

There are several ways that domains can help or hinder creativity. Three major dimensions are particularly relevant: the clarity of structure, the centrality within the culture, and accessibility. Say that pharmaceutical companies A and B are competing in the same market. The amount of money they devote to research and development, as well as the creative potential of their researchers, is equal. Now we want to predict whether company A or B will come up with the most effective new drugs, basing our prediction solely on domain characteristics. The questions we would ask are the following: Which company has the more detailed data about pharmaceuticals? Where are the data better organized? Which company puts more emphasis in its culture on research, relative to other areas such as production and marketing? Where does pharmaceutical knowledge earn more respect? Which company disseminates knowledge better among its staff? Where is it easier to test a hypothesis? The company where knowledge is better structured, more central, and more accessible is likely to be the one where—other things still being equal—creative innovations are going to happen.

It has been often remarked that superior ability in some domains—such as mathematics or music—shows itself earlier in life than in other domains—such as painting or philosophy. Similarly, it has been suggested that the most creative performances in some domains are the work of young people, while in other domains older persons have the edge. The most creative lyric verse is believed to be that written by the young, while epics tend to be written by more mature poets. Mathematical genius peaks in the twenties, physics in the thirties, but great philosophical works are usually achieved later in life.

The most likely explanation for these differences lies in the different ways these domains are structured. The symbolic system of mathematics is organized relatively tightly; the internal logic is strict; the system maximizes clarity and lack of redundancy. Therefore, it is easy for a young person to assimilate the rules quickly and jump to the cutting edge of the domain in a few years. For the same structural reasons, when a novelty is proposed—like the long-awaited proof of Fermat's last theorem presented by a relatively young mathematician in 1993—it is immediately recognized and, if viable, accepted. By contrast, it takes decades for social scientists or philosophers to master their domains, and if they produce a new idea, it takes the field many years to assess whether it is an improvement worth adding to the knowledge base.

Heinz Maier-Leibnitz tells the story of a small physics seminar he taught in Munich, which was interrupted one day by a graduate student who suggested a new way to represent on the blackboard the behavior of a subatomic particle. The professor agreed that the new formulation was an improvement and praised the student for having thought of it. By the end of the week, Maier-Leibnitz says, he started getting calls from physicists at other German universities, asking in effect, "Is it true that one of your students came up with such and such an idea?" The next week, calls began to come in from American universities on the East Coast. In two weeks, colleagues from Cal Tech, Berkeley, and Stanford were asking the same question.

This story could never have been told about my branch of psychology. If a student stood up in a psychology seminar at any school in the world and uttered the most profound ideas, he or she would not create a ripple beyond the walls of the classroom. Not because psychology students are less intelligent or original than the ones in physics. Nor because my colleagues and I are less alert to our students' new ideas. But because with the exception of a few highly structured subdomains, psychology is so diffuse a system of thought that it takes years of intense writing for any person to say something that others recognize as new and important. The young student in Maier-Leibnitz's class was eventually awarded the Nobel Prize in physics, something that could never happen to a psychologist.

Does this mean that a domain that is better structured—where creativity is easier to determine—is in some sense "better" than one that is more diffuse? That it is more important, more advanced, more serious? Not at all. If that were true, then chess, microeconomics, or computer programming, which are very clearly structured domains, would have to be considered more advanced than morality or wisdom.

But it is certainly true that nowadays a quantifiable domain with sharp boundaries and well-defined rules is taken more seriously. In a typical university it is much easier to get funding for such a department. It is also easier to justify promotion for a teacher in a narrowly defined domain: Ten colleagues will willingly write letters of recommendation stating that professor X should be promoted because she is the world's authority on the mating habits of the kangaroo rat or on the use of the subjunctive in Dravidic languages. It is much less likely that ten scholars would agree on who is a world authority on personality development. From this it is easy to make the regrettable mistake of inferring that personality development is a scientifically less respectable domain than the one that studies the mating practices of the kangaroo rat.

In the current historical climate, a domain where quantifiable measurement is possible takes precedence over one where it does not. We believe that things that can be measured are real, and we ignore those that we don't know how to measure. So people take intelligence very seriously, because the mental ability we call by that name can be measured by tests; whereas few bother about how sensitive, altruistic, or helpful someone is, because as yet there is no good way to measure such qualities. Sometimes this bias has profound consequences—for instance, in how we define social progress and achievement. One of futurist Hazel Henderson's life goals is to convince world governments to start computing less easily measured trends in their Gross Natural Product. As long as the costs of pollution, depredation of natural resources, decline in the quality of life, and various other human costs are left out of the reckoning of the GNP, she claims, entirely distorted pictures of reality result. A country may pride itself on all its new highways while the resulting auto emissions are causing widespread emphysema.

FIELDS OF ACCOMPLISHMENT

If a symbolic domain is necessary for a person to innovate in, a field is necessary to determine whether the innovation is worth making a fuss about. Only a very small percentage of the great number of novelties produced will eventually become part of the culture. For instance, about one hundred thousand new books are published every year in the United States. How many of these will be remembered ten years from now? Similarly, about five hundred thousand people in this country state on their census forms that they are artists. If each of them painted only one picture a year, it would amount to about fifteen *million* new paintings per generation. How many of these will end up in museums or in textbooks on art? One in a million, ten in a million, one in ten thousand? One?

George Stigler, the Nobel laureate in economics, made the same point about new ideas produced in his domain, and what he says can be applied to any other field of science:

> The profession is too busy to read much. I keep telling my colleagues at the *Journal of Political Economy* that anytime we get an article that fifteen of our profession, of the seven thousand subscribers, read carefully, that must be truly a major article of the year.

These numbers suggest that the competition between memes, or units of cultural information, is as fierce as the competition between the units of chemical information we call genes. In order to survive, cultures must eliminate most of the new ideas their members produce. Cultures are conservative, and for good reason. No culture could assimilate all the novelty people produce without dissolving into chaos. Suppose you had to pay equal attention to the fifteen million paintings—how much time would you have left free to eat, sleep, work, or listen to music? In other words, no person can afford to pay attention to more than a very small fraction of new things produced. Yet a culture could not survive long unless all of its members paid attention to at least a few of the same things. In fact it could be said that a culture exists when the majority of people agree that painting X deserves more attention than painting Y, or idea X deserves more thought than idea Y.

Because of the scarcity of attention, we must be selective: We remember and recognize only a few of the works of art produced, we read only a few of the new books written, we buy only a few of the new appliances busily being invented. Usually it is the various fields that act as filters to help us select among the flood of new information those memes worth paying attention to. A field is made up of experts in a given domain whose job involves passing judgment on performance in that domain. Members of the field choose from among the novelties those that deserve to be included in the canon.

This competition also means that a creative person must convince the field that he or she has made a valuable innovation. This is never an easy task. Stigler emphasizes the necessity of this difficult struggle for recognition:

> I think you have to accept the judgment of others. Because if one were allowed to judge his own case, every one of us should have been president of the United States and received all the medals and so forth. And so I guess I am most proud of the things in which I succeeded in impressing other people with what I have done. And those would be things like the two areas of work in which I received the Nobel Prize, and things like that. So those and certain other works that my profession has liked would be, as far as my professional life goes, the things of which I'm most proud.
>
> I have always looked upon the task of a scientist as bearing the responsibility for persuading his contemporaries of the cogency and validity of his thinking. He isn't entitled to a warm reception. He has to earn it, whether by the skill of his exposition, the novelty of his ideas, or what. I've written on subjects which I thought had promise which haven't amounted to much. That's all right. That may well mean that my judgment wasn't good, because I

don't think any one person's judgment is as good as that of a collection of his better colleagues.

Fields vary greatly in terms of how specialized versus how inclu-
sive they are. For some domains, the field is as broad as society itself.
It took the entire population of the United States to decide whether
the recipe for New Coke was an innovation worth keeping. On the
other hand, it has been said that only four or five people in the world
initially understood Einstein's theory of relativity, but their opinion
had enough weight to make his name a household word. But even in
Einstein's case, the broader society had a voice in deciding that his
work deserved a central place in our culture. To what extent, for
instance, did his fame depend on the fact that he looked like a scientist from Hollywood central casting? That he was persecuted by our
enemies, the Nazis? That many interpreted his discoveries as supportive of the relativity of values, and thus offering a refreshing alternative to binding social norms and beliefs? That while yearning to
overthrow old beliefs, we also thirst for new certainties, and Einstein
was said to have come up with an important new truth? Although
none of these considerations bears in the least on the theory of relativity, they were all very much part of how the media portrayed Einstein—and it is these traits rather than the profundity of his theory
that presumably convinced most people that he was worth including
in the cultural pantheon.

Fields can affect the rate of creativity in at least three ways. The
first way is by being either reactive or proactive. A reactive field does
not solicit or stimulate novelty, while a proactive field does. One of
the major reasons the Renaissance was so bountiful in Florence is
that the patrons actively demanded novelty from artists. In the
United States, we make some effort to be proactive in terms of stimulating scientific creativity in the young: science fairs and prestigious
prizes like the Westinghouse, which goes to the one hundred best
high school science projects each year, are some examples. But of
course much more could be done to stimulate novel thinking in science early on. Similarly, some companies like Motorola take seriously
the idea that one way to increase creativity is for the field to be
proactive.

The second way for the field to influence the rate of novelty is by
choosing either a narrow or a broad filter in the selection of novelty.

Some fields are conservative and allow only a few new items to enter the domain at any given time. They reject most novelty and select only what they consider best. Others are more liberal in allowing new ideas into their domains, and as a result these change more rapidly. At the extremes, both strategies can be dangerous: It is possible to wreck a domain either by starving it of novelty or by admitting too much unassimilated novelty into it.

Finally, fields can encourage novelty if they are well connected to the rest of the social system and are able to channel support into their own domain. For instance, after World War II it was easy for nuclear physicists to get all sorts of money to build new laboratories, research centers, experimental reactors, and to train new physicists, because politicians and voters were still enormously impressed by the atomic bomb and the future possibilities it represented. During a few years in the 1950s, the number of students in theoretical physics at the University of Rome went from seven to two hundred; the proportions were not so far off elsewhere around the world.

There are several ways that domains and fields can affect each other. Sometimes domains determine to a large extent what the field can or cannot do; this is probably more usual in the sciences, where the knowledge base severely restricts what the scientific establishment can or cannot claim. No matter how much a group of scientists would like their pet theory accepted, it won't be if it runs against the previously accumulated consensus. In the arts, on the other hand, it is often the field that takes precedence: The artistic establishment decides, without firm guidelines anchored in the past, which new works of art are worthy of inclusion in the domain.

Sometimes fields that are not competent in the domain take control over it. The church interfered in Galileo's astronomical findings; the Communist party for a while directed not only Soviet genetics but art and music as well; and fundamentalists in the United States are trying to have a voice in teaching evolutionary history. In more subtle ways, economic and political forces always influence, whether intentionally or not, the development of domains. Our knowledge of foreign languages would be even less if the U.S. government stopped subsidizing Title IV programs. Opera and ballet would virtually disappear without massive outside support. The Japanese government is heavily invested in stimulating new ideas and applications in microcircuitry, while the Dutch government, understandably enough, encourages pioneering work in the building of dams and hydraulic devices. The Romanian government was actively involved in the destruction of the art forms of its ethnic minorities in order to maintain the purity of Dacian culture; the Nazis tried to destroy what they considered "degenerate" Jewish art.

At times fields become unable to represent well a particular domain. A leading philosopher in our study maintains that if a young person wants to learn philosophy these days, he or she would be better advised to become immersed in the domain directly and avoid the field altogether: "I'd tell him to read the great books of philosophy. And I would tell him not to do graduate study at any university. I think all philosophy departments are no good. They are all terrible." By and large, however, jurisdiction over a given domain is officially left in the hands of a field of experts. These may range from grade school teachers to university professors and include anyone who has a right to decide whether a new idea or product is "good" or "bad." It is impossible to understand creativity without understanding how fields operate, how they decide whether something new should or should not be added to the domain.

THE CONTRIBUTIONS OF THE PERSON

Finally we get to the individual responsible for generating novelty. Most investigations focus on the creative person, believing that by understanding how his or her mind works, the key to creativity will be found. But this is not necessarily the case. For though it is true that behind every new idea or product there is a person, it does not follow that such persons have a single characteristic responsible for the novelty.

Perhaps being creative is more like being involved in an automobile accident. There are some traits that make one more likely to be in an accident—being young and male, for instance—but usually we cannot explain car accidents on the basis of the driver's characteristics alone. There are too many other variables involved: the condition of the road, the other driver, the type of traffic, the weather, and so on. Accidents, like creativity, are properties of systems rather than of individuals.

Nor can we say that it is the person who starts the creative process. In the case of the Florentine Renaissance one could just as well say that it was started by the rediscovery of Roman art, or by the stimulation provided by the city's bankers. Brunelleschi and his friends found themselves in a stream of thought and action that started before they were born, and then they stepped into the middle of it. At first it appears that they initiated the great works that made the epoch famous, but in reality they were only catalysts for a much more complex process with many participants and many inputs.

When we asked creative persons what explains their success, one of the most frequent answers—perhaps the most frequent one—was that they were lucky. Being in the right place at the right time is an almost universal explanation. Several scientists who were in graduate school in the late 1920s or 1930s remember being among the first cohorts to be exposed to quantum theory. Inspired by the work of Max Planck and Niels Bohr, they applied quantum mechanics to chemistry, to biology, to astrophysics, to electrodynamics. Some of them, like Linus Pauling, John Bardeen, Manfred Eigen, Subrahmanyan Chandrasekhar, were awarded Nobel Prizes for extending the theory to new domains. Many women scientists who entered graduate school in the 1940s mention that they wouldn't have been accepted by the schools, and certainly they wouldn't have been given fellowships and special attention from supervisors, except for the fact that there were so few male students left to compete against, most of them having gone to war.

Luck is without doubt an important ingredient in creative discoveries. A very successful artist, whose work sells well and hangs in the best museums and who can afford a large estate with horses and a swimming pool, once admitted ruefully that there could be at least a thousand artists as good as he is—yet they are unknown and their work is unappreciated. The one difference between him and the rest, he said, was that years back he met at a party a man with whom he had a few drinks. They hit it off and became friends. The man eventually became a successful art dealer who did his best to push his friend's work. One thing led to another: A rich collector began to buy the artist's work, critics started paying attention, a large museum added one of his works to its permanent collection. And once the artist became successful, the field discovered his creativity.

It is important to point out the tenuousness of the individual contribution to creativity, because it is usually so often overrated. Yet one can also fall in the opposite error and deny the individual any credit. Certain sociologists and social psychologists claim that creativity is all a matter of attribution. The creative person is like a blank screen on which social consensus projects exceptional qualities. Because we need to believe that creative people exist, we endow some individuals with this illusory quality. This, too, is an oversimplification. For while the individual is not as important as it is commonly supposed, neither is it true that novelty could come about without the contribution of individuals, and that all individuals have the same likelihood of producing novelty.

Luck, although a favorite explanation of creative individuals, is also easy to overstate. Many young scientists in Linus Pauling's generation were exposed to the arrival of quantum theory from Europe. Why didn't they see what this theory implied for chemistry, the way he saw it? Many women would have liked to become scientists in the 1940s. Why did so few take the opportunity when the doors to graduate training were opened to them? Being in the right place at the right time is clearly important. But many people never realize that they are standing in a propitious space/time convergence, and even fewer know what to do when the realization hits them.

INTERNALIZING THE SYSTEM

A person who wants to make a creative contribution not only must work within a creative system but must also reproduce that system within his or her mind. In other words, the person must learn the rules and the content of the domain, as well as the criteria of selection, the preferences of the field. In science, it is practically impossible to make a creative contribution without internalizing the fundamental knowledge of the domain. All scientists would agree with the words of Frank Offner, a scientist and inventor: "The important thing is that you must have a good, a very solid grounding in the physical sciences, before you can make any progress in understanding." The same conclusions are voiced in every other discipline. Artists agree that a painter cannot make a creative contribution without looking, and looking, and looking at previous art, and without knowing what other artists and critics consider good and bad art. Writers say that you have to read, read, and read some more, and know what the critics' criteria for good writing are, before you can write creatively yourself.

An extremely lucid example of how the internalization of the system works is given by the inventor Jacob Rabinow. At first, he talks about the importance of what I have called the *domain*:

> So you need three things to be an original thinker. First, you have to have a tremendous amount of information—a big database if you like to be fancy. If you're a musician, you should know a lot about music, that is, you've heard music, you remember music, you could repeat a song if you have to. In other words, if you were born on a desert island and never heard music, you're not likely to be a Beethoven. You might, but it's not likely. You may imitate birds but you're not going to write the Fifth Symphony. So you're brought up in an atmosphere where you store a lot of information.
>
> So you have to have the kind of memory that you need for the kind of things you want to do. And you do those things which are easy and you don't do those things which are hard, so you get better and better by doing the things you do well, and eventually you become either a great tennis player or a good inventor or whatever, because you tend to do those things which you do well and the more you do, the easier it gets, and the easier it gets, the better you do it, and eventually you become very one-sided but you're very good at it and you're lousy at everything else because you don't do it well. This is what engineers call positive feedback. So the small differences at the beginning of life become enormous differences by the time you've done it for forty, fifty, eighty years as I've done it. So anyway, first you have to have the big database.

Next Rabinow brings up what the *person* must contribute, which is mainly a question of motivation, or the enjoyment one feels when playing (or working?) with the contents of the domain:

> Then you have to be willing to pull the ideas, because you're interested. Now, some people could do it, but they don't bother. They're interested in doing something else. So if you ask them, they'll, as a favor to you, say: "Yeah, I can think of something." But there are people like myself who *like* to do it. It's fun to come up with an idea, and if nobody wants it, I don't give a damn. It's just fun to come up with something strange and different.

Finally he focuses on how important it is to reproduce in one's mind the criteria of judgment that the *field* uses:

> And then you must have the ability to get rid of the trash which you think of. You cannot think only of good ideas, or write only beautiful music. You must think of a lot of music, a lot of ideas, a lot of poetry, a lot of whatever. And if you're good, you must be able to throw out the junk immediately without even saying it. In other words, you get many ideas appearing and you discard them because you're well trained and you say, "that's junk." And when you see the good one, you say, "Oops, this sounds interesting. Let me pursue that a little further." And you start developing it. Now, people don't like this explanation. They say, "What? You think of junk?" I say, "Yup. You must." You cannot a priori think only of good ideas. You cannot think only of great symphonies. Some people do it very rapidly. And this is a matter of training. And by the way, if you're not well trained, but you've got ideas, and you don't know if they're good or bad, then you send them to the Bureau of Standards, National Institute of Standards, where I work, and *we* evaluate them. And *we* throw them out.

He was asked what constitutes "junk." Is it something that doesn't work, or—

> It doesn't work, or it's old, or you know that it will not gel. You suddenly realize it's not good. It's too complicated. It's not what mathematicians call "elegant." You know, it's not good poetry. And this is a matter of training. If you're well trained in technology, you see an idea and say, "Oh, God, this is terrible." First of all, it's too complicated. Secondly, it's been tried before. Thirdly, he could have done it in three different easier ways. In other words, you can evaluate the thing. That doesn't mean that he wasn't original. But he simply didn't do enough. If he were well trained, if he had the experience I had, and had good bosses and worked with great people, he could say this is not really a good idea. It's an idea, but it's not a *good* idea. And you have arguments with people. And you say, "Look, this is not a good way. Look at the number of parts you're gluing together. Look at the amount of energy it'll take. This is really not good." And the guy says, "But to me it's new." I

say, "Yup. To you it's new. It may be new to the world. But it's still not good."

To say what is beautiful you have to take a sophisticated group of people, people who know that particular art and have seen a lot of it, and say this is good art, or this is good music, or this is a good invention. And that doesn't mean everybody can vote on it; they don't know enough. But if a group of engineers who work on new stuff look at it and say, "That's pretty nice," that's because they know. They know because they've been trained in it.

And a good creative person is well trained. So he has first of all an enormous amount of knowledge in that field. Secondly, he tries to combine ideas, because he enjoys writing music or enjoys inventing. And finally, he has the judgment to say, "This is good, I'll pursue this further."

It would be very difficult to improve on this description of how the systems model works after it is internalized. Drawing on over eighty years of varied experience, Rabinow has distilled with great insight what is involved in being a creative inventor. And as his words suggest, the same process holds for other domains, whether poetry, music, or physics.

THREE

THE CREATIVE PERSONALITY

To be creative, a person has to internalize the entire system that makes creativity possible. So what sort of person is likely to do that? This question is very difficult to answer. Creative individuals are remarkable for their ability to adapt to almost any situation and to make do with whatever is at hand to reach their goals. If nothing else, this distinguishes them from the rest of us. But there does not seem to be a particular set of traits that a person must have in order to come up with a valuable novelty. What John Reed, the CEO of Citicorp, who has thought quite a lot about such things, says about businesspeople could be applied to creative persons in other domains as well:

> Well, because of my job, I tend to know the guys who run the top fifty, one hundred companies in the country, and there's quite a range. It has little to do with the industry. It's funny, there is a consistency in what people look at in businesspeople, but there's no consistency in style and approach, personality, and so forth. There is not a consistent norm with regard to anything other than business performance.

Personality type, style. There are guys who drink too much, there are guys who chase girls; there are guys who are conservative, do none of the above; there are guys who are very serious and workaholics; there are guys who—it's quite amazing, the range of styles. You're paid to run companies, they watch quite carefully as to results. But there's an amazing lack of consistency on any other dimension. How you do it seems to be a wide-open variable. There isn't a clear pattern, tremendously different personality types. And it doesn't seem to run by industry either.

The same is true for scientists: What leads to an important discovery doesn't matter as long as you play by the rules. Or for artists: You can be a happy extrovert like Raphael, or a surly introvert like Michelangelo—the only thing that matters is how good your paintings are judged to be. This is all well and true; yet at the same time it is somewhat disappointing. After all, to say that what makes a person creative is his or her creativity is a tautology. Can we do any better? We don't really have very sound evidence, let alone proof, but we can venture some rather robust and credible suggestions.

Perhaps the first trait that facilitates creativity is a *genetic predisposition* for a given domain. It makes sense that a person whose nervous system is more sensitive to color and light will have an advantage in becoming a painter, while someone born with a perfect pitch will do well in music. And being better at their respective domains, they will become more deeply interested in sounds and colors, will learn more about them, and thus are in a position to innovate in music or art with greater ease.

On the other hand, a sensory advantage is certainly not necessary. El Greco seems to have suffered from a disease of the optic nerve, and Beethoven was functionally deaf when he composed some of his greatest work. Although most great scientists seem to have been attracted to numbers and experimentation early in life, how creative they eventually became bears little relationship to how talented they were as children.

But a special sensory advantage may be responsible for developing an early *interest in the domain*, which is certainly an important ingredient of creativity. The physicist John Wheeler remembers being interested in "toy mechanisms, things that would shoot rubber bands, Tinkertoys, toy railroads, electric light bulbs, switches, buzzers." His father, who was a librarian, used to take him to New York State University, where he left John in the library office while he lectured. John was fascinated by the typewriters and other machines, especially hand calculators: "You pushed a button down and turned a crank, and how the thing worked, that intrigued me immensely." When he was twelve, he built a primitive calculator that had gears whittled out of wood.

Without a good dose of curiosity, wonder, and interest in what things are like and in how they work, it is difficult to recognize an interesting problem. Openness to experience, a fluid attention that constantly processes events in the environment, is a great advantage for recognizing potential novelty. Every creative person is more than amply endowed with these traits. Here is how the historian Natalie Davis selects what historical projects to focus on:

> Well, I just get really curious about some problem. It just hooks in very deeply. At the time I don't know why necessarily it is that I invest so much curiosity and eros into some project. At the time, it just seems terribly interesting and important for the field. I may not know what is personally invested in it, other than my curiosity and my delight.

Without such interest it is difficult to become involved in a domain deeply enough to reach its boundaries and then push them farther. True, it is possible to make one creative discovery, even a very important one, by accident and without any great interest in the topic. But contributions that require a lifetime of struggle are impossible without curiosity and love for the subject.

A person also needs *access to a domain*. This depends to a great extent on luck. Being born to an affluent family, or close to good schools, mentors, and coaches obviously is a great advantage. It does no good to be extremely intelligent and curious if I cannot learn what it takes to operate in a given symbolic system. The ownership of what sociologist Pierre Bourdieu calls "cultural capital" is a great resource. Those who have it provide their children with the advantage of an environment full of interesting books, stimulating conversation, expectations for educational advancement, role models, tutors, useful connections, and so on.

But here too, luck is not everything. Some children fight their way to the right schools while their peers stay behind. Manfred Eigen was captured by Russian troops at age seventeen and taken to a prisoner-of-war camp at the end of World War II, because he had been drafted to serve in an antiaircraft unit two years earlier. But he was determined to get back to studying science, even though he had had to leave high school at fifteen and never finished his studies. He escaped from the POW camp, walked back across half of Europe, and made a beeline for Göttingen, for he had heard that the best faculty in physics was reassembling there after the ravages of the war. He reached the city before the university actually had a chance to open but was admitted later with the first cohort of students, even though he lacked a high school diploma. Caught up in the ascetic postwar dedication to scholarship, led by the most knowledgeable teachers, surrounded by other equally dedicated students, he made quick progress. A few years later he received his doctorate and in 1967 the Nobel Prize. It is true that in early childhood Eigen could draw on substantial cultural capital, because his family had been musical and intellectually ambitious. Nevertheless, few people tossed by fate so far outside the circle of knowledge found their way back to its center as quickly and surely as he did.

Access to a field is equally important. Some people are terribly knowledgeable but are so unable to communicate with those who matter among their peers that they are ignored or shunned in the formative years of their careers. Michelangelo was reclusive, but in his youth was able to interact with leading members of the Medici court long enough to impress them with his skill and dedication. Isaac Newton was equally solitary and cantankerous, but somehow convinced his tutor at Cambridge that he deserved a lifetime tenured fellowship at the university, and so was able to continue his work undisturbed by human contact for many years. Someone who is not known and appreciated by the relevant people has a very difficult time accomplishing something that will be seen as creative. Such a person may not have a chance to learn the latest information, may not be given the opportunity to work, and if he or she does manage to accomplish something novel, that novelty is likely to be ignored or ridiculed.

In the sciences, being at the right university—the one where the most state-of-the-art research is being done in the best equipped labs by the most visible scientists—is extremely important. George Stigler describes this as a snowballing process, where an outstanding scientist gets funded to do exciting research, attracts other faculty, then the best students—until a critical mass is formed that has an irresistible appeal to any young person entering the field. In the arts, the attraction is more to the centers of distribution, now primarily New York City, where the major galleries and collectors are located. Just as a century ago aspiring young artists felt they had to go to Paris if they wanted to be recognized, now they feel that unless they run the gauntlet of Manhattan they don't have a chance. One can paint beautiful pictures in Alabama or North Dakota, but they are likely to be misplaced, ignored, and forgotten unless they get the stamp of approval of critics, collectors, and other gatekeepers of the field. Eva Zeisel's work received the imprimatur of the art establishment after her ceramics were shown by the Museum of Modern Art. The same is true of the other arts: Michael Snow spent ten years in New York City to catch up with the field of jazz music, and writers have to make connections with the agents and publishers there.

Access to fields is usually severely restricted. There are many gates to pass, and bottlenecks form in front of them. Writers who want to catch the attention of an editor long enough to have their work read have to compete with thousands of similarly hopeful writers who have also submitted their manuscripts. The editor typically has only a few minutes to dedicate to each writer's work, assuming he or she even glances at the submission in the first place. Getting a literary agent to sell the manuscript is no solution either, since a good agent's attention is as difficult to get as that of an editor.

Because of these bottlenecks, access to a field is often determined by chance or by irrelevant factors, such as having good connections. Students applying to good universities in some disciplines are so many and have such excellent credentials that it is difficult to rank them in any meaningful way. Yet the openings are few, so a selection must be made. Hence the joke that the admissions committee throws all the application folders down a long stairway, and the students whose files travel farthest get admitted.

THE TEN DIMENSIONS OF COMPLEXITY

Access to the domain and access to the field are all well and good, but when are we going to deal with the *real* characteristics of creative persons? When do we get to the interesting part—the tortured souls, the impossible dreams, the agony and the ecstasy of creation? The reason I hesitate to write about the deep personality of creative individuals is that I am not sure that there is much to write about, since creativity is the property of a complex system, and none of its components alone can explain it. The personality of an individual who is to do something creative must adapt itself to the particular domain, to the conditions of a particular field, which vary at different times and from domain to domain.

Giorgio Vasari in 1550 noted with chagrin that the new generations of Italian painters and sculptors seemed to be very different from their predecessors of the early Renaissance. They tended to be savage and mad, wrote the good Vasari, whereas their elders and betters had been tame and sensible. Perhaps Vasari was reacting to the artists who had embraced the ideology of Mannerism, the style ushered in by Michelangelo near the end of his long career, which relied on interesting distortions of figures and on grand gestures. This style would have been considered ugly a hundred years earlier, and the painters who used it would have been shunned. But a few centuries later, at the height of the Romantic period, an artist who was not more than a little savage and mad would not have been taken very seriously, because these qualities were de rigueur for creative souls.

In the 1960s, when abstract expressionism was the reigning style, those art students who tended to be sullen, brooding, and antisocial were thought by their teachers to be very creative. They were encouraged, and they won the prizes and fellowships. Unfortunately, when these students left school and tried to establish careers in the art world, they found that being antisocial did not get them very far. To get the attention of dealers and critics they had to throw wild parties and be constantly seen and talked about. Hence a hecatomb of introverted artists ensued: Most were selected out, ending up as art teachers in the Midwest or as car salesmen in New Jersey. Then the Warhol cohort replaced the abstract expressionists, and it was young artists with cool, clever, flip personalities who projected the aura of creativity. This, too, was a transient mask. The point is that you cannot assume the mantle of creativity just by assuming a certain personality style. One can be creative by living like a monk, or by burning the candle at both ends. Michelangelo was not greatly fond of women, while Picasso couldn't get enough of them. Both changed the domain of painting, even though their personalities had little in common.

Are there then no traits that distinguish creative people? If I had to express in one word what makes their personalities different from others, it would be *complexity*. By this I mean that they show tendencies of thought and action that in most people are segregated. They contain contradictory extremes—instead of being an "individual," each of them is a "multitude." Like the color white that includes all the hues in the spectrum, they tend to bring together the entire range of human possibilities within themselves.

These qualities are present in all of us, but usually we are trained to develop only one pole of the dialectic. We might grow up cultivating the aggressive, competitive side of our nature, and disdain or repress the nurturant, cooperative side. A creative individual is more likely to be both aggressive and cooperative, either at the same time or at different times, depending on the situation. Having a complex personality means being able to express the full range of traits that are potentially present in the human repertoire but usually atrophy because we think that one or the other pole is "good," whereas the other extreme is "bad."

This kind of person has many traits in common with what the Swiss analytic psychologist Carl Jung considered a mature personality. He also thought that every one of our strong points has a repressed shadow side that most of us refuse to acknowledge. The very orderly person may long to be spontaneous, the submissive person wishes to be dominant. As long as we disown these shadows, we can never be whole or satisfied. Yet that is what we usually do, and so we keep on struggling against ourselves, trying to live up to an image that distorts our true being.

A complex personality does not imply neutrality, or the average. It is not some position at the midpoint between two poles. It does not imply, for instance, being wishy-washy, so that one is never very competitive or very cooperative. Rather it involves the ability to move from one extreme to the other as the occasion requires. Perhaps a central position, a golden mean, is the place of choice, what software writers call the default condition. But creative persons definitely know both extremes and experience both with equal intensity and without inner conflict. It might be easier to illustrate this conclusion in terms of ten pairs of apparently antithetical traits that are often both present in such individuals and integrated with each other in a dialectical tension.

1. Creative individuals have a great deal of physical energy, but they are also often quiet and at rest. They work long hours, with great concentration, while projecting an aura of freshness and enthusiasm. This suggests a superior physical endowment, a genetic advantage. Yet it is surprising how often individuals who in their seventies and eighties exude energy and health remember a childhood plagued by illness. Heinz Maier-Leibnitz was bedridden for months in the Swiss mountains recovering from a lung ailment; György Faludy was often ill as a child, and so was the psychologist Donald Campbell. Public opinion analyst Elisabeth Noelle-Neumann was given no hope of survival by her physicians, but a homeopathic cure so improved her health that thirty years later she works harder than any four persons half her age. It seems that the energy of these people is internally generated and is due more to their focused minds than to the superiority of their genes. (Although it must be said that some respondents, such as Linus Pauling, answered "good genes," when asked to explain what accounted for their achievements.)

This does not mean that creative persons are hyperactive, always "on," constantly churning away. In fact, they often take rests and sleep a lot. The important thing is that the energy is under their own control—it is not controlled by the calendar, the clock, an external schedule. When necessary they can focus it like a laser beam; when it is not, they immediately start recharging their batteries. They consider the rhythm of activity followed by idleness or reflection very important for the success of their work. And this is not a biorhythm they inherited with their genes; it was learned by trial and error, as a strategy for achieving their goals. A humorous example is given by Robertson Davies:

> Well, you know, that leads me to something which I think has been very important in my life, and it sounds foolish and rather trivial. But I've always insisted on having a nap after lunch, and I inherited this from my father. And one time I said to him, "You know, you've done awfully well in the world. You came to Canada as an immigrant boy without anything and you

have done very well. What do you attribute it to?" And he said, "Well, what drove me on to be my own boss was that the thing that I wanted most was to be able to have a nap every day after lunch." And I thought, What an extraordinary impulse to drive a man on! But it did, and he always had a twenty-minute sleep after lunch. And I'm the same. I think it is very important. If you will not permit yourself to be driven and flogged through life, you'll probably enjoy it more.

One manifestation of energy is sexuality. Creative people are paradoxical in this respect also. They seem to have quite a strong dose of eros, or generalized libidinal energy, which some express directly into sexuality. At the same time, a certain spartan celibacy is also a part of their makeup; continence tends to accompany superior achievement. Without eros, it would be difficult to take life on with vigor; without restraint, the energy could easily dissipate.

2. Creative individuals tend to be smart, yet also naive at the same time. How smart they actually are is open to question. It is probably true that what psychologists call the g factor—meaning a core of general intelligence—is high among people who make important creative contributions. But we should not take seriously the lists that used to be printed on the sidebars of psychology textbooks, according to which John Stuart Mills must have had an IQ of 170 and Mozart an IQ of 135. Had they been tested at the time, perhaps they would have scored high. Perhaps not. And how many children in the eighteenth century would have scored even higher but never did anything memorable?

The earliest longitudinal study of superior mental abilities, initiated at Stanford University by the psychologist Lewis Terman in 1921, shows rather conclusively that children with very high IQs do well in life, but after a certain point IQ does not seem to be correlated any longer with superior performance in real life. Later studies suggest that the cutoff point is around 120; it might be difficult to do creative work with a lower IQ, but beyond 120 an increment in IQ does not necessarily imply higher creativity.

Why a low intelligence interferes with creative accomplishment is quite obvious. But being intellectually brilliant can also be detrimental to creativity. Some people with high IQs get complacent, and, secure in their mental superiority, they lose the curiosity essential to achieving anything new. Learning facts, playing by the existing rules of domains, may come so easily to a high-IQ person that he or she never has any incentive to question, doubt, and improve on existing knowledge. This is probably why Goethe, among others, said that naïveté is the most important attribute of genius.

Another way of expressing this dialectic is by the contrasting poles of wisdom and childishness. As Howard Gardner remarked in his study of the major creative geniuses of this century, a certain immaturity, both emotional and mental, can go hand in hand with deepest insights. Mozart comes immediately to mind.

Furthermore, people who bring about an acceptable novelty in a domain seem able to use well two opposite ways of thinking: the *convergent* and the *divergent*. Convergent thinking is measured by IQ tests, and it involves solving well-defined, rational problems that have one correct answer. Divergent thinking leads to no agreed-upon solution. It involves fluency, or the ability to generate a great quantity of ideas; flexibility, or the ability to switch from one perspective to another; and originality in picking unusual associations of ideas. These are the dimensions of thinking that most creativity tests measure and that most workshops try to enhance.

It is probably true that in a system that is conducive to creativity, a person whose thinking is fluent, flexible, and original is more likely to come up with novel ideas. Therefore, it makes sense to cultivate divergent thinking in laboratories and corporations—especially if management is able to pick out and implement the most appropriate ideas from the many that are generated. Yet there remains the nagging suspicion that at the highest levels of creative achievement the generation of novelty is not the main issue. A Galileo or a Darwin did not have that many new ideas, but the ones they fastened upon were so central that they changed the entire culture. Similarly, the individuals in our study often claimed to have had only two or three good ideas in their entire career, but each idea was so generative that it kept them busy for a lifetime of testing, filling out, elaborating, and applying.

Divergent thinking is not much use without the ability to tell a good idea from a bad one—and this selectivity involves convergent thinking. Manfred Eigen is one of several scientists who claim that the only difference between them and their less creative colleagues is that they can tell whether a problem is soluble or not, and this saves enormous amounts of time and many false starts. George Stigler stresses the importance of fluidity, that is, divergent thinking on the one hand, and good judgment in recognizing a viable problem on the other:

> I consider that I have good intuition and good judgment on what problems are worth pursuing and what lines of work are worth doing. I used to say (and I think this was bragging) that whereas most scholars have ideas which do not pan out more than, say, 4 percent of the time, mine come through maybe 80 percent of the time.

3. A third paradoxical trait refers to the related combination of playfulness and discipline, or responsibility and irresponsibility. There is no question that a playfully light attitude is typical of creative individuals. John Wheeler says that the most important thing in a young physicist is "this bounce, which I always associate with fun in science, kicking things around. It's not quite joking, but it has some of the lightness of joking. It's exploring ideas." David Riesman, in describing the attitude of "detached attachment" that makes him an astute observer of the social scene, stresses the fact that he always "wanted at the same time to be irresponsible and responsible."

But this playfulness doesn't go very far without its antithesis, a quality of doggedness, endurance, perseverance. Much hard work is necessary to bring a novel idea to completion and to surmount the obstacles a creative person inevitably encounters. When asked what enabled him to solve the physics problems that made him famous, Hans Bethe answered with a smile: "Two things are required. One is a brain. And second is the willingness to spend long times in thinking, with a definite possibility that you come out with nothing."

Nina Holton, whose playfully wild germs of ideas are the genesis of her sculpture, is very firm about the importance of hard work:

Tell anybody you're are a sculptor and they'll say, "Oh, how exciting, how wonderful." And I tend to say, "What's so wonderful?" I mean, it's like being a mason, or being a carpenter, half the time. But they don't wish to hear that because they really only imagine the first part, the exciting part. But, as Khrushchev once said, that doesn't fry pancakes, you see. That germ of an idea does not make a sculpture which stands up. It just sits there. So the next stage, of course, is the hard work. Can you really translate it into a piece of sculpture? Or will it be a wild thing which only seemed exciting while you were sitting in the studio alone? Will it look like something? Can you actually do it physically? Can you, personally, do it physically? What do you have by way of materials? So the second part is a lot of hard work. And sculpture is that, you see. It is the combination of wonderful wild ideas and then a lot of hard work.

Jacob Rabinow uses an interesting mental technique to slow himself down when work on an invention requires more endurance than intuition:

Yeah, there's a trick I pull for this. When I have a job to do like that, where you have to do something that takes a lot of effort, slowly, I pretend I'm in jail. Don't laugh. And if I'm in jail, time is of no consequence. In other words, if it takes a week to cut this, it'll take a week. What else have I got to do? I'm going to be here for twenty years. See? This is a kind of mental trick. Because otherwise you say, "My God, it's not working," and then you make mistakes. But the other way, you say time is of absolutely no consequence. People start saying how much will it cost me in time? If I work with somebody else it's fifty bucks an hour, a hundred dollars an hour. Nonsense. You just forget everything except that it's got to be built. And I have no trouble doing this. I work fast, normally. But if something will take a day gluing and then next day I glue the other side—it'll take two days—it doesn't bother me at all.

Despite the carefree air that many creative people affect, most of them work late into the night and persist when less driven individuals would not. Vasari wrote in 1550 that when the Renaissance painter Paolo Uccello was working out the laws of visual perspective, he would walk back and forth all night, muttering to himself: "What a beautiful thing is this perspective!" while his wife kept calling him back to bed with no success. Close to five hundred years later, physicist and inventor Frank Offner describes the time he was trying to understand how the membrane of the ear works:

> Ah, the answer may come to me in the middle of the night. My wife, when I was first into this membrane stuff, would kick me in the middle of the night and say, "Now get your mind off of membranes and get to sleep."

4. Creative individuals alternate between imagination and fantasy at one end, and a rooted sense of reality at the other. Both are needed to break away from the present without losing touch with the past. Albert Einstein once wrote that art and science are two of the greatest forms of escape from reality that humans have devised. In a sense he was right: Great art and great science involve a leap of imagination into a world that is different from the present. The rest of society often views these new ideas as fantasies without relevance to current reality. And they are right. But the whole point of art and science is to go beyond what we now consider real, and create a new reality. At the same time, this "escape" is not into a never-never land. What makes a novel idea creative is that once we see it, sooner or later we recognize that, strange as it is, it is true.

This dialectic is reflected by the way that, many years ago, the artists we studied responded to so-called projective tests, like the Rorschach or the Thematic Apperception Test. These require you to make up a story about some ambiguous stimuli, such as inkblots or drawings, that could represent almost anything. The more creative artists gave responses that were definitely more original, with unusual, colorful, detailed elements. But they never gave "bizarre" responses, which normal people occasionally do. A bizarre response is one that, with all the goodwill in the world, one could not see in the stimulus. For instance if an inkblot looks vaguely like a butterfly, and you say that it looks like a submarine without being able to give a sensible clue as to what in the inkblot made you say so, the response would be scored as bizarre. Normal people are rarely original, but they are sometimes bizarre. Creative people, it seems, are original without being bizarre. The novelty they see is rooted in reality.

Most of us assume that artists—musicians, writers, poets, painters—are strong on the fantasy side, whereas scientists, politicians, and businesspeople are realists. This may be true in terms of day-to-day routine activities. But when a person begins to work creatively, all bets are off—the artist may be as much a realist as the physicist, and the physicist as imaginative as the artist.

We certainly think of bankers, for example, as having a rather pedestrian, commonsense view of what is real and what is not. Yet a financial leader such as John Reed has much to say that dispels that notion. In his interview, he returns again and again to the theme that reality is relative and constantly changing, a perspective that he thinks is essential to confronting the future creatively:

I don't think there is such a thing as reality. There are widely varying descriptions of reality, and you've got to be alert to when they change and what's really going on. No one is going to truly grasp it, but you have to stay truly active on that end. That implies you have to have a multifaceted perspective.

There is a set of realities that exist at any moment in time. I always have some kind of a model in my mind as to what I think is going on in the world. I'm always tuning that [model] and trying to get different insights as I look at things, and I try to relate it back to what it means to our business, to how one behaves, if you will.

I don't mean to say there isn't anything in the center. I just think we can look at it [reality] in so many different ways. Right now, in my business, banks are deemed to be successful based on capital ratios. Ten years ago there was no concept of the "capital ratio." I failed totally to understand the impact of the savings and loan crisis on Congress, the regulators, and the industry. The world I'm living in today bears little resemblance to the world I lived in ten years ago, with regard to what was thought to be important. So we have defined a reality, which as I say is not empty, but it's close to being empty.

Like anybody else, I was slow to recognize the new reality. Knowing these kinds of things turns out to be awfully relevant, because your degrees of freedom get taken away if you're off base. I went through a massive adjustment to play a game that was different from the one you saw before. But it's a changing reality. I know goddamn well that these capital ratios are not sufficiently robust to be long-term, decent leading indicators of things, and five years from now the people who worry about how to price bank stocks are not going to be focusing on those. I describe success as *evolutionary* success.

What Einstein implied about art and science reappears in this account of banking: It is an *evolutionary* process, where current reality becomes rapidly obsolete, and one must be on the alert for the shape of things to come. At the same time, the emerging reality is not a fanciful conceit but something inherent in the here and now. It would be easy to dismiss Reed's visionary view as the romancing of a businessman who has had one too many encounters with reality. But apparently his unorthodox approach works: A recent issue of *Newsweek* announced: "John Reed might be excused a little gloating. . . . Since his darkest days three years ago he's quietly produced a stunning 425 percent return for investors who bought Citicorp shares." And one commentator adds that the overseas investments Reed made were considered junk five years ago, whereas now they are seen as a hot stock. "Nothing's changed but the perception," the financial expert says, echoing Reed's take on the reality of the market.

5. Creative people seem to harbor opposite tendencies on the continuum between extroversion and introversion. Usually each of us tends to be one or the other, either preferring to be in the thick of crowds or sitting on the sidelines and observing the passing show. In fact, in current psychological research, extroversion and introversion are considered the most stable personality traits that differentiate people from each other and that can be reliably measured. Creative individuals, on the other hand, seem to express both traits at the same time.

The stereotype of the "solitary genius" is strong and gets ample support also from our interviews. After all, one must generally be alone in order to write, paint, or do experiments in a laboratory. As we know from studies of young talented people, teenagers who cannot stand being alone tend not to develop their skills because practicing music or studying math requires a solitude they dread.

Only those teens who can tolerate being alone are able to master the symbolic content of a domain.

Yet over and over again, the importance of seeing people, hearing people, exchanging ideas, and getting to know another person's work and mind are stressed by creative individuals. The physicist John Wheeler expresses this point with his usual directness: "If you don't kick things around with people, you are out of it. Nobody, I always say, can be anybody without somebody being around."

Physicist Freeman Dyson expresses with a fine nuance the opposite phases of this dichotomy in his work. He points to the door of his office and says:

Science is a very gregarious business. It is essentially the difference between having this door open and having it shut. When I am doing science I have the door open. I mean, that is kind of symbolic, but it is true. You want to be, all the time, talking with people. Up to a point you welcome being interrupted because it is only by interacting with other people that you get anything interesting done. It is essentially a communal enterprise. There are new things happening all the time, and you should keep abreast and keep yourself aware of what is going on. You must be constantly talking. But, of course, writing is different. When I am writing I have the door shut, and even then too much sound comes through, so very often when I am writing I go and hide in the library. It is a solitary game. So, I suppose that is the main difference. But then, afterward, of course the feedback is very strong, and you get a tremendous enrichment of contacts as a result. Lots and lots of people write me letters simply because I have written books which address a general public, so I get into touch with a much wider circle of friends. It's broadened my horizons very much. But that is only after the writing is finished and not while it is going on.

John Reed builds the alternation between inner-directed reflection and intense social interaction into his daily routine:

I'm an early morning guy. I get up at five always, get out of the shower about 5:30, and I typically try to work either at home or at the office, and that's when I do a good bit of my thinking and priority setting. I'm a great lister. I have twenty lists of things to do all the time. If I ever have five free minutes I sit and make lists of things that I should be worrying about or doing. Typically I get to the office about 6:30. I try to keep a reasonably quiet time until 9:30 or 10:00. Then you get involved in lots of transactions. If you are chairman of the company it's like being a tribal chieftain. People come into your office and talk to you.

Even in the very private realm of the arts the ability to interact is essential. Nina Holton describes well the role of sociability in art:

You really can't work entirely alone in your place. You want to have a fellow artist come and talk things over with you— "How does that strike you?" You have to have some sort of feedback. You can't be sitting there entirely by yourself and never show it. And then eventually, you know, when you begin to show, you have to have a whole network. You have to get to know gallery people, you have to get to know people who work in your field who are involved. And you may want to find out whether you wish to be part of it or not be part of it, but you cannot help being part of a fellowship, you know?

Jacob Rabinow again puts into clear words the dilemma that many creative individuals face:

I remember once we had a big party and Gladys [his wife] said that I sometimes walk to a different drummer. In other words, I'm so involved in an idea I'm working on, I get so carried away, that I'm all by myself. I'm not listening to what anybody says. This sometimes happens. That you've got a new idea and you feel that it's very good and you're so involved that you're not paying attention to anybody. And you tend to drift away from people. It's very hard for me to be objective. I don't know. I'm social, I like people, I like to tell jokes, I like to go to the theater. But it's probably true that there are times when Gladys would have liked me to pay more attention to her and to the family. I love my children, they love me, and we have a wonderful relationship. But it could be that if I were not an inventor but had a routine job, I'd spend more time at home and I'd pay more attention to them, and the job would be something that I wouldn't like to do. So maybe people who don't like their jobs love their home more. It's quite possible.

6. Creative individuals are also remarkably humble and proud at the same time. It is remarkable to meet a famous person whom you expect to be arrogant or supercilious, only to encounter self-deprecation and shyness instead. Yet there are good reasons why this should be so. In the first place, these individuals are well aware that they stand, in Newton's words, "on the shoulders of giants." Their respect for the domain in which they work makes them aware of the long line of previous contributions to it, which puts their own into perspective. Second, they also are aware of the role that luck played in their own achievements. And third, they are usually so focused on future projects and current challenges that their past accomplishments, no matter how outstanding, are no longer very interesting to them. Elisabeth Noelle-Neumann's answer to the question "Looking back on all your accomplishments, which one would you say you are most proud of?" is typical:

I never think of what I am proud about. I never look back, except to find out about mistakes. Because mistakes are hard to remember and to draw conclusions from. But I only see danger in thinking back about things you are proud of. When people ask me if I am proud of something, I just shrug and hope to get away as soon as possible. I should explain that my way is always to look ahead, all my pleasant thoughts are about the future. It has been this way since I was twenty years old. I start every day fresh. The most important thing for me is to keep up the research institute, to keep up empirical research.

Despite her great accomplishments and reputation in the field, neuropsychologist Brenda Milner tells of being very self-critical and of having enormous self-doubts about being creative. The Canadian artist Michael Snow attributes the restless experimentation that led him to so many successes to a sense of confusion and insecurity he has been trying to dispell.

Another indication of modesty is how often this question was answered in terms of the family rather than the accomplishments that made a person famous. For instance, Freeman Dyson's answer was: "I suppose it is just to have raised six kids, and brought them up, as far as one can see, all to be interesting people. I think that's what I am most proud of, really." And John Reed's: "Oh, God. That's real . . . I suppose being a parent. I have four kids. If you had to say what has both surprised and given you a lot of pleasure, I'd say that I'm close to my kids and I enjoy them, and I never would have guessed that that would be as much fun as it's turned out to be."

At the same time, of course, no matter how modest these individuals are, they know that in comparison with others they have accomplished a great deal. And this knowledge provides a sense of security, even pride. This is often expressed as a sense of self-assurance. For instance, medical physicist Rosalyn Yalow mentioned repeatedly that all through her life she never had any doubts about succeeding in what she started out to do. Jacob Rabinow concurs: "There's one other thing that you do when you invent. And that is what I call the Existence Proof. This means that you have to assume that it can be done. If you don't assume that, you won't even try. And I always assume that not only it can be done, but *I* can do it." Some individuals stress humility, others self-assurance, but in actuality all of the people we interviewed seemed to have a good dose of both.

Another way of expressing this duality is to see it as a contrast between *ambition* and *selflessness*, or competition and cooperation. It is often necessary for creative individuals to be ambitious and aggressive. Yet at the same time, they are often willing to subordinate their own personal comfort and advancement to the success of whatever project they are working on. Aggressiveness is required especially in fields where competition is acute, or in domains where it is difficult to introduce novelty. In George Stigler's words:

> Every scholar, I think, is aggressive in some sense. He has to be aggressive if he wants to change his discipline. Now, if you get a Keynes or a Friedman, they are also aggressive in that they want to change the world, and so they become splendid public figures as well. But that's a very hard game to play.

Brenda Milner claims that the she has always been very aggressive verbally. John Gardner, statesman and founder of several national grassroots political organizations, describes well both the peaceful and aggressive instincts that coexist within the same person:

> I was the president of the Carnegie Corporation. I had a very interesting life, but not a lot of new challenges, not a tumultuous life. I was well protected. When I went to Washington I discovered a lot of things about myself that I didn't know. I discovered that I liked politicians. I got along well with them. I enjoyed dealing with the press, as much as anyone can enjoy dealing with the press. And then I discovered that I enjoyed a political fight, which was about as far away from my self-image as you can get. I'm a very peaceful person. But these things come out. Life pulls them out of you, and as I say, I'm a slow learner, but in my midfifties I learned some interesting things.

Several persons mention that in the course of their careers motivation has shifted from self-centered goals to more altruistic interests. For instance, Sarah LeVine, who started out as an anthropologist and then became a fiction writer, has this to say:

> Up until quite recently, I used to think of production only for the greater glory of myself, really. I don't see it that way at all anymore. I mean, it's nice if one gets recognition for what one does, but much more important is to leave something that other people can learn about, and I suppose that comes with middle age.

7. In all cultures, men are brought up to be "masculine" and to disregard and repress those aspects of their temperament that the culture regards as "feminine," whereas women are expected to do the opposite. Creative individuals to a certain extent escape this rigid gender role stereotyping. When tests of masculinity/femininity are given to young people, over and over one finds that creative and talented girls are more dominant and tough than other girls, and creative boys are more sensitive and less aggressive than their male peers.

This tendency toward androgyny is sometimes understood in

purely sexual terms, and therefore it gets confused with homosexuality. But psychological androgyny is a much wider concept, referring to a person's ability to be at the same time aggressive and nurturant, sensitive and rigid, dominant and submissive, regardless of gender. A psychologically androgynous person in effect doubles his or her repertoire of responses and can interact with the world in terms of a much richer and varied spectrum of opportunities. It is not surprising that creative individuals are more likely to have not only the strengths of their own gender but those of the other one, too.

Among the people we interviewed, this form of androgyny was difficult to detect—no doubt in part because we did not use any standard test to measure its presence. Nevertheless, it was obvious that the women artists and scientists tended to be much more assertive, self-confident, and openly aggressive than women are generally brought up to be in our society. Perhaps the most noticeable evidence for the "femininity" of the men in the sample was their great preoccupation with their family and their sensitivity to subtle aspects of the environment that other men are inclined to dismiss as unimportant. But despite having these traits that are not usual to their gender, they retained the usual gender-specific traits as well. In general, the women were perfectly "feminine" and the men thoroughly "masculine," in addition to having cross-gender traits.

8. Generally, creative people are thought to be rebellious and independent. Yet it is impossible to be creative without having first internalized a domain of culture. And a person must believe in the importance of such a domain in order to learn its rules; hence, he or she must be to a certain extent a traditionalist. So it is difficult to see how a person can be creative without being both *traditional and conservative* and at the same time *rebellious and iconoclastic*. Being only traditional leaves the domain unchanged; constantly taking chances without regard to what has been valued in the past rarely leads to novelty that is accepted as an improvement. The artist Eva Zeisel, who says that the folk tradition in which she works is "her home," nevertheless produces ceramics that were recognized by the Museum of Modern Art as masterpieces of contemporary design. This is what she says about innovation for its own sake:

This idea to create something different is not my aim, and shouldn't be anybody's aim. Because, first of all, if you are a designer or a playful person in any of these crafts, you have to be able to function a long life, and you can't always try to be different. I mean different from different from different. Secondly, wanting to be different can't be the motive of your work. Besides—if I talk too much let me know—to be different is a negative motive, and no creative thought or created thing grows out of a negative impulse. A negative impulse is always frustrating. And to be different means not like this and not like that. And the "not like"—that's why postmodernism, with the prefix of "post" couldn't work. No negative impulse can work, can produce any happy creation. Only a positive one.

But the willingness to take risks, to break with the safety of tradition, is also necessary. The economist George Stigler is very emphatic in this regard:

> I'd say one of the most common failures of able people is a lack of nerve. They'll play safe games. They'll take whatever the literature's doing and add a little bit to it. In our field, for example, we study duopoly, which is a situation in which there are two sellers. Then why not try three and see what that does. So there's a safe game to play. In innovation, you have to play a less safe game, if it's going to be interesting. It's not predictable that it'll go well.

9. Most creative persons are very *passionate* about their work, yet they can be extremely *objective* about it as well. The energy generated by this conflict between attachment and detachment has been mentioned by many as being an important part of their work. Why this is the case is relatively clear. Without the passion, we soon lose interest in a difficult task. Yet without being objective about it, our work is not very good and lacks credibility. So the creative process tends to be what some respondents called a yin-yang alternation between these two extremes. Here is how the historian Natalie Davis puts it:

> I am sometimes like a mother trying to bring the past to life again. I love what I am doing and I love to write. I just have a

great deal of affect invested in bringing these people to life again, in some way. It doesn't mean that I love my characters, necessarily, these people from the past. But I love to find out about them and re-create them or their situation. I think it is very important to find a way to be detached from what you write, so that you can't be so identified with your work that you can't accept criticism and response, and that is the danger of having as much affect as I do. But I am aware of that and of when I think it is particularly important to detach oneself from the work, and that is something where age really does help.

10. Finally, the openness and sensitivity of creative individuals often exposes them to *suffering and pain yet also a great deal of enjoyment.* The suffering is easy to understand. The greater sensitivity can cause slights and anxieties that are not usually felt by the rest of us. Most would agree with Rabinow's words: "Inventors have a low threshold of pain. Things bother them." A badly designed machine causes pain to an inventive engineer, just as the creative writer is hurt when reading bad prose. Being alone at the forefront of a discipline also makes you exposed and vulnerable. Eminence invites criticism and often vicious attacks. When an artist has invested years in making a sculpture, or a scientist in developing a theory, it is devastating if nobody cares.

Ever since the Romantic movement gained ascendance a few centuries ago, artists have been expected to suffer in order to demonstrate the sensitivity of their souls. In fact, research shows that artists and writers do have unusually high rates of psychopathology and addictions. But what is the cause, what is the effect? The poet Mark Strand comments:

There have been a lot of unfortunate cases of writers, painters, who have been melancholic, depressed, taken their own lives. I don't think it goes with the territory. I think those people would have been depressed, or alcoholic, suicidal, whatever, even if they weren't writing. I just think it's their characterological makeup. Whether that characterological makeup drove them to write or to paint, as well as to alcohol or to suicide, I don't know. I know there are an awful lot of healthy writers and painters who have no thoughts of suicide. I think it's a myth, by and large. It creates a special aura, a frailty, around the artist to say that he lives so close to the edge. He's so responsive to the world around him, so sensitive, so driven to respond to it, it's almost unbearable. That he must escape either through drugs or alcohol, finally suicide, the burden of consciousness is so great. But the burden of consciousness is great for people who don't—you know—want to kill themselves.

It is also true that deep interest and involvement in obscure subjects often goes unrewarded, or even brings on ridicule. Divergent thinking is often perceived as deviant by the majority, and so the creative person may feel isolated and misunderstood. These occupational hazards do come with the territory, so to speak, and it is difficult to see how a person could be creative and at the same time insensitive to them.

Perhaps the most difficult thing for a creative individual to bear is the sense of loss and emptiness experienced when, for some reason or another, he or she cannot work. This is especially painful when a person feels one's creativity drying out; then the whole self-concept is jeopardized, as Mark Strand suggests:

Yeah, there's a momentary sereneness, a sense of satisfaction, when you come up with an idea that you think is worth pursuing. Another form of that is when you have completed, where you've done as much as you can with an idea that you thought was worth working on. Then you sort of bask in the glow of completion for a day, maybe. You know, have a glass or two more of wine at night because you don't feel you have to go upstairs and look at anything again.

And then you're beginning again. You hope. Sometimes the hiatus will last not overnight but for weeks, months, and years. And the longer the hiatus is between books that you're committed to finishing, the more painful and frustrating life becomes. When I say "painful," that's probably too grandiose a term for the petty frustration one feels. But if it goes on, and on, and you develop what people call a writer's block, it's painful, because your identity's at stake. If you're not writing, and you're a writer and known as a writer, what are you?

Yet when the person is working in the area of his or her expertise, worries and cares fall away, replaced by a sense of bliss. Perhaps the most important quality, the one that is most consistently present in all creative individuals, is the ability to enjoy the process of creation for its own sake. Without this trait poets would give up striving for perfection and would write commercial jingles, economists would work for banks where they would earn at least twice as much as they do at the university, physicists would stop doing basic research and join industrial laboratories where the conditions are better and the expectations more predictable. In fact, enjoyment is such an important part of creativity that we devote chapter 5 to the connection. Here I report a single illustration, just as a place marker, to make sure that we don't lose sight of this essential component:

Margaret Butler is a computer scientist and mathematician, the first woman elected a fellow of the American Nuclear Society. In describing her work, like most of our respondents, she keeps stressing this element of fun, of enjoyment. In answer to the question "Of your accomplishments at work, what are you most proud of?" she answers:

> Well, in my work I think that the most interesting and exciting things that I have done were in the early days at Argonne when we were building computers. We worked on a team to design one of the first computers. We developed image analysis software with the people in the biology division for scanning chromosomes and trying to do automatic karyotyping, and I think that was the most fun that I had in all of my forty-plus years at the lab.

It is interesting that this response, stressing fun and excitement, came in answer to a question about what she is most proud of in her work. Later on, she says:

> I worked and worked. You work hard. You try to do your best. When we were working on the chromosome project, Jim [her husband] and I spent sometimes the whole night over there working. We would come out in the morning and the sun would be coming up. Science is very much fun. And I think women should have the opportunity to have fun.

I may work as hard as Butler did out of ambition or a desire to make money. But unless I also enjoy the task, my mind is not fully concentrated. My attention keeps shifting to the clock, to daydreams of better things to do, to resenting the job and wishing it was over. This kind of split attention, of halfhearted involvement, is incompatible with creativity. And creative people usually enjoy not only their work but also the many other activities in their lives. Margaret Butler, in describing what she does after her formal retirement, uses the word *enjoy* in reference to everything she does: helping her husband to continue his mathematical research, writing a careers-for-women guide for the American Nuclear Society, working with teachers to get women students interested in science, organizing support groups for women scientists, reading, and being involved in local politics.

These ten pairs of contrasting personality traits might be the most telling characteristic of creative people. Of course, this list is to a certain extent arbitrary. It could be argued that many other important traits have been left out. But what is important to keep in mind is that these conflicting traits—or any conflicting traits—are usually difficult to find in the same person. Yet without the second pole, new ideas will not be recognized. And without the first, they will not be developed to the point of acceptance. Therefore, the novelty that survives to change a domain is usually the work of someone who can operate at both ends of these polarities—and that is the kind of person we call "creative."

FOUR

THE WORK OF CREATIVITY

I s there a single series of mental steps that leads to novelties that result in changing a domain? Or, to put it differently, is every creative product the result of a single "creative process"? Many individuals and business training programs claim that they know what "creative thinking" consists of and that they can teach it. Creative individuals usually have their own theories—often quite different from one another. Robert Galvin says that creativity consists of anticipation and commitment. Anticipation involves having a vision of something that will become important in the future before anybody else has it; commitment is the belief that keeps one working to realize the vision despite doubt and discouragement.

On the other hand, in his letter of refusal, the management guru Peter Drucker lists four reasons that account for his accomplishments (in addition to the fifth, never participate in studies such as this):

(a) I have been able to produce because I have always been a loner and have not had to spend time on keeping subordinates, assistants, secretaries, and other time-wasters; because (b) I never set foot in my university office—I do my teaching; and if students want to see me I give them lunch; because (c) I have been a workaholic since I was 20; and (d) because I thrive on stress and begin to pine if there is no deadline. Otherwise—if I may be presumptuous: I was born like the sentry in Goethe's *Faust II*:

> *Zum Sehen geboren*
> *Zum Schauen bestellt*

("Born to see, my task is to watch")

Given how different domains are from one another, however, and given the variety of tasks and the different strengths and weaknesses of individuals, we should not expect a great deal of similarity in how people arrive at a novel idea or product. Yet some common threads do seem to run across boundaries of domains and individual idiosyncrasies, and these might well constitute the core characteristics of what it takes to approach a problem in a way likely to lead to an outcome the field will perceive as creative. Let's illustrate this process with a description of how the Italian author Grazia Livi wrote one of her short stories.

THE WRITING OF A STORY

One day Livi went to her bank to talk to a financial adviser who managed her portfolio of investments. The adviser was a woman Livi had met before; she seemed to her the epitome of a contemporary career woman bent on success and not much else, immaculately groomed, cold, hard, impatient. A person without a private life, with no dreams except money and advancement. This particular day the appointment started in the usual key: the adviser looking distant and frigid, asking questions in a dry, uninterested voice. Then a ringing phone interrupted the conversation. To Livi's surprise, as the woman turned away to take the call, her face changed—the chiseled features softened, even the hard helmet of hair became velvety—her posture relaxed, her voice became low and caressing. Livi had an immediate visual image of the person at the other end of the phone: a handsome, tanned, laid-back architect who drove a Maserati. After returning from the bank, she made a few notes to herself in a log she keeps for this purpose and then apparently forgot the incident. Some months later, rereading the log, she saw a connection between the entry she had made of the episode at the bank and entries she had written about a dressed-for-success woman sitting for hours in a beauty shop and other similar types she had met in the course of the past years. She was seized with a strong feeling of emotional discovery: Here was an insight about the current predicament of women—torn between contrasting demands—that could yield a true story. True not in the sense of representing what she had seen—the woman at the bank may have been talking to her mother or her child—but true to a widespread condition of our times, where many women feel that they have to be aggressive and cold to compete in the business world yet at the same time cannot give up what they think of as their femininity. So she sat down to write about a career woman grooming herself all day for a date that never comes off—and it was a terrific story. Not because of the plot, which is as old as the hills, but because the emotional currents of her character reflected so achingly and accurately the experience of our time.

Livi's story may not change the domain of literature, and hence it is not an example of the highest order of creativity. But it may well be included in future collections of short stories, because it is an excellent example of a contemporary genre. And to the extent that it expands the domain, it qualifies as a creative achievement. Is there a way to analyze what Livi did, to see more clearly what her mental processes were as she wrote the story?

The creative process has traditionally been described as taking five steps. The first is a period of preparation, becoming immersed, consciously or not, in a set of problematic issues that are interesting and arouse curiosity. In the case of Grazia Livi, the emotional quandary of modern women was something she experienced personally, as a writer trying to compete for prizes, reviews, and publications, and also as a woman trying to balance the responsibilities of motherhood with her writing.

The second phase of the creative process is a period of incubation, during which ideas churn around below the threshold of consciousness. It is during this time that unusual connections are likely to be made. When we intend to solve a problem consciously, we process information in a linear, logical fashion. But when ideas call to each other on their own, without our leading them down a straight and narrow path, unexpected combinations may come into being.

The third component of the creative process is insight, sometimes called the "Aha!" moment, the instant when Archimedes cried out "Eureka!" as he stepped into the bath, when the pieces of the puzzle fall together. In real life, there may be several insights interspersed with periods of incubation, evaluation, and elaboration. For instance, in the case of Livi's short story, there are at least two moments of significant insight: when she saw the investment adviser transformed by the phone call, and when she saw the connection between the similar entries in the log.

The fourth component is evaluation, when the person must decide whether the insight is valuable and worth pursuing. This is often the most emotionally trying part of the process, when one feels most uncertain and insecure. This is also when the internalized criteria of the domain, and the internalized opinion of the field, usually become prominent. Is this idea really novel, or is it obvious? What will my colleagues think of it? It is the period of self-criticism, of soul-searching. For Grazia Livi, much of this sifting took place as she read through her log and decided which ideas to develop.

The fifth and last component of the process is elaboration. It is probably the one that takes up the most time and involves the hardest work. This is what Edison was referring to when he said that creativity consists of 1 percent inspiration and 99 percent perspiration. In Livi's case, elaboration consisted in selecting the characters of the story, deciding on a plot, and then translating the emotions she had intuited into strings of words.

But this classical analytic framework leading from preparation to elaboration gives a severely distorted picture of the creative process if it is taken too literally. A person who makes a creative contribution never just slogs through the long last stage of elaboration. This part of the process is constantly interrupted by periods of incubation and is punctuated by small epiphanies. Many fresh insights emerge as one is presumably just putting finishing touches on the initial insight. As Grazia Livi was struggling to find words to describe her character, the words themselves suggested new emotions that were sometimes more "right" to the personality she was trying to create than the ones she had initially envisioned. These new feelings in turn suggested actions, turns of the plot she had not thought of before. The character became more complex, more nuanced, as the writing progressed; the plot became more subtle and intriguing.

Thus the creative process is less linear than recursive. How many iterations it goes through, how many loops are involved, how many insights are needed, depends on the depth and breadth of the issues dealt with. Sometimes incubation lasts for years; sometimes it takes a few hours. Sometimes the creative idea includes one deep insight and innumerable small ones. In some cases, as with Darwin's formulation of the theory of evolution, the basic insight may appear slowly, in separate disconnected flashes that take years to coalesce into a coherent idea. By the time Darwin clearly understood what his theory implied, it was hardly an insight any longer, because its components had all emerged in his thought at different times in the past and had slowly connected with one another along the way. It was a thunderous "Aha!" built up over a lifetime, made up of a chorus of little "Eurekas."

A more linear account is Freeman Dyson's description of the creative process that brought him scientific fame. Dyson had been a student of Richard Feynman, who in the late 1940s was trying to make electrodynamics understandable in terms of the principles of quantum mechanics. Success in this task would mean translating the laws of electricity so that they conformed to the more basic laws of subatomic behavior. It would be a great simplification, a welcome ordering of the domain of physics. Unfortunately, while most colleagues felt that Feynman was onto something deep and important, not many could follow the few scribbles and sketches he used to prove his points, especially since he usually went from A directly to Z with no stops in between. At the same time, another physicist, Julian Schwinger, also was working on the unification of quantum and electrodynamic principles. Schwinger was in many ways Feynman's opposite: He worked slowly and methodically and was such a perfectionist that he never felt ready to claim a solution to the problem he was working on. Freeman Dyson, working in Feynman's orbit at Cornell University, was exposed to a series of lectures by Schwinger. It gave him the idea of bringing together Feynman's leaps of intuition with Schwinger's painstaking calculations and to resolve once and for all the puzzle of how the behavior of quanta related to electrical phenomena. After Dyson finished his work, Feynman's and Schwinger's theories became understandable, and the two received the Nobel Prize in physics. Several colleagues felt that if anyone deserved the prize, it was Dyson. Here is how he describes the process that led to his accomplishment:

It was the summer of 1948, so I was then twenty-four. There was a big problem which essentially the whole community of physicists was concentrated on. Physics is usually like that—there is some particularly fascinating problem that everybody is working on and it tends to be sort of one thing at a time. And at that time the big problem was called quantum electrodynamics, which was a theory of radiation and atoms, and the theory was in a mess and nobody knew how to calculate with it. It was sort of a logjam for all kinds of further developments. So somebody had to learn how to calculate with this theory. It wasn't a question of the theory being wrong, but it was somehow not decently organized, so that people tried to calculate and always got silly answers, like zero or infinity, or something. Anyhow, at that moment there appeared two great ideas which were associated with two people, Schwinger and Feynman, both of them about five years older than I was. Each of them produced a new theory of radiation, which looked as though it was going to work, although there were difficulties with both of them. I was in this happy position of being familiar with both of them and I got to know both of them and I got to work.

I spent six months working very hard to understand both of them clearly, and that meant simply hard, hard work of calculating. I would sit down for days and days with large stacks of papers doing calculations so that I could understand precisely what Feynman was saying. And at the end of six months, I went off on a vacation. I took a Greyhound bus to California and spent a couple of weeks just bumming around. This was soon after I had arrived from England, so I had never been to the West before. After two weeks in California, where I wasn't doing any work, I was just sight-seeing, I got on the bus to come back to Princeton, and suddenly in the middle of the night when we were going through Kansas, the whole sort of suddenly became crystal clear, and so that was sort of the big revelation for me, it was the Eureka experience or whatever you call it. Suddenly the whole picture became clear, and Schwinger fit into it beautifully and Feynman fit into it beautifully and the result was a theory that actually was useful. That was the big creative moment of my life. Then I had to spend another six months working out the details and writing it all up and so forth. It finally ended up with two long papers in the *Physical Review,* and that was my passport to the world of science.

It would be difficult to imagine a clearer example of the classical version of the creative process. It starts with Dyson, immersed in the field of physics, sensing from his teachers and colleagues where the next opportunity for adding something important to the domain lies. He has a privileged access to both the domain and the field—he is personally acquainted with the two central individuals involved. Having found his problem—to reconcile the two leading theories in the domain—he goes through a six-month period of consciously directed, hard preparation. Then he spends two weeks relaxing, a period during which the ideas marshaled up during the past half year have a chance to incubate, to sort out and shake together. This is followed by the sudden insight that occurs unbidden during a night bus ride. And finally another half year of hard work evaluating and elaborating the insight. The idea having been accepted by the field—in this case, the editors of *Physical Review*—it is then added to the domain. As is often the case, most of the credit for the accomplishment does not go directly to the author, but to those whose work he has built upon.

The five stage view of the creative process may be too simplified, and it can be misleading, but it does offer a relatively valid and simple way to organize the complexities involved. Therefore, I use these categories to describe how creative people work, starting with the beginning phase, that of preparation. It is essential to remember in what follows, however, that the five stages in reality are not exclusive but typically overlap and recur several times before the process is completed.

THE EMERGENCE OF PROBLEMS

Occasionally it is possible to arrive at a creative discovery without any preparation. The fortunate person simply stumbles into a wholly unpredictable situation, as Roentgen did when he tried to find out why his photographic plates were being ruined and discovered radiation in the process. But usually insights tend to come to prepared minds, that is, to those who have thought long and hard about a given set of problematic issues. There are three main sources from which problems typically arise: personal experiences, requirements of the domain, and social pressures. While these three sources of inspiration are usually synergistic and intertwined, it is easier to consider them separately, as if they acted independently, which in reality is not the case.

Life as a Source of Problems

We have seen that Grazia Livi's idea for a story about the conflict between career and femininity was influenced by her own experiences as a woman. From the time she was a little girl, her parents expected her two brothers to be educated and successful while Grazia and her sister were expected to grow up to be traditional housewives. Throughout her life Livi rebelled against the role cut out for her. Even though she married and had children, she resolved to become successful on her own. It is this direct experience in her own life that made her sensitive to the episodes involving career women that she jotted down in her diary.

The origins of problematic elements in life experience are easiest to see in the work of artists, poets, and humanists in general. Eva Zeisel, who was considered the "dumb one" in a family that eventually included two Nobel laureates and many other outstanding male scientists, also resolved to prove herself by breaking away from traditional family interests and becoming an independent artist. Most of the creative ideas for her pottery come from a tension between two contrasting, self-imposed requirements: to make pots that conform to the human hand and are steeped in tradition, and yet can be mass-produced inexpensively by modern technology.

Poets like Anthony Hecht, György Faludy, and Hilde Domin write down daily impressions, events, and especially feelings on index cards or in notebooks, and these caches of experience are the raw material out of which their work evolves. "I had a friend, a poet called Radnòty, who wrote poems I considered atrocious," says Faludy. "And then after suffering in the concentration camps it changed him totally and he wrote wonderful verse. Suffering is not bad: It helps you very much. Do you know a novel about happiness? Or a film about happy people? We are a perverse race, only suffering interests us." He then relates how once when he was sitting in a cabin on beautiful Vancouver Island, trying to find inspiration to start a poem, he could think of nothing interesting. Finally, a set of strong images occurred to him: Five secret policemen arrive in a boat, break into the cabin, throw his books out of the window into the sea, take him five thousand miles to Siberia, and beat him mercilessly—a great scenario for a poem, one with which the poet was unfortunately all too familiar.

The historian Natalie Davis describes the project she is working on, a book about three women of the seventeenth century, one Jewish, one Catholic, one Protestant, exploring the "sources of adventuresomeness for women":

> They were all sort of me in the sense that they were all middle-aged mothers, although in one case a grandmother—which I am—and so I keep thinking that it is no coincidence that I got started on this completely different project.

The painter Ed Paschke tears off dozens of arresting images each day from magazines and newspapers and keeps these strange or funny cutouts in boxes to which he returns occasionally for inspiration. Rummaging through these icons of the times he may find one that he projects on the wall and uses as a starting point for a sardonic pictorial commentary. Another painter, Lee Nading, tears off newspaper headlines that have to do with the conflict of nature and technology—DAM ENDANGERS RARE FISH or TRAIN FULL OF GARBAGE DERAILS IN IOWA—and eventually uses one of them to inspire a canvas. To understand why Nading is particularly sensitive to this kind of event, it helps to know that he had a beloved elder brother who committed suicide just as his career was becoming successful. This brother worked at one of the most prestigious scientific research laboratories but became disillusioned with the competitiveness and the lack of concern for human consequences that he felt around him. Nading never quite forgave science for having contributed to his brother's death, and he finds in the threats posed by the fruits of science the source for his artistic problems.

Artists find inspiration in "real" life—emotions like love and anxiety, events like birth and death, the horrors of war, and a peaceful afternoon in the country. We shall see in a little while that artists are also influenced in the choice of their problems by the domain and the field. It has been said that every painting is a response to all previous paintings, and every poem reflects the history of poetry. Yet paintings and poems are also very clearly inspired by the artist's experiences.

The experiences of scientists are relevant to the problems they deal with in a much more general, but perhaps not less important way. This has to do with the fundamental interest and curiosity the scientist brings to the task. One of the very first studies of creative scientists, conducted by Ann Roe, concluded that the chemists and physicists in her sample became interested as children in the properties of matter because the normal interests of childhood were not accessible to them. Their parents were emotionally distant, they had few friends, they were not very athletic. Perhaps this kind of generalization is drawn with too thick a brush, but the basic idea underlying it—that early experience predisposes a young person to be interested in a certain range of problems—is probably sound.

For instance, the physicist Viktor Weisskopf describes with great emotion the sense of awe and wonder he felt when, as a young man, he and a friend used to climb in the Austrian Alps. Many of the great physicists of his generation, like Max Planck, Werner Heisenberg, and Hans Bethe, claim that what inspired them to try to understand the movement of atoms and stars was the exhilaration they felt at the sight of tall peaks and the night sky.

Linus Pauling became interested in chemistry when his father, a pharmacist in turn-of-the-century Portland, let him mix powders and potions in the back of the drugstore. The young Pauling was fascinated by the fact that two different substances could turn into a third entirely different one. He experienced a godlike sense of being able to create something entirely new. By the age of seven he had read and practically memorized the enormous *Pharmacopoeia* containing the knowledge of basic elements and mixtures a pharmacist was expected to know. It was this early curiosity about how matter could be transformed that fueled Pauling's career for the next eighty years. The psychologist Donald Campbell makes the point that the difference between a scholar who comes up with new ideas and one who does not is often a difference in curiosity:

So many of my professor friends who know that they should be continuing to do research look around and find no problem that fascinates them. Whereas I have a scattered dilettante backlog of problems that I would love to work on and I feel are within reach of a solution. Many talented people can't think of anything to do that they feel is worth doing. Now, I think that I am blessed that there are trivial problems that can excite me.

Without a burning curiosity, a lively interest, we are unlikely to persevere long to make a significant new contribution. This kind of interest is rarely only intellectual in nature. It is usually rooted in deep feelings, in memorable experiences that need some sort of resolution—a resolution that can be achieved only by a new artistic expression or a new way of understanding. Someone who is motivated solely by the desire to become rich and famous might struggle hard to get ahead but will rarely have enough inducement to work beyond what is necessary, to venture beyond what is already known.

The Influence of Past Knowledge

The other main source of problems is the domain itself. Just as personal experiences produce tensions that cannot be resolved in terms of ordinary solutions, so does working within a symbolic system. Over and over, both in the arts and the sciences, the inspiration for a creative solution comes from a conflict suggested by the "state of the art." Every domain has its own internal logic, its pattern of development, and those who work within it must respond to this logic. A young painter in the 1960s had two choices: Either paint in the fashionable abstract expressionist style or discover a viable way of rebelling against it. Natural scientists in the early part of this century were confronted by the development of quantum theory in physics: Many of the most challenging problems in chemistry, biology, astronomy, as well as physics, were generated by the possibility of applying quantum theory to these new realms. Freeman Dyson's concern with quantum electrodynamics is only one example.

Gerald Holton, a physicist who later turned to the history of science, gives a lucid account of how a problematic issue in the domain can fuse with a personally felt conflict to suggest the theme for a person's lifework. As a graduate student at Harvard, Holton was immersed in the heady atmosphere of logical positivism. His teachers and fellow students were bent on demonstrating that science could be reduced to an absolutely logical enterprise. Nothing intuitive or metaphysical was admitted to this new domain. But Holton, who read about the way Kepler and Einstein had worked, started to feel that the kind of science everyone around him took for granted did not apply to some of the most celebrated scientific breakthroughs.

I discovered that these models don't quite work, that you do not in fact have built into usual accounts of the scientific process the kind of presuppositions that these people were very fond of. It was not true, for example, that the way to think about science is to think in terms of protocol sentences, and verification theory of meaning, and all of those things that were very dear to them. But these presuppositions were the things that the best of them were willing to put their money on, their reputation, their time, their very life, and stick with it even against the evidence for a while. They were enchanted with an idea for which there was in fact no proof. I had to really struggle with that.

And it is at that point that I found the idea of a *thematic proposition,* that some people are imbued with prior thematic ideas which would survive a period of disconfirmation. And that was not part of the logic of positivism or empiricism at all.

Holton describes the genesis of his own intellectual problem as a conjunction of personal interest and a sense that something was askew in the intellectual environment:

Your research project gets defined partly by some internal fascination for which one cannot account in any detail, preparation that is unique because of the life history of that person, luck, and something to work against. That is, something that you are dissatisfied with that other people are doing.

An intellectual problem is not restricted to a particular domain. Indeed, some of the most creative breakthroughs occur when an idea that works well in one domain gets grafted to another and revitalizes it. This was certainly the case with the widespread applications of physics' quantum theory to neighboring disciplines like chemistry and astronomy. Creative people are ever alert to what colleagues across the fence are doing. Manfred Eigen, whose recent work involves the attempt to replicate inorganic evolution in the laboratory, is bringing together concepts and experimental procedures from physics, chemistry, and biology. The ideas coalesced in part from conversations over the years with colleagues from different disciplines—whom he invited to informal winter meetings in Switzerland.

A large majority of our respondents were inspired by a tension in their domain that became obvious when looked at from the perspective of another domain. Even though they do not think of themselves as interdisciplinary, their best work bridges realms of ideas. Their histories tend to cast doubt on the wisdom of overspecialization, where bright young people are trained to become exclusive experts in one field and shun breadth like the plague.

And then there are people who sense problems in "real" life that cannot be accommodated within the symbolic system of any existing domain. Barry Commoner, trained as a biophysicist, decided to step out of the formalities of the academic approach and confront such issues as the quality of water and the disposal of garbage. His problems are defined by real-life concerns, not disciplines.

Well, I established a pretty good reputation in biochemistry and biophysics. In the beginning all of the papers were published in academic journals. But in various ways and for various reasons I moved more and more in the direction of doing work that was relevant to real world problems. And every now and then a paper of mine will appear in an academic journal, but that's just by accident.

As the generation of World War II scientists began to get older, the academic world became very isolated from the real world. Academic work was discipline dictated and discipline oriented, which is really pretty dull, I think.

The prevailing philosophy in academic life is reductionism, which is exactly the reverse of my approach to things, and I'm not interested in doing it.

This is a typical reaction against a domain becoming too confining and its members mistaking the symbolic system in which they operate for the broader reality of which it is a part. Commoner's feelings may be similar to those that young scholars in Byzantium must have felt when the church councils spent so much time debating how many angels could dance on the head of a pin. When a field becomes too self-referential and cut off from reality, it runs the risk of becoming irrelevant. It is often dissatisfaction with the rigidity of domains that makes great creative advances possible.

Of course, a person cannot be inspired by a domain unless he or she learns its rules. That is why everyone we talked to, whether artist or scientist, emphasized over and over the importance of basic knowledge, of thorough familiarity with the symbolic information and the basic procedures of the discipline. György Faludy can recite long stretches of verse by Catullus that he memorized in Latin sixty years ago; he has read all the Greek, Chinese, Arabic, and European poetry that he has been able to find. He translated more than fourteen hundred poems from around the world to master his craft, even though his own powerful poems are simple, discursive, and based on personal experience. In science, mastery of the basic symbolic tools is equally important. Practically everyone echoes what Margaret Butler tells high school students:

> The message that we were trying to get [across] is that if you do not know what you want to be, at least take science and math. Especially math, so that when you get into college if you change your mind and you like science or math more, or you find that you want to get into it, then you will have the background that is needed. Many women find later on that they do not have the background [mathematics] because they copped out early on.

You cannot transform a domain unless you first thoroughly understand how it works. Which means that one has to acquire the tools of mathematics, learn the basic principles of physics, and become aware of the current state of knowledge. But the old Italian saying seems to apply: *Impara l'arte, e mettila da parte* (learn the craft, and then set it aside). One cannot be creative without learning what others know, but then one cannot be creative without becoming dissatisfied with that knowledge and rejecting it (or some of it) for a better way.

The Pressures of the Human Environment

The third source of ideas and problems is the field one works in. All through life, a creative person is exposed to the influence of teachers, mentors, fellow students, and coworkers, and later in life to the ideas of one's own students and followers. Moreover, the institutions one works for and the events of the wider society in which one lives provide powerful influences that can redirect one's career and channel a person's thinking in new directions.

Indeed, if we look at creativity from this perspective, personal experience and domain knowledge may pale in comparison with the contribution of the social context to determine which problems one tackles. What an artist paints is a response not only to the classic canon of art but also to what others are painting right now. Scientists don't learn only from books or experiments they conduct but also from seminars, meetings, workshops, and journal articles reporting what is happening, or about to happen elsewhere. Whether one follows the crowd or takes a different path, it is usually impossible to ignore what takes place in the field.

Many people are introduced to the wonders of a domain by a teacher. There is often a particular teacher who recognizes the child's curiosity or ability and starts cultivating his or her mind in the discipline. Some creative persons have a long list of such teachers. The critic and rhetorician Wayne Booth says that each year in school he idealized a different one and tried to live up to that teacher's expectations. In his case, as in several others, the changes from one career direction to the other—from engineering to English—occurred in response to the quality of the teachers encountered.

For some, the introduction to the domain comes later. John Gardner started college intending to become a writer but found in the psychology departments of Berkeley and then Stanford an intellectual community that satisfied his curiosity as well as his desire for congenial company.

The field is paramount for individuals who work primarily in an organizational context. John Reed of Citicorp must constantly interact with several groups in order to assimilate the information that he needs to make difficult decisions. About twice a year he meets for a few days with the half-dozen heads of the national banks of Germany, Japan, and so on to exchange ideas about future trends in the world economy. At more frequent intervals he has similar meetings with the CEOs of General Motors, General Electric, or IBM. Even more often, he meets with the key executives of his own corporation. His inner network consists of about thirty people whom he trusts to provide the input he needs to navigate a multibillion-dollar corporation through constantly changing times. Reed spends at least half of his mornings talking on the phone or in person with members of this network and never makes a major decision involving the company without conferring with at least some of them.

Another organizational approach is represented by Robert Galvin, president of Motorola. Galvin sees his company as a gigantic creative enterprise, with more than twenty thousand engineers anticipating trends, reacting to them with new ideas, creating new products and processes. He sees his own job as orchestrating all this effort, being a role model for everyone else. In cases when the responsibility is to lead a group of people in novel directions, work is usually dictated not by a symbolic domain but by the requirements of the organization itself. It could be said for them, to borrow Marshall McLuhan's phrase, that the medium is the message; what they accomplish within their organizational structure *is* their creative accomplishment.

Scientists also mention the importance of particular research institutions. The Bell Labs, the Rockefeller Institute, and the Argonne National Laboratories are some of the places that have allowed young scientists to pursue their interests in a stimulating and supportive environment. Not surprisingly, many of them feel strong loyalties to such institutions and are more than willing to follow their research policies. Many a Nobel Prize was won by tackling problems that arose out of such institutional contexts.

New ideas are also generated when someone attempts to create a new organization or perhaps a new field. Manfred Eigen founded an interdisciplinary Max Planck Institute in Göttingen to replicate experimentally evolutionary forces in the laboratory. George Klein built up the tumor biology research center at the Karolinska Institute in Stockholm, and employs a large cadre of Ph.D.'s. Initiatives of this sort not only allow the principal investigator to pursue his or her research but also make it possible for a new discipline to emerge. If the lab is successful, entirely new sets of problems are opened up for investigation, and with time a new symbolic system—or domain—may develop.

Finally, some creative individuals attempt to form entirely new organizations outside the pale of accepted scientific, academic, or business institutions. Hazel Henderson dedicates most of her time to developing groups that will further her vision; she sees herself as the progenitor of innumerable special interest groups united in their ecological consciousness. Similarly, Barry Commoner has purposefully positioned his center in a no-man's-land where he can move unfettered by the pressures of academic or political conformity. When John Gardner founded Common Cause, he insisted on financing it only through small independent contributions so as to avoid the major influences that come with large donations. By creating new forms of association, these individuals hope to see new problems emerge, leading to solutions that couldn't be attempted through old ways of thinking.

But organizations are embedded in larger human groups and broader historical processes. An economic depression or a change in political priorities will stimulate one line of research and send another into oblivion. According to George Stigler, the Great Depression is what sent him and many of his colleagues to study economics in graduate school. The availability of nuclear reactors built to support World War II projects stimulated many bright students to major in physics. György Faludy spent many years in concentration camps for writing one poem critical of Joseph Stalin.

Wars are notorious for affecting the direction of science, and, indirectly, of the arts as well. Let's take psychology as an example. The domain of mental testing, including the whole concept of the IQ test and its uses, owes much of its success to the U.S. Army's need to have a way of selecting recruits for World War I. Afterward the testing technology was transported into the field of education, where it has achieved a prominence that many educators find disturbing. Creativity testing owes its existence to World War II, when the air force commissioned J. P. Guilford, a psychologist at the University of Southern California, to study the subject. The air force wanted to select pilots who in an emergency—the unexpected failure of a gear or instrument—would respond with appropriately original behavior, saving themselves and the plane. The usual IQ tests were not designed to tap originality, and hence Guilford was funded to develop what later became known as the tests for divergent thinking.

As mentioned earlier, World War II was especially beneficial for women scientists. Several said that they probably would not have been admitted to graduate school if so many men had not been drafted and the graduate departments had not been looking desperately for qualified students. After graduating, these same women found jobs in government-sponsored research labs involved with the war effort, or the later attempts to keep up scientific superiority fueled by the Cold War. Margaret Butler fondly recalls the early postwar years at Argonne, where she became involved with the birth and the infancy of computer science. Those were exciting times, when outside historical events, technological advances, and new scientific discoveries fused into a single stimulus to work hard and tackle important problems.

The influence of historical events on the arts is less direct but probably not less important. It could be argued, for instance, that the breakaway from classical literary, musical, and artistic styles that is so characteristic of the twentieth century was an indirect reaction to the disillusion people felt at the inability of Western civilization to avoid the bloodshed of World War I. It is no coincidence that Einstein's theory of relativity, Freud's theory of the unconscious, Eliot's free-form poetry, Stravinsky's twelve-tone music, Martha Graham's abstract choreography, Picasso's deformed figures, James Joyce's stream of consciousness prose were all created—and were accepted by the public—in the same period in which empires collapsed and belief systems rejected old certainties.

The Egyptian writer Naguib Mahfouz has spent many decades chronicling imaginatively the forces that are tearing apart the ancient fabric of his culture: colonialism, shifting of values, social mobility that creates new wealth and new poverty, and the changing roles of men and women. His ideas originate:

> in the process of living. We learn to get on with life even before we think of writing about it. There are particular events that sink deeper into our heart than others. My concerns were always political. Politics attracts me very much. Politics, interpersonal relationships, and love. The oppressed people in society. These were the sort of things that attracted me most.

For Nina Gruenenberg, associate editor and editorial columnist for the elite opinion-making weekly *Die Zeit,* unfolding world events provide a constant stream of problematic issues. Her challenge is to grasp the essential elements of the human conflicts involved, the sociocultural context in which the drama is played out, and then to report concisely her personal impression of the events. In the weeks prior to being interviewed, she had been in Texas covering the World Economic Summit, in London for the NATO summit, and in Russia for a meeting between German chancellor Helmut Kohl and Russian president Mikhail Gorbachev.

You know, I run a weekly newspaper, and normally I am very proud Wednesday mornings after the newspaper is out of the machinery, and it's ready and fresh, and I am satisfied with the piece I did. The last time I was very satisfied was after Chancellor Kohl went to the Caucasus and talked with President Gorbachev. This was on Monday, and we returned on Monday evening. I came back here to Hamburg on Tuesday morning, and by that evening the article had to be written. It was the end, it was the event of the week, and so I had to do an article which seemed to me and to all of my colleagues very important. But I was very tired and exhausted. And so I had really some difficulty in getting it done my way and in concentrating. And after that, the next morning, I was very happy!

The creative process starts with a sense that there is a puzzle somewhere, or a task to be accomplished. Perhaps something is not right, somewhere there is a conflict, a tension, a need to be satisfied. The problematic issue can be triggered by a personal experience, by a lack of fit in the symbolic system, by the stimulation of colleagues, or by public needs. In any case, without such a felt tension that attracts the psychic energy of the person, there is no need for a new response. Therefore, without a stimulus of this sort, the creative process is unlikely to start.

PRESENTED AND DISCOVERED PROBLEMS

Problems are not all alike in the way they come to a person's attention. Most problems are already formulated; everybody knows what is to be done and only the solution is missing. The person is expected by employers, patrons, or some other external pressure to apply his or her mind to the solution of a puzzle. These are "presented" problems. But there are also situations in which nobody has asked the question yet, nobody even knows that there *is* a problem. In this case the creative person identifies both the problem and the solution. Here we have a "discovered" problem. Einstein, among others, believed that the really important breakthroughs in science come as a result of reformulating old problems or discovering new ones, rather than by just solving existing problems. Or as Freeman Dyson said: "It is characteristic of scientific life that it is easy when you have a problem to work on. The hard part is finding your problem."

Frank Offner illustrates a presented problem-solving process:

> When I first was getting into aircraft, I had a best friend who introduced me to Hamilton Standard, who made propellers, now part of United Technology. He suggested that I go see them and see if I could help them, and the chief of the vibration group said to me, "Now, Frank, we have had this problem for months, we cannot figure how to get the maximum positive and the maximum negative value of the voltage and take the sum of them and figure out the total stress. We don't know how to choose a resistor. You have to have a capacitor that has to agree with the resistor, because if the resistor is too high it's too sluggish and if it's too low you lose one before you get the other." Well, before he was finished talking I knew the answer. I said, "Don't use a resistor, use a little relay and you short the capacitor . . . "

In contrast, Robert Galvin describes a problem that is discovered. His father had founded Motorola early in the century to make car radios. For several decades the business was a small one-room operation, with perhaps a dozen engineers and no large contracts, so Galvin's father worked very hard to make ends meet. In 1936 he felt that he finally could afford to take a vacation. He took his wife and young Robert on a European tour. As they traveled across Germany, the elder Galvin became convinced that sooner or later Hitler would start a war. Upon his return home, he followed up his hunch by sending Don Mitchell, one of his assistants, to Camp McCoy in Wisconsin to find out how the army passed on information among its various units.

Mitchell drove to Wisconsin, rang a bell at the gate of the camp, sat down with the major in charge, and in a short time found out that, as far as communications were concerned, the army hadn't changed at all since World War I: A phone wire was run from the front line to the back trenches. Upon being told this, Galvin's ears perked up. "Don," he is supposed to have said, "if we can make a radio that fits in a car and receives signals, can't we marry a little transmitter with it, and could we add some kind of power unit and put it into a box so someone could hold it, and he could talk from the front trench to the back trench with radios instead of stringing out the wire?" They figured it was a good idea and went to work. By the time Hitler invaded Poland, Motorola was ready to produce what became the SCR 536, the walkie-talkie of World War II. Robert Galvin uses this story to illustrate what he means by anticipation and commitment: on the one hand, having the foresight to realize how you could contribute to the future and thereby profit from it, and on the other, to have faith in your intuition and work hard to actualize it.

Presented problems usually take a much shorter time to prepare for and to solve than discovered problems. Sometimes the solution appears with the immediacy of Offner's example. Although it may require little time and effort, a novel solution to a presented problem could change the domain in significant ways and therefore be judged creative. Even in the arts, some of the most enduring paintings of the Middle Ages and the Renaissance were ordered by patrons who specified the size of the canvas, how many figures of what kind, the amount of expensive ground lapis lazuli pigment to be used, the weight of gold foil to be used in the frame, down to the smallest detail. Bach turned out a new cantata every few weeks to satisfy his patron's demands for religious hymns. Such cases show that, when approached with a desire to come up with the best solution, even the most rigidly predefined problems can result in creative outcomes.

Nevertheless, discovered problems have a chance to make a larger difference in the way we see the world. An example is Darwin's slow development of the theory of evolution. Darwin was commissioned to travel with the *Beagle* around the coast of South America and describe the largely unrecorded plant and animal life he encountered there. This was not an assignment that required a creative solution, and Darwin did what he was expected to do. But at the same time, he became more and more interested in and then puzzled by subtle differences in otherwise similar species living in what we now would call different ecological niches. He saw the connection between specific physical traits and corresponding environmental opportunities, such as the shape of a bird's beak and the kind of food available. These observations led to the concept of differential adaptation, which in turn, after many more detailed observations, led to the idea of natural selection and finally to the concept of the evolution of species.

The theory of evolution answered a great number of questions, ranging from why do animals look so different from each other to where do men and women come from. But perhaps the most remarkable feature of Darwin's accomplishment was that these questions had not been stated in an answerable form before, and he had to formulate the problem as well as propose a solution to it. Most great changes in a domain share this feature of Darwin's work: They tend to fall toward the discovered rather than the presented end on the continuum of problematic situations.

THE MYSTERIOUS TIME

After a creative person senses that on the horizon of his or her expertise there is something that does not fit, some problem that might be worth tackling, the process of creativity usually goes underground for a while. The evidence for incubation comes from reports of discoveries in which the creator becomes puzzled by an issue and remembers coming to a sudden insight into the nature of a problem, but does not remember any intermediate conscious mental steps. Because of this empty space in between sensing a problem and intuiting its solution, it has been assumed that an indispensable stage of incubation must take place in an interval of the conscious process.

Because of its mysterious quality, incubation has often been thought the most creative part of the entire process. The conscious sequences can be analyzed, to a certain extent, by the rules of logic and rationality. But what happens in the "dark" spaces defies ordinary analysis and evokes the original mystery shrouding the work of genius: One feels almost the need to turn to mysticism, to invoke the voice of the Muse as an explanation.

Our respondents unanimously agree that it is important to let problems simmer below the threshold of consciousness for a time. One of the most eloquent accounts of the importance of this stage comes again from the physicist Freeman Dyson. In describing his current work he has this to say:

> I am fooling around not doing anything, which probably means that this is a creative period, although of course you don't know until afterward. I think that it is very important to be idle. I mean, they always say that Shakespeare was idle between plays. I am not

comparing myself to Shakespeare, but people who keep themselves busy all of the time are generally not creative. So I am not ashamed of being idle.

Frank Offner is equally strong in his belief in the importance of not always thinking about one's problem:

I will tell you one thing that I found in both science and technology: If you have a problem, don't sit down and try to solve it. Because I will never solve it if I am just sitting down and thinking about it. It will hit me maybe in the middle of the night, while I am driving my car or taking a shower, or something like that.

How long a period of incubation is needed varies depending on the nature of the problem. It may range from a few hours to several weeks and even longer. Manfred Eigen says that he goes to sleep every night mulling some unresolved problem in his mind, some experimental procedure that does not work, some laboratory process that is not quite right. Miraculously, when he wakes up in the morning he has the solution clearly in mind. Hazel Henderson jogs or does gardening when she runs dry of ideas, and when she returns to the computer they usually flow freely again. Elisabeth Noelle-Neumann needs plenty of sleep, otherwise she feels that her thoughts become routine and predictable. Donald Campbell is very clear about the importance of letting ideas make connections with each other without external distractions:

One of the values in walking to work is mental meandering. Or if driving, not to have the car radio on. Now I don't think of myself as necessarily especially creative, but this creativity has to be a profoundly wasteful process. And that mental meandering, mind wandering and so on, is an essential process. If you are allowing that mentation to be driven by the radio or the television or other people's conversations, you are just cutting down on your exploratory, your intellectual exploratory time.

These short periods of incubation, usually having to do with a "presented" problem, tend to result in minuscule, perhaps imperceptible, changes in the domain. Examples of somewhat longer periods of incubation are the few weeks Freeman Dyson spent in California sight-seeing and not thinking consciously at all about how to reconcile Feynman and Schwinger's theories. In general, it seems that the more thorough the revolution brought about by the novelty, the longer it was working its way underground. But this hypothesis is difficult to verify. How long did Einstein's theory of relativity incubate? Or Darwin's theory of evolution? Or Beethoven's ideas for the Fifth Symphony? Because it is impossible to determine with precision when the first germs of these great works appeared in the minds of their authors, it is also impossible to know how long the process of incubation lasted.

The Functions of Idle Time

But what happens during this mysterious idle time, when the mind is not consciously preoccupied with the problem? There are several competing explanations of why incubation helps the creative process. Perhaps the best known is an offshoot of psychoanalytic theory. According to Freud, the curiosity at the roots of the creative process—especially in the arts—is triggered by a childhood experience of sexual origin, a memory so devastating that it had to be repressed. The creative person is one who succeeds in displacing the quest for the forbidden knowledge into a permissible curiosity. The artist's zeal in trying to find new forms of representation and the scientist's urge to strip away the veils of nature are really disguised attempts to understand the confusing impressions the child felt when witnessing his parents having sex, or the ambivalently erotic emotions toward one of the parents.

But if the secondary creative process is to drain effectively the repressed primary interest, it has to dip occasionally below the threshold of consciousness, where it can connect again with its original libidinal source. This is presumably what happens during the period of incubation. The content of the conscious line of thought is taken up by the subconscious, and there, out of reach of the censorship of awareness, the abstract scientific problem has a chance to reveal itself for what it is—an attempt to come to terms with a very personal conflict. Refreshed by having been able to commune with its true source, the subconscious thought can then reemerge in consciousness, its disguise back in place, and the scientist can continue his or her research with renewed vigor.

Many creative people use a watered-down version of this account to explain their own work and often drop hints as to the probable libidinal origin of their interests. It is difficult to know what to make of such intelligence. Often it turns out that the artists or scientists who are most convinced that in their works they are attempting to resolve a childhood trauma are those who have spent many years in therapy and have been well socialized into Freudian ideology. It could be that analysis helped them uncover the repressed sources of their curiosity. Or it could be that it helped them come up with an interesting explanation for what is mysterious about their experiences—an explanation, however, that may have little basis in reality.

In any case, although a psychoanalytic approach might explain some of the motivation for a person to engage in the process of discovery, it provides very little guidance as to why a vacation in California yielded Dyson the key to quantum electrodynamics. The transformation of libido in such a case is so spectacularly implausible as to lack credibility.

Cognitive accounts of what happens during incubation assume, like the psychoanalytic ones, that some kind of information processing keeps going on in the mind even when we are not aware of it, even when we are asleep. The difference is that cognitive theories do not posit any direction to subconscious thought. There is no trauma at the center of the unconscious, seeking resolution through disguised curiosity. Cognitive theorists believe that ideas, when deprived of conscious direction, follow simple laws of association. They combine more or less randomly, although seemingly irrelevant associations between ideas may occur as a result of a prior connection: For example, the German chemist August Kekulé had the insight that the benzene molecule might be shaped like a ring after he fell asleep while watching sparks in the fireplace make circles in the air. If he had stayed awake, Kekulé would have presumably rejected as ridiculous the thought that there might be a connection between the sparks and the shape of the molecule. But in the subconscious, rationality could not censor the connection, and so when he woke up he was no longer able to ignore its possibility. According to this perspective, truly irrelevant connections dissolve and disappear from memory, while the ones that are robust survive long enough to emerge eventually into consciousness.

The distinction between serial and parallel processing of information may also explain what happens during incubation. In a serial system like that of an old-fashioned calculator, a complex numerical problem must be solved in a sequence, one step at a time. In a parallel system such as in advanced computer software, a problem is broken up into its component steps, the partial computations are carried out simultaneously, and then these are reconstituted into a single final solution.

Something similar to parallel processing may be taking place when the elements of a problem are said to be incubating. When we think consciously about an issue, our previous training and the effort to arrive at a solution push our ideas in a linear direction, usually along predictable or familiar lines. But intentionality does not work in the subconscious. Free from rational direction, ideas can combine and pursue each other every which way. Because of this freedom, original connections that would be at first rejected by the rational mind have a chance to become established.

The Field, the Domain, and the Unconscious

At first sight, incubation seems to occur exclusively within the mind; what's more, within the mind's hidden recesses where consciousness is unable to reach. But after a closer look, we must admit that even in the unconscious the symbol system and the social environment play important roles. In the first place, it is obvious that incubation cannot work for a person who has not mastered a domain or been involved in a field. A new solution to quantum electrodynamics doesn't occur to a person unfamiliar with this branch of physics, no matter how long he or she sleeps.

Even though subconscious thinking may not follow rational lines, it still follows patterns that were established during conscious learning. We internalize the knowledge of the domain, the concerns of the field, and they become part of the way our minds are organized. It is often not necessary to perform an experiment to know that something won't work: Theoretical knowledge can predict the outcome. Similarly, we can predict what our colleagues will say if we express publicly certain ideas. When we sit alone in our study and say that an idea won't work, what we may be saying is that none of the people whose opinions matter will accept it. These internalized criteria of the domain and the field do not disappear when the thought process goes underground. They are probably less insistent than when we are aware of what we are doing, but they still shape and control how combinations of ideas are evaluated and selected.

But just as one must take the concerns of the discipline seriously, one must also be willing to take a stand against received wisdom, if the conditions warrant it. Otherwise no advance is possible. The all-important tension between trusting domain knowledge yet being ready to reject it is well illustrated by Frank Offner's description of what went on in his mind as he was trying to develop the first electronic controls that eventually made possible the commercial use of jet engines:

If you understand science and a question comes up and you want to do something, then you can work out a good solution very easily. If you don't have a good scientific background, you can't. If I had looked at what other people had done before, like in the jet engines, I would have been lost. Everybody attacked it exactly the wrong way. They thought the way that I did it was impossible. [Norbert] Weiner, the mathematician—I read his book on cybernetics—that said it was impossible. But I used rate acceleration feedback, and it worked.

What Offner points out here is that a creative solution often requires using knowledge from one part of the domain to correct the accepted beliefs of the field—which are based on different conclusions derived from other parts of the same domain. In this case, cybernetic theory seemed to exclude the possibility of controls that would keep the speed of the jet engine exactly constant. But before ever seeing a jet engine, by thinking about what the controls were supposed to accomplish and then going back to basic physics, Offner came up with a design that worked and was implemented.

Creative thoughts evolve in this gap filled with tension—holding on to what is known and accepted while tending toward a still ill-defined truth that is barely glimpsed on the other side of the chasm. Even when thoughts incubate below the threshold of consciousness, this tension is present.

THE "AHA!" EXPERIENCE

Most of the people in our sample—but not all—recall with great intensity and precision a particular moment when some major problem crystallized in their minds in such a way that a solution became all but inevitable, requiring only a matter of time and hard work. For presented problems, the insight might even include the particulars of the solution. Here are two examples from Frank Offner:

> It will hit me maybe in the middle of the night. It turns around somehow inside your brain. I can tell you where I was when I got the answer how to stabilize the jet control with a feedback. I was sitting on a sofa, I guess this was before I was married, at some friend's house and a little bit bored and the answer hit me, "Ah!" and I put in the derivative term.
>
> And another one. I was going to do my Ph.D. thesis on nerve excitation. There were two sets of equations describing nerve excitation. I was going to make some experiments to see which was the right one, one made at the University of Chicago, the other in England, and I was going to see which was the more accurate. And I tried to work out the mathematics to see what kind of experiment would [be decisive]. I remember I was taking a shower when I saw how to solve that problem. I sat down to solve that problem and I found that the equations were just two ways of saying the same thing. So I had to do something else [for the thesis].

The insight presumably occurs when a subconscious connection between ideas fits so well that it is forced to pop out into awareness, like a cork held underwater breaking out into the air after it is released.

THE 99 PERCENT PERSPIRATION

After an insight occurs, one must check it out to see if the connections genuinely make sense. The painter steps back from the canvas to see whether the composition works, the poet rereads the verse with a more critical eye, the scientist sits down to do the calculations or run the experiments. Most lovely insights never go any farther, because under the cold light of reason fatal flaws appear. But if everything checks out, the slow and often routine work of elaboration begins.

There are four main conditions that are important during this stage of the process. First of all, the person must pay attention to the developing work, to notice when new ideas, new problems, and new insights arise out of the interaction with the medium. Keeping the mind open and flexible is an important aspect of the way creative persons carry on their work. Next, one must pay attention to one's goals and feelings, to know whether the work is indeed proceeding as intended. The third condition is to keep in touch with domain knowledge, to use the most effective techniques, the fullest information, and the best theories as one proceeds. And finally, especially in the later stages of the process, it is important to listen to colleagues in the field. By interacting with others involved with similar problems, it is possible to correct a line of solution that is going in the wrong direction, to refine and focus one's ideas, and to find the most convincing mode of presenting them, the one that has the best chance of being accepted.

The historian Natalie Davis describes how she feels during the last stage of the creative process, when all that is left is the writing up of the results of her research:

> If I didn't have affect in a project, if I had lost it or maybe it didn't last too long, it would lose its spark. I mean, I don't want to do something that I have lost my love for. I think that everybody is perhaps that way, but I am very much that way. It is hard to be creative if you are just doing something doggedly. If I didn't have curiosity, if I felt that my curiosity was limited, then the novelty part of it would be gone. Because it is the curiosity that has often pushed me to think of ways of finding out about something that people thought you could never find out about. Or ways of looking at a subject that have never been looked at before. That's what keeps me running back and forth to the library, and just thinking, and thinking, and thinking.

Barry Commoner describes the last phases of his work, when he has to write things down, or communicate them to an audience:

> Some of the work is extremely hard from the point of view of creating a clear statement. For example, in one of my books I wrote a chapter on thermodynamics designed for the lay public. That probably went through fifteen drafts. It was the most difficult writing I ever had to do, because it's a very difficult subject to put

into ordinary lay terms. And that's one of the things I've done I'm most proud of. I've had engineers tell me that for the first time they had a clear picture of thermodynamics from it. So I enjoy that a great deal. I enjoy communicating. Same with speaking. I do a lot of speaking. And I really enjoy seeing the audience paying attention—listening, understanding it.

One thing about creative work is that it's never done. In different words, every person we interviewed said that it was equally true that they had worked every minute of their careers, and that they had never worked a day in all their lives. They experienced even the most focused immersion in extremely difficult tasks as a lark, an exhilarating and playful adventure.

It is easy to resent this attitude and see the inner freedom of the creative person as an elite privilege. While the rest of us are struggling at boring jobs, they have the luxury of doing what they love to do, not knowing whether it is work or play. There might be an element of truth in this. But far more important, in my opinion, is the message that the creative person is sending us: You, too, can spend your life doing what you love to do. After all, most of the people we interviewed were not born with a silver spoon in their mouth; many came from humble origins and struggled to create a career that allowed them to keep exploring their interests. Even if we don't have the good fortune to discover a new chemical element or write a great story, the love of the creative process for its own sake is available to all. It is difficult to imagine a richer life.

FIVE

THE FLOW OF CREATIVITY

Creative persons differ from one another in a variety of ways, but in one respect they are unanimous: They all love what they do. It is not the hope of achieving fame or making money that drives them; rather, it is the opportunity to do the work that they enjoy doing. Jacob Rabinow explains: "You invent for the hell of it. I don't start with the idea, 'What will make money?' This is a rough world, money's important. But if I have to trade between what's fun for me and what's money-making, I'll take what's fun." The novelist Naguib Mahfouz concurs in more genteel tones: "I love my work more than I love what it produces. I am dedicated to the work regardless of its consequences." We found the same sentiments in every single interview.

What is extraordinary in this case is that we talked to engineers and chemists, writers and musicians, businesspersons and social reformers, historians and architects, sociologists and physicians—and they all agree that they do what they do primarily because it's fun. Yet many others in the same occupations don't enjoy what they do. So we have to assume that it is not *what* these people do that counts but *how* they do it. Being an engineer or a carpenter is not in itself enjoyable. But if one does these things a certain way, then they become intrinsically rewarding, worth doing for their own sake. What is the secret of transforming activities so that they are rewarding in and of themselves?

PROGRAMMED FOR CREATIVITY

When people are asked to choose from a list the best description of how they feel when doing whatever they enjoy doing most—reading, climbing mountains, playing chess, whatever—the answer most frequently chosen is "designing or discovering something new." At first, it seems strange that dancers, rock climbers, and composers all agree that their most enjoyable experiences resemble a process of discovery. But when we think about it some more, it seems perfectly reasonable that at least some people should enjoy discovering and creating above all else.

To see the logic of this, try a simple thought experiment. Suppose that you want to build an organism, an artificial life form, that will have the best chance of surviving in a complex and unpredictable environment, such as that on Earth. You want to build into this organism some mechanism that will prepare it to confront as many of the sudden dangers and to take advantage of as many of the opportunities that arise as possible. How would you go about doing this? Certainly you would want to design an organism that is basically conservative, one that learns the best solutions from the past and keeps repeating them, trying to save energy, to be cautious and go with the tried-and-true patterns of behavior.

But the best solution would also include a relay system in a few organisms that would give a positive reinforcement every time they discovered something new or came up with a novel idea or behavior, whether or not it was immediately useful. It is especially important to make sure that the organism was not rewarded only for useful discoveries, otherwise it would be severely handicapped in meeting the future. For no earthly builder could anticipate the kind of situations the species of new organisms might encounter tomorrow, next year, or in the next decade. So the best program is one that makes the organism feel good whenever something new is discovered, regardless of its present usefulness. And this is what seems to have happened with our race through evolution.

By random mutations, some individuals must have developed a nervous system in which the discovery of novelty stimulates the pleasure centers in the brain. Just as some individuals derive a keener pleasure from sex and others from food, so some must have been born who derived a keener pleasure from learning something new. It is possible that children who were more curious ran more risks and so were more likely to die early than their more stolid companions. But it is also probable that those human groups that learned to appreciate the curious children among them, and helped to protect and reward them so that they could grow to maturity and have children of their own, were more successful than groups that ignored the potentially creative in their midst.

If this is true, we are the descendants of ancestors who recognized the importance of novelty, protected those individuals who enjoyed being creative, and learned from them. Because they had among them individuals who enjoyed exploring and inventing, they were better prepared to face the unpredictable conditions that threatened their survival. So we too share this propensity for enjoying whatever we do, provided we can do it in a new way, provided we can discover or design something new in doing it. This is why creativity, no matter in what domain it takes place, is so enjoyable. This is why Brenda Milner, among many others, said: "I would say that I am impartial about what is important or great, because every new little discovery, even a tiny one, is exciting at the moment of discovery."

But this is only part of the story. Another force motivates us, and it is more primitive and more powerful than the urge to create: the force of entropy. This too is a survival mechanism built into our genes by evolution. It gives us pleasure when we are comfortable, when we relax, when we can get away with feeling good without expending energy. If we didn't have this built-in regulator, we could easily kill ourselves by running ragged and then not having enough reserves of strength, body fat, or nervous energy to face the unexpected.

This is the reason why the urge to relax, to curl up comfortably on the sofa whenever we can get away with it, is so strong. Because this conservative urge is so powerful, for most people "free time" means a chance to wind down, to park the mind in neutral. When there are no external demands, entropy kicks in, and unless we understand what is happening, it takes over our body and our mind.

We are generally torn between two opposite sets of instructions programmed into the brain: the least-effort imperative on one side, and the claims of creativity on the other.

In most individuals entropy seems to be stronger, and they enjoy comfort more than the challenge of discovery. A few, like the ones who tell their stories in this book, are more responsive to the rewards of discovery. But we all respond to both of these rewards; the tendencies toward conserving energy as well as using it constructively are simultaneously part of our inheritance. Which one wins depends not only on our genetic makeup but also presumably on our early experiences. However, unless enough people are motivated by the enjoyment that comes from confronting challenges, by discovering new ways of being and doing, there is no evolution of culture, no progress in thought or feeling. It is important, therefore, to understand better what enjoyment consists of and how creativity can produce it.

WHAT IS ENJOYMENT?

In order to answer that question, many years ago I started to study people who seemed to be doing things that they enjoyed but were not rewarded for with money or fame. Chess players, rock climbers, dancers, and composers devoted many hours a week to their avocations. Why were they doing it? It was clear from talking to them that what kept them motivated was the quality of experince they felt when they were involved with the activity. This feeling didn't come when they were relaxing, when they were taking drugs or alcohol, or when they were consuming the expensive privileges of wealth. Rather, it often involved painful, risky, difficult activities that stretched the person's capacity and involved an element of novelty and discovery. This optimal experience is what I have called *flow,* because many of the respondents described the feeling when things were going well as an almost automatic, effortless, yet highly focused state of consciousness.

The flow experience was described in almost identical terms regardless of the activity that produced it. Athletes, artists, religious mystics, scientists, and ordinary working people described their most rewarding experiences with very similar words. And the description did not vary much by culture, gender, or age; old and young, rich and poor, men and women, Americans and Japanese seem to experience enjoyment in the same way, even though they may be doing very different things to attain it. Nine main elements were mentioned over and over again to describe how it feels when an experience is enjoyable.

1. *There are clear goals every step of the way.* In contrast to what happens in everyday life, on the job or at home, where often there are contradictory demands and our purpose is unsure, in flow we always know what needs to be done. The musician knows what notes to play next, the rock climber knows the next moves to make. When a job is enjoyable, it also has clear goals: The surgeon is aware how the incision should proceed moment by moment; the farmer has a plan for how to carry out the planting.

2. *There is immediate feedback to one's actions.* Again, in contrast to the usual state of affairs, in a flow experience we know how well we are doing. The musician hears right away whether the note played is the one. The rock climber finds out immediately whether the move was correct because he or she is still hanging in there and hasn't fallen to the bottom of the valley. The surgeon sees there is no blood in the cavity, and the farmer sees the furrows lining up neatly in the field.

3. *There is a balance between challenges and skills.* In flow, we feel that our abilities are well matched to the opportunities for action. In everyday life we sometimes feel that the challenges are too high in relation to our skills, and then we feel frustrated and anxious. Or we feel that our potential is greater than the opportunities to express it, and then we feel bored. Playing tennis or chess against a much better opponent leads to frustration; against a much weaker opponent, to boredom. In a really enjoyable game, the players are balanced on the fine line between boredom and anxiety. The same is true when work, or a conversation, or a relationship is going well.

4. *Action and awareness are merged.* It is typical of everyday experience that our minds are disjointed from what we do. Sitting in class, students may appear to be paying attention to the teacher, but

they are actually thinking about lunch, or last night's date. The worker thinks about the weekend; the mother cleaning house is worried about her child; the golfer's mind is preoccupied with how his swing looks to his friends. In flow, however, our concentration is focused on what we do. One-pointedness of mind is required by the close match between challenges and skills, and it is made possible by the clarity of goals and the constant availability of feedback.

5. *Distractions are excluded from consciousness.* Another typical element of flow is that we are aware only of what is relevant here and now. If the musician thinks of his health or tax problems when playing, he is likely to hit a wrong note. If the surgeon's mind wanders during an operation, the patient's life is in danger. Flow is the result of intense concentration on the present, which relieves us of the usual fears that cause depression and anxiety in everyday life.

6. *There is no worry of failure.* While in flow, we are too involved to be concerned with failure. Some people describe it as a feeling of total control; but actually we are not in control, it's just that the issue does not even come up. If it did, we would not be concentrating totally, because our attention would be split between what we did and the feeling of control. The reason that failure is not an issue is that in flow it is clear what has to be done, and our skills are potentially adequate to the challenges.

7. *Self-consciousness disappears.* In everyday life, we are always monitoring how we appear to other people; we are on the alert to defend ourselves from potential slights and anxious to make a favorable impression. Typically this awareness of self is a burden. In flow we are too involved in what we are doing to care about protecting the ego. Yet after an episode of flow is over, we generally emerge from it with a stronger self-concept; we know that we have succeeded in meeting a difficult challenge. We might even feel that we have stepped out of the boundaries of the ego and have become part, at least temporarily, of a larger entity. The musician feels at one with the harmony of the cosmos, the athlete moves at

one with the team, the reader of a novel lives for a few hours in a different reality. Paradoxically, the self expands through acts of self-forgetfulness.

8. *The sense of time becomes distorted.* Generally in flow we forget time, and hours may pass by in what seem like a few minutes. Or the opposite happens: A figure skater may report that a quick turn that in real time takes only a second seems to stretch out for ten times as long. In other words, clock time no longer marks equal lengths of experienced time; our sense of how much time passes depends on what we are doing.

9. *The activity becomes autotelic.* Whenever most of these conditions are present, we begin to enjoy whatever it is that produces such an experience. I may be scared of using a computer and learn to do it only because my job depends on it. But as my skills increase, and I recognize what the computer allows me to do, I may begin to enjoy using the computer for its own sake as well. At this point the activity becomes *autotelic*, which is Greek for something that is an end in itself. Some activities such as art, music, and sports are usually autotelic: There is no reason for doing them except to feel the experience they provide. Most things in life are *exotelic*: We do them not because we enjoy them but in order to get at some later goal. And some activities are both: The violinist gets paid for playing, and the surgeon gets status and good money for operating, as well as getting enjoyment from doing what they do. In many ways, the secret to a happy life is to learn to get flow from as many of the things we have to do as possible. If work and family life become autotelic, then there is nothing wasted in life, and everything we do is worth doing for its own sake.

THE CONDITIONS FOR FLOW IN CREATIVITY

Creativity involves the production of novelty. The process of discovery involved in creating something new appears to be one of the most enjoyable activities any human can be involved in. In fact, it is easy to recognize the conditions of flow in the accounts of our respondents, as they describe how it feels to do the sort of things they do.

The Clarity of Goals

In certain conditions, the creative process begins with the goal of solving a problem that is given to the person by someone else or is suggested by the state of the art in the domain. Moreover, anything that does not work as well as it could can provide a clear goal to the inventor. This is what Frank Offner describes:

> Oh, I love to solve problems. If it is why our dishwasher does not work, or why the automobile does not work, or how the nerve works, or anything. Now I am working on how the hair cells work, and ah . . . it is so very interesting. I don't care what kind of problem it is. If I can solve it, it is fun. It is really a lot of fun to solve problems, isn't it? Isn't that what is interesting in life? Especially if people say one thing and you show that they have been wrong for twenty years and you can solve it in five minutes.

Or the goal may emerge as a problem in the domain—a gap in the network of knowledge, a contradiction among the findings, a puzzling result. Here the goal is to restore harmony in the system by reconciling the apparent disparities. The physicist Viktor' Weisskopf describes the enjoyment involved in this process:

> Well, in science, obviously, if I understand something, you know, a new discovery, it need not be my own, a discovery of somebody else, where I say, "Aha, now I understand natural processes that I did not understand before," that is the joy of insight.
>
> In music it is the insight into what the piece means. What it tells you, what the composer wanted to tell you, the beauty or expression or religious feelings, things like that.

For artists the goal of the activity is not so easily found. In fact, the more creative the problem, the less clear it is what needs to be done. Discovered problems, the ones that generate the greatest changes in the domain, are also the most difficult to enjoy working on because of their elusiveness. In such cases, the creative person somehow must develop an unconscious mechanism that tells him or her what to do. The poet György Faludy usually does not start writing until a "voice" tells him, often in the middle of the night, "György, it's time to start writing." He adds ruefully: "That voice has my number, but I don't have his." The ancients called that voice the Muse. Or it can be a vision, as it is for Robertson Davies:

> You are always writing, and you're always fantasizing. What I find very much in my own work, though I don't know if it applies to the work of other people, is that an idea for a novel seizes me and will not let me go until I have given it careful consideration. And that is not to say that a complete story appears in my head, but very often what appears is a picture which seems somehow significant and which must be considered. Now, a great many years ago, I found that whenever I stopped thinking about something in particular, a picture kept coming up in my head. It was a picture of a street, and I knew what street it was; it was the street on which I was born in a small Ontario village. And there were two boys playing in the snow, and one threw a snowball at the other.

Readers of Davies's oeuvre will recognize in this picture the opening scene of *Fifth Business,* the first volume of his famous Deptford trilogy. In many ways, the writing of the book consisted in finding out what that image, charged with emotion and nostalgia, portended. The goal was to find out what were the consequences of throwing that snowball. Probably if Davies had told himself rationally that this is what the book would be about he would have thought it a trivial goal, not worth all the time and effort. But fortunately the goal presented itself as a vision, a mysterious call that he felt impelled to follow. Very often this is how the Muse communicates—through a glass darkly, as it were. It is a splendid arrangement, for if the artist were not tricked by the mystery, he or she might never venture into the unexplored territory.

Knowing How Well One Is Doing

Games are designed so that we can keep score and know how well we are doing. Most jobs give some sort of information about performance: The salesman can add up daily sales, the assembly worker can count pieces produced. If all else fails, the boss may tell you how well you are doing. But the artist, the scientist, and the inventor are moving on very different timelines. How do they know, day in and day out, whether they are wasting their time or actually accomplishing something?

This is indeed a difficult problem. Many artists give up because it is just too excruciating to wait until critics or galleries take notice and pass judgment on their canvases. Research scientists drift away from pure science because they cannot tolerate the long cycles of insecurity before reviewers and editors evaluate their results. So how can they experience flow without external information about their performance?

The solution seems to be that those individuals who keep doing creative work are those who succeed in internalizing the field's criteria of judgment to the extent that they can give feedback to themselves, without having to wait to hear from experts. The poet who keeps enjoying writing verse is the one who knows how good each line is, how appropriate is each word chosen. The scientist who enjoys her work is the one who has a sense of what a good experiment is like and who appreciates it when a test is well run or when a report is clearly written. Then she need not wait until October to see if her name is on the Nobel Prize list.

Many creative scientists say that the difference between them and their less creative peers is the ability to separate bad ideas from good ones, so that they don't waste much time exploring blind alleys. Everyone has both bad and good ideas all the time, they say. But some people can't tell them apart until it's too late, until they have already invested a great deal of time in the unprofitable hunches. This is another form of the ability to give oneself feedback: to know in advance what is feasible and what will work, without having to suffer the consequences of bad judgment. At Linus Pauling's sixtieth birthday celebration, a student asked him, "Dr. Pauling, how does one go about having good ideas?" He replied, "You have a lot of ideas and throw away the bad ones." To do that, of course, one has to have a very well internalized picture of what the domain is like and what constitutes "good" and "bad" ideas according to the field.

Balancing Challenges and Skills

The pursuit of a creative problem is rarely easy. In fact, in order to be enjoyable it should be hard, and of course so it is, almost by definition. It is never easy to break new ground, to venture into the unknown. When one starts out, the difficulties may seem almost overwhelming. Here is how Freeman Dyson describes this aspect of the process:

Well, I think that you have to describe it as sort of a struggle. I have to always force myself to write, and also to work harder at a science problem. You have to put blood, sweat, and tears into it first. And it is awfully hard to get started. I think most writers have this problem. I mean, it's part of the business. You may work very hard for a week producing the first page. That's really blood, tears, and sweat, and there is nothing else to describe it. You have to force yourself to push and push and push with the hope that something good will come out. And you have to go through that process before it really starts to flow easily, and without that preliminary forcing and pushing probably nothing would ever happen. So, I think that is what distinguishes it from just having a good time— you have a good time once you are really in the flowing phase, but you have to overcome some sort of barrier to get there. That is why I say it is unconscious, because you don't know actually whether you are really getting anywhere or not. In that phase it just seems to be unadulterated torture.

The creative person is not immune to the conflict between the two programs we all carry in our genetic inheritance. As Dyson knows, even the most creative persons must overcome the barrier of entropy. It is impossible to accomplish something that is truly new and worthwhile without struggling with it. It isn't just in competitive sports that the saying "no pain, no gain" applies. The less well defined the problem, the more ambitious it is, and the harder it is for the creative person to get a handle on it. Barry Commoner points out:

I enjoy doing things that other people won't do. Because what are they? They're usually things that are difficult and important— and that people shy away from. I have a general approach to thinking of the way in which issues develop. I'm interested in the origins of problems. And so I have a pretty good idea of where things are going, and what's important and what isn't important. And I try very hard to be at the cutting edge of problems. Very often that puts me so far out in front that people are upset about it, but that's OK.

To be able to cope with such problems, the creative person has to have a great many personality traits that are conducive to discovery and hard work, including the ability to internalize the rules of the domain and the judgments of the field. Commoner also gives a hint of another skill that creative individuals develop: a personal approach, an internal model that allows them to put the problem into a manageable context. The same idea is expressed by Linus Pauling:

I think one thing that I do is to bring ideas from one field of knowledge into another field of knowledge. And, I've often said I don't think that I'm smarter than a lot of other scientists, but perhaps I think more about the problems. I have a picture, a sort of general theory of the universe in my mind that I've built up over the decades. If I read an article, or hear someone give a seminar talk, or in some other way get some piece of information about science that I hadn't had before, I ask myself, "How does that fit into my picture of the universe?" and if it doesn't fit, I ask, "Why doesn't it fit in?"

The strategies creative individuals develop are not always successful. They take risks, and what is risk without an occasional failure? When the challenges become too great for the person to cope with, a sense of frustration rather than joy creeps in—at least for a while. Our interview with John Reed took place a few years after Citicorp was bloodied in the market; its shares lost a great deal of their value almost overnight. Reed blamed himself for not foreseeing the contingency that caused the loss. As a result, at the time he felt that some of the fun had gone out of his job. What used to be spontaneous turned into hard work; he had to force himself to be more of an accountant than a builder and leader; and the new skill he had to acquire required unfamiliar discipline.

The Merging of Action and Awareness
But when the challenges are just right, the creative process begins to hum, and all other concerns are temporarily shelved in the deep involvement with the activity. Here is Dyson again, describing how it feels after the initial struggle is over:

I always find that when I am writing, it is really the fingers that are doing it and not the brain. Somehow the writing takes charge. And the same thing happens of course with equations. You don't really think of what you are going to write. You just scribble, the equations lead the way, and what you are doing is sort of architectural. You have to have a design in view, in which you design a chapter, or a proof of a theorem, as the case may be. Then you have to put it together out of words or out of symbols as the case may be, but if you don't have a clear architecture in mind then the thing won't end up being any good. The trick is to start from both ends and to meet in the middle, which is essentially like building a bridge. That seems to me the way that I think, anyhow. So the original design is somehow accidental and you don't know how it comes into your head. It just sort of happens, maybe when you are shaving or taking a walk, then you sit down and actually work through and that is when the hard work is done. And that is very largely a matter of putting pieces together, finding out what works and what doesn't.

Barry Commoner uses similar terms to describe the almost automatic quality of the flow experience when writing, expressing the feeling of merging action and awareness through the image of the flowing ink and the flowing of ideas:

I write with this pen [he removes a fountain pen from his breast pocket and holds it up]. And it's very clear to me that my ability to think and write at the same time depends on the flow of ink. The thing I enjoy most is the flow of my own ideas and getting them down on paper. I will not write with a ballpoint pen, because it doesn't really flow. That's why I use a fountain pen. And only a fountain pen that really works very well.

The novelist Richard Stern gives a classic description of how it feels to become lost in the process of writing and to feel the rightness of one's actions in terms of what is happening in that special world of one's own creation:

At your best you're not thinking, How am I making my way ahead in the world by doing this? No. You're concentrated on your characters, on the situation, on the form of the book, on the words which are coming out. And their shape. You've lost . . . you're not an ego at that point. It's not competitive. It's . . . I would use the word *pure*. You know that this is right. I don't mean that it works in the world, or that it adds up, but that it's right in this place. In this story. It belongs to it. It's right for that person, that character.

Avoiding Distractions

Many of the peculiarities attributed to creative persons are really just ways to protect the focus of concentration so that they may lose themselves in the creative process. Distractions interrupt flow, and it may take hours to recover the peace of mind one needs to get on with the work. The more ambitious the task, the longer it takes to lose oneself in it, and the easier it is to get distracted. A scientist working on an arcane problem must detach himself from the "normal" world and roam with his mind in a world of disembodied symbols that now you see, now you don't. Any intrusion from the solid world of everyday reality can make that world disappear in an instant. It is for this reason that Freeman Dyson "hides" in the library when he's writing and why Marcel Proust used to seclude himself in a windowless room lined with cork when he sat down to write *À la recherche du temps perdu*. Even the slightest noise could break the thread of his teetering imagination.

More serious health, family, or financial problems could occupy the mind of a person so insistently that he or she is no longer able to devote enough attention to work. Then a long period of drought may follow, a writer's block, a burnout, which may even end a creative career. It is this kind of distraction that Jacob Rabinow talks about:

> Freedom from worry is one thing—that you don't have any problem of health or sickness in the family or something that occupies your mind. Or financial worries, that you're going crazy about how you're going to pay the next bill. Or children's worries, or drugs or something. No, it's nice to be free of responsibility. That doesn't mean you have no responsibility to the project, but to be free of other things. And you're not likely to be an inventor if you're very sick. You're too busy with your problems, too many pains.

Many of our respondents were thankful to their spouses for providing a buffer from exactly these kinds of distractions. This was

especially true of the men; the women sometimes mentioned point-
edly that they also would have liked to have had a wife to spare them
from worries that interfered with their concentration on work.

Forgetting Self, Time, and Surroundings

When distractions are out of the way and the other conditions for
flow are in place, the creative process acquires all the dimensions of
flow. Here it is described by the poet Mark Strand:

> Well, you're right in the work, you lose your sense of time,
> you're completely enraptured, you're completely caught up in
> what you're doing, and you're sort of swayed by the possibilities
> you see in this work. If that becomes too powerful, then you get
> up, because the excitement is too great. You can't continue to
> work or continue to see the end of the work because you're
> jumping ahead of yourself all the time. The idea is to be so . . . so
> *saturated* with it that there's no future or past, it's just an extended
> present in which you're, uh, making meaning. And dismantling
> meaning, and remaking it. Without undue regard for the words
> you're using. It's meaning carried to a high order. It's not just
> essential communication, daily communication; it's a *total* com-
> munication. When you're working on something and you're
> working well, you have the feeling that there's no other way of
> saying what you're saying.

He captures precisely the sense of flowing along this extended present and the powerful sense of doing exactly the right thing the only
way it could be done. It may not happen often, but when it does the
beauty of it justifies all the hard work.

Creativity as Autotelic Experience

This then brings us back to where we started this chapter and the
observation that all of the respondents placed the joy of working
ahead of any extrinsic rewards they may receive from it. Like most of
the others, the psychologist Donald Campbell gives unambiguous
advice to young people entering the field:

> I would say: "Don't go into science if you are interested in
> money. Don't go into science if you will not enjoy it even if you

do not become famous. Let fame be something that you accept graciously if you get it, but make sure that it is a career that you can enjoy. That requires intrinsic motivation. And try to pick a setting in which you can work on the problems that intrinsically motivate you even if they are not exciting to others. Try to have the situational setting so that you can enjoy that work intrinsically, even if you are out of step with the time."

Scientists often describe the autotelic aspects of their work as the exhilaration that comes from the pursuit of truth and of beauty. What they seem to describe, however, is the joy of discovery, of solving a problem, of being able to express an observed relationship in a simple and elegant form. So what is rewarding is not a mysterious and ineffable external goal but the activity of science itself. It is the *pursuit* that counts, not the attainment. Of course this distinction is to a certain extent misleading, because without occasional successes the scientist might become discouraged. But what makes science intrinsically rewarding is the everyday practice, not the rare success. This is how Subrahmanyan Chandrasekhar, the Nobel laureate physicist, describes his own motivation:

There are two things about me which people generally don't know. I've never worked in anything which is glamorous in any sense. That's point number one. Point number two: I have always worked in areas which, during the time I have worked on them, did not attract attention.

The word *success* is an ambiguous word. Success with respect to the outside? Or success with respect to oneself? And if it is a success with respect to the outside, then how do you evaluate it? Very often outside success is irrelevant, wrong, and misplaced. So how can one talk about it? Externally, you may think I am successful because people write about some aspects of my work. But that is an external judgment. And I have no idea as to how to value that judgment.

Success is not one of my motives. Because success stands in contrast to failure. But no worthwhile effort in one's life is either a success or a failure. What do you mean by success? You take a problem and you want to solve it. Well, if you solve it, in a limited sense it is a success. But it may be a trivial problem. So a judgment about success is not something about which I've ever been serious about in any sense whatever.

Certainly all of these people seem to have heeded their own advice. None pursued money and fame. Some became comfortably wealthy from their inventions or their books, but none of them felt fortunate because of it. What they felt fortunate about was that they could get paid for something they had such fun doing and that in the bargain they could feel that what they did might help the human condition along. It is indeed lucky to be able to justify one's life activity with words such as those of C. Vann Woodward, who explains why he writes history:

> It interests me. It is a source of satisfaction. Achieving something that one thinks is important. Without such a consciousness or motivation it seems to me that life could be rather dull and purposeless, and I wouldn't want to attempt that kind of life. Of complete leisure, say, of having absolutely nothing to do that one felt was worth doing—that strikes me as a rather desperate situation to be in.

FLOW AND HAPPINESS

What is the relation between flow and happiness? This is a very interesting and delicate question. At first, it is easy to conclude that the two must be the same thing. But actually the connection is a bit more complex. First of all, when we are in flow, we do not usually feel happy—for the simple reason that in flow we feel only what is relevant to the activity. Happiness is a distraction. The poet in the middle of writing or the scientist working out equations does not feel happy, at least not without losing the thread of his or her thought.

It is only after we get out of flow, at the end of a session or in moments of distraction within it, that we might indulge in feeling happy. And then there is the rush of well-being, of satisfaction that comes when the poem is completed or the theorem is proved. In the long run, the more flow we experience in daily life, the more likely we are to feel happy overall. But this also depends on what activity provides flow. Unfortunately, many people find the only challenges they can respond to are violence, gambling, random sex, or drugs. Some of these experiences can be enjoyable, but these episodes of flow do not add up to a sense of satisfaction and happiness over time. Pleasure does not lead to creativity, but soon turns into addiction— the thrall of entropy.

So the link between flow and happiness depends on whether the flow-producing activity is complex, whether it leads to new challenges and hence to personal as well as cultural growth. Thus we might conclude that all our respondents must be happy, because they do enjoy their work, and their work is certainly complex. But there are further complications to consider. For instance, what if a person enjoyed being a physicist for thirty years, and then found out that his work resulted in a nuclear device that killed millions of people? How would Jonas Salk have felt if his vaccine, instead of saving lives, had been used by others for biological warfare? Certainly these are not idle questions in today's world, and they suggest that it is possible for complex activities that produce flow to cause long-range unhappiness. Yet when all is said and done, it is much easier to be happy when one's life has been enjoyable.

FLOW AND THE EVOLUTION OF CONSCIOUSNESS

There are many things that people enjoy: the pleasures of the body, power and fame, material possessions. Some enjoy collecting different beer bottles, and a few even enjoy causing pain to themselves or to others. Strangely enough, even though the means to obtain it are widely different, the resulting feeling of well-being is very much the same. Does that mean that all forms of enjoyment are equally worth pursuing?

Twenty-five centuries ago, Plato wrote that the most important task for a society was to teach the young to find pleasure in the right objects. Now Plato was conservative even for his times, so he had rather definite ideas about what those "right things" were that young people should learn to enjoy. We are much too sophisticated in this day and age to have strong feelings in the matter. Yet we probably agree that we would feel better if our children learned to enjoy cooperation rather than violence; reading rather than stealing; chess rather than dice; hiking rather than watching television. In other words, no matter how relativistic and tolerant we have become, we still have priorities. And we do want the next generation to share those priorities. Finally, many of us suspect that the next generation will not preserve what we value unless they now enjoy it to some extent.

The problem is that it is easier to find pleasure in things that *are* easier, in activities like sex and violence that are already programmed into our genes. Hunting, fishing, eating, and mating have privileged places in our nervous system. It is also easy to enjoy making money, or discovering new lands, or conquering new territories, or building elaborate palaces, temples, or tombs because these projects are in synchrony with survival strategies established long ago in our physiological makeup. It is much more difficult to learn to enjoy doing things that were discovered recently in our evolution, like manipulating symbolic systems by doing math or science or writing poetry or music, and learning from doing these things about the world and about ourselves.

Children grow up believing that football players and rock singers must be happy and envy the stars of the entertainment world for what they think must be fabulous, fulfilling lives. Asked what they would like to do when they grow up, most of them would choose to be athletes and entertainers. They don't realize until much later, if at all, that the glamour of those lives is vulgar tinsel, that to be like them leads anywhere but to happiness.

Neither parents nor schools are very effective at teaching the young to find pleasure in the right things. Adults, themselves often deluded by infatuation with fatuous models, conspire in the deception. They make serious tasks seem dull and hard, and frivolous ones exciting and easy. Schools generally fail to teach how exciting, how mesmerizingly beautiful science or mathematics can be; they teach the routine of literature or history rather than the adventure.

It is in this sense that creative individuals live exemplary lives. They show how joyful and interesting complex symbolic activity is. They have struggled through marshes of ignorance, deserts of disinterest, and with the help of parents and a few visionary teachers they have found themselves on the other side of the known. They have become pioneers of culture, models for what men and women of the future will be—if there is to be a future at all. It is by following their example that human consciousness will grow beyond the limitations of the past, the programs that genes and cultures have wired into our brains. Perhaps our children, or their children, will feel more joy in writing poetry and solving theorems than in being passively entertained. The lives of these creative individuals reassure us that it is not impossible.

SIX

CREATIVE SURROUNDINGS

Even the most abstract mind is affected by the surroundings of the body. No one is immune to the impressions that impinge on the senses from the outside. Creative individuals may seem to disregard their environment and work happily in even the most dismal surroundings: Michelangelo contorted on his scaffold below the Sistine ceiling, the Curies freezing in their shabby Parisian lab, and an infinitude of poets scribbling away in dingy rented rooms. But in reality, the spatiotemporal context in which creative persons live has consequences that often go unnoticed. The right milieu is important in more ways than one. It can affect the production of novelty as well as its acceptance; therefore, it is not surprising that creative individuals tend to gravitate toward centers of vital activity, where their work has the chance of succeeding. From time immemorial artists, poets, scholars, and scientists have sought out places of natural beauty expecting to be inspired by the majestic peaks or the thundering sea. But in the last analysis, what sets creative individuals apart is that regardless of whether the conditions in which they find themselves are luxurious or miserable, they manage to give their surroundings a personal pattern that echoes the rhythm of their thoughts and habits of action. Within this environment of their own making, they can forget the rest of the world and concentrate on pursuing the Muse.

BEING IN THE RIGHT PLACE

The great centers of learning and commerce have always acted as magnets for ambitious individuals who wanted to leave their mark on the culture. From the Middle Ages onward, master craftsmen traveled all over Europe to build cathedrals and palaces, attracted now by the wealth of one city, then by that of another. Milanese stonemasons built fortresses for Teutonic knights in Poland; Venetian architects and painters went to decorate the courts of the tsars of Russia. Even Leonardo, that paragon of creativity, kept serving one master after another depending on whether duke, pope, or king could best finance his dreams.

The place where one lives is important for three main reasons. The first is that one must be in a position to access the domain in which one plans to work. Information is not distributed evenly in space but is clumped in different geographical nodes. In the past, when the diffusion of information was slower, one went to Göttingen to study some branches of physics, to Cambridge or Heidelberg for others. Even with our dazzling electronic means for exchanging information, New York is still the best place for an aspiring artist to find out firsthand what's happening in the art world, what future trends other artists are talking about now. But New York is not the best place to learn oceanography, or economics, or astronomy. Iowa might be the place to learn creative writing or etching, and one can learn things about neural networks in Pittsburgh that one cannot learn anywhere else.

People in our sample often moved to places where information of interest was stored: Subrahmanyan Chandrasekhar took a boat from India to study physics at Cambridge; Freeman Dyson joined Richard Feynman at Cornell; Nina Holton went to Rome to learn bronze casting techniques. Sometimes it is not the person who chooses the place to further his or her knowledge: The opportunities for learning that a place offers capture the person's interest, and involvement with the domain follows. Brenda Milner happened to be in Montreal when the neurophysiologist D. O. Hebb started to teach at McGill University. She was so impressed by his seminars that both she and her husband changed the direction of their research, and she became one of the pioneers of the field. Margaret Butler found herself at the Argonne National Laboratories when computers were first put to use in biochemical research, and her lifelong interest in this domain was started by the opportunity to be a pioneer in this area. Rosalyn Yalow became interested in nuclear medicine because she happened to be where the instruments that made such studies possible were available. Of course, it is not that knowledge is stored in the place; rather it resides in an institution, a local tradition, or a particular person who happens to live in that place. To learn to cast bronze it helps to see how the old Italian craftsmen do it, and if one wanted to learn psychology from Hebb, one just had to go to Montreal.

The second reason why a place may help creativity is that novel stimulation is not evenly distributed. Certain environments have a greater density of interaction and provide more excitement and a greater effervescence of ideas; therefore, they prompt the person who is already inclined to break away from conventions to experiment with novelty more readily than if he or she had stayed in a more conservative, more repressive setting. The young artists who were drawn to Paris from all over the world at the end of the last century lived in a heady atmosphere where new ideas, new expressions, and new ways of living constantly jostled one another and called forth further novelty. The novelist Richard Stern describes how an artist may depend on such variety for his inspiration:

I yearned to go abroad when I was young, reading Hemingway, Fitzgerald, and so on. And once I went there it was extremely exciting for me to become a new personality, to be detached from everything that bound me, noticing everything that was different. That noticing of difference was very important. The languages, even though I was no good at them, were very important. How things were said that were different, the different formulas. Extremely exciting to me. The first time I went abroad, I was twenty-one, I began to keep the journal which I've still kept. I would keep it mostly not to go a little nuts—because there's so much that comes in. If I can get it down, then I don't have to worry about it. So being abroad has been very important in that way too.

For a theoretical physicist like Freeman Dyson, the stimulation of colleagues in neighboring offices is indispensable. Science, even more than art, is a collective enterprise where information grows much faster in "hot spots" where the thought of one person builds on that of many others. And then there are places that inhibit the generation of novelty. According to some, universities are too committed to their primary function, which is the preservation of knowledge, to be very good at stimulating creativity. Here Anthony Hecht comments on the pros and cons of this position from a poet's perspective, but his argument applies to other domains as well:

> There have been a number of poets in modern times who've said poets who teach in the academies end up being dry as dust, unimaginative and without daring and all that sort of stuff. I don't think that's true. The academy is neutral; it can, if you want to let it, curtail your imagination, but it doesn't have to. It's a place where you do a certain kind of work and live with certain kinds of people. The kind of people that you live with are pretty good on the whole. They're interesting, quirky, imaginative, idiosyncratic, lively, controversial. And I find that pleasant. I know this would not be the case if I were in a business organization where everybody's trying much more eagerly to conform.

Finally, access to the field is not evenly distributed in space. The centers that facilitate the realization of novel ideas are not necessarily the ones where the information is stored or where the stimulation is greatest. Often sudden availability of money at a certain place attracts artists or scientists to an otherwise barren environment, and that place becomes, at least for a while, one of the centers of the field. When in the 1890s William R. Harper was able to convince John D. Rockefeller, flush with dollars made in the oil fields, to part with a few million to start a university in the cornfields south of Chicago, he almost immediately attracted a number of leading scholars from the Northeast who flocked to the wilderness and established a great center of research and scholarship. Eighty years later the same phenomenon repeated itself farther west, when oil money made it possible for the University of Texas to attract a new generation of intellectual leaders to Austin. Oil is just one source of financial lure that greases the movement of academic fields from one place to another. After luminaries settle down in a particular place, it becomes difficult for young people with similar interests to resist their attraction. George Stigler, member of a department that has collected more Nobel Prizes in economics than any other in the world, explains some of the reasons why this is so:

The intellectual atmosphere in which you are determines a lot how you work. And Chicago in economics has been a virile, challenging, aggressive, and political environment. You're surrounded by able colleagues who are quite willing to embarrass you a little if you're doing something that's foolish or wrong but are quite willing to help you, too, on things that have promise, so that it's an extremely helpful environment.

The career of John Bardeen is typical. He went to graduate school at Princeton, where he became the second doctoral student of Eugene Wigner, a distinguished theoretical physicist who was awarded the Nobel Prize in 1963. Not surprisingly, many of Wigner's students also became leaders in the field. Bardeen then went to work at the Bell Research Laboratories, where many of the bright young physicists were being hired. This is how he describes the atmosphere there:

Bell Labs had a really outstanding group in solid-state theory. The way the organization was designed, they didn't have a theoretical group as such, but the theorists had their offices in close proximity so that they could talk readily with one another but they'd report to different experimental groups. So there was very close interaction between theory and experiment, and most papers were coauthored jointly by theorists and experimentalists. And that was a very exciting time to be there because there was a great enthusiasm for applying quantum theory to make new materials for the telephone system.

While working at Bell Labs, Bardeen developed the theory of semiconductors, which eventually led to the revolutionary invention of transistors. (For this work, he and two colleagues received the Nobel Prize in 1956.) Then Bardeen left for the University of Illinois, where he became fascinated by superconductivity, which promised to fulfill the medieval dream of the *perpetuum mobile,* the frictionless machine that in principle might go on working forever. In 1957 he contributed to a theory that became the benchmark in that domain, and for that he shared the 1972 Nobel Prize with two new colleagues. This is how he explains why he moved from Bell Labs:

> One reason I left to come to the University of Illinois in 1951 is that I thought that superconductivity was a just purely theoretical thing with no practical applications and it would be better to work on it in an academic environment. And Fred Seitz, who was the first student of Eugene Wigner's in Princeton, a good friend of mine for many years, had come from Carnegie Tech, now Carnegie-Mellon, with some of his coworkers to establish a group in solid-state physics at the University of Illinois. And I thought, if I came here with the group that was already present, they'd have a very strong effort in solid-state physics here. And that was true. It would attract the outstanding graduate students from places like Cal Tech and MIT; if they wanted to study solid-state physics their professors would send them out here as the best place to go.

In sciences and in the arts, in business and in politics, location matters almost as much as in buying real estate. The closer one is to the major research laboratories, journals, departments, institutes, and conference centers, the easier it is for a new voice to be heard and appreciated. At the same time, there is a downside to being near the centers of power. No one is more aware of this than Donald Campbell, whose warnings about the dangers young scholars run by being immersed too soon in a competitive, high-pressure environment are relevant beyond the confines of academia:

> I do think that environments make a difference. And the assistant professorships at Big Ten universities in psychology, where you have to produce five papers a year for five years to make tenure, are far less ideal than the British system in which a Francis Crick need not publish for years and years, yet still be kept in the system on the basis of interpersonal esteem. So much less pressure and much

greater freedom to explore and try out things without fear of failing.

People are responding to these conditions adaptively, and they are getting out the five papers a year for five years. But their freedom to be creative is being reduced by the pressure for quickness and number, and so is their ability to write a whole manuscript.

Look, you have two job offers, both of them have reasonable teaching loads. In one job you are going to be under high publish-or-perish pressure. In the other job you are going to feel adequate and under less pressure. Obviously the two universities have different national esteem levels. Which job would you take? I say clearly take the one where you will be free of tenure anxiety and be free to explore intellectually.

As with so many other things we have learned from these people's lives, there is no recipe for deciding, once and for all, which place is most suitable for the development of creativity. Certainly moving to the center of information and action makes sense; occasionally, it may even be indispensable. In certain domains there is really only one place in the world where one can learn and practice. But there might be disadvantages to being where the action, and therefore the pressure, is most intense. Where is the right place to be? Unfortunately, there is no single answer. Creativity is not determined by outside factors but by the person's hard resolution to do what must be done. Which place is best depends on the total configuration of a person's characteristics and those of the task he or she is involved in. Someone who is relatively more introverted may wish to perfect his act before stepping before the limelight. A more extroverted person may enjoy competitive pressures from the very beginning of her career. In either case, however, choosing the wrong environment will probably hinder the unfolding of creativity.

INSPIRING ENVIRONMENTS

I wrote the first draft of this chapter in a small stone cell, seven feet square, with two French windows looking out over the eastern branch of Lake Como, in northern Italy, near the foothills of the Alps. The cell was inhabited by hermit monks about five hundred years ago, and it is built over a chapel dedicated to Our Lady of

Monserrat. An earlier version of the chapel slid into the lake a long time ago. Now, from its windows, between the dense branches of laurel, oak, cedar, and beech trees, I can see, below the rocks on which the chapel is perched, the huge body of the lake rippling toward the south, like a fabulous dragon straining to break its chains.

The walls of the cell are covered with graffiti left by earlier occupants of this secluded haven. They too had the good fortune of having been selected by the Rockefeller Foundation to spend a month at the Villa Serbelloni, in the hope that the grand views, the panoramic paths through the forests, and the romantic ruins would inspire in them fresh bursts of scholarship. "Hundreds of trails, / Thousands of pines, / Limitless are the views" goes a haikulike verse scratched by a Harvard visitor. "Generations of guests, / Ten thousand experiences, / Attainment of resonant harmony." "Sun on the waters" begins an entry from UCLA, "the waves aglitter, / birds in the branches, / the trees atwitter; / bells of Bellagio—a new day's birth. Scholars in the Chapel: Heaven on earth!" Another verse, this time from Sussex University in England, ends: ". . . our graffiti, / Make grateful, / if grotesque entreaty, / That in this tree-encircled chapel, / We taste the tree of learning's apple."

There is ample precedent for such hopes. After all, the village of Bellagio, where the Villa Serbelloni stands, has been visited through the centuries by the likes of Pliny the Younger, Leonardo da Vinci, and the poets Giuseppe Parini and Ippolito Nievo—who once wrote from Sicily that he "would gladly exchange a month in Palermo for twenty-four hours in Bellagio"—all of whom sought to refresh their creativity in its magic atmosphere. "I feel that all the various features of Nature around me . . . provoked an emotional reaction in the depth of my soul, which I have tried to transcribe in music" wrote Franz Liszt during his stay here.

And from the highest points of the villa one can see at least three other similar enclaves across the lake: the Villa Monastero, formerly a convent for nuns from good families, where Italian physicists now repair to meditate and discuss quarks and neutrinos; the Villa Collina, once the private retreat of German chancellor Konrad Adenauer, now a place for German politicians to congregate; and the Villa Vigoni, built by a patriotic count of the Napoleonic era, now used for conferences that bring together Italian and German scientists. The air of these mountains, the smell of the azaleas, the shimmering reflection of old church spires in the fjordlike branches of the lake, are supposedly conducive to the creation of beautiful paintings, gorgeous music, and deep thoughts.

Nietzsche chose to write *Thus Spake Zarathustra* in the coolness of the nearby Engadine; Wagner loved to write his music in a villa in Ravello overlooking the hypnotic blue Tyrrhenian Sea; Petrarch was inspired to write his poetry in the Alps and in his villa near the Adriatic; the European physicists of the early part of this century seem to have had their most profound ideas while climbing mountains or looking at the stars from the peaks.

The belief that the physical environment deeply affects our thoughts and feelings is held in many cultures. The Chinese sages chose to write their poetry on dainty island pavilions or craggy gazebos. The Hindu Brahmins retreated to the forest to discover the reality hidden behind illusory appearances. Christian monks were so good at selecting the most beautiful natural spots that in many European countries it is a foregone conclusion that a hill or plain particularly worth seeing must have a convent or monastery built upon it.

A similar pattern exists in the United States. The Institute for Advanced Studies in the physical sciences at Princeton and its twin for the behavioral sciences in Palo Alto are situated in especially beautiful settings. Deer tiptoe through the immaculate grounds of the Educational Testing Services headquarters, and the research and development center of any corporation worth its salt will be situated among rolling meadows or within hearing range of thundering surf. The Aspen conferences unfold in the heady, thin air of the Rockies, and the Salk Institute sparkles over the cliffs of La Jolla like a Minoan temple; the idea is that such a setting will stimulate thought and refresh the mind, and thus bring forth novel and creative ideas.

Unfortunately, there is no evidence—and probably there never will be—to prove that a delightful setting induces creativity. Certainly a great number of creative works of music, art, philosophy, and science were composed in unusually beautiful sites. But wouldn't the same works have issued forth even if their authors had been confined to a steamy urban alley or a sterile suburban spread? One cannot answer that question without a controlled experiment, and given the fact that creative works are by definition unique, it is difficult to see how a controlled experiment could ever be performed.

However, accounts by creative individuals strongly suggest that their thought processes are not indifferent to the physical environment. But the relationship is not one of simple causality. A great view does not act like a silver bullet, embedding a new idea in the mind. Rather, what seems to happen is that when persons with prepared minds find themselves in beautiful settings, they are more likely to find new connections among ideas, new perspectives on issues they are dealing with. But it is essential to have a "prepared mind." What this means is that unless one enters the situation with some deeply felt question and the symbolic skills necessary to answer it, nothing much is likely to happen.

For instance, John Reed, of Citicorp, remembers two instances in his professional life, separated in time by several years, when he had been especially creative. Both of these involved recognizing the main problem his company was facing and sketching out possible solutions. As with most creative moments, it was the formulation rather than the solution of the problem that mattered most. In both cases, Reed wrote letters to himself, more than thirty pages in length, detailing the issues his company was confronting, the dangers and the opportunities of the next years, and the steps that could be taken to make the most of them. The interesting thing is that both letters were written when Reed was far away from the office, ostensibly free to relax: the first on a beach in the Caribbean, the second on a park bench in Florence. He describes how the second "letter" came about:

I write myself lots of letters. And I keep some of them. In September before the third quarter I had been kind of tired, working Saturdays and Sundays, and I had gone to Italy for a week, just to get away. I went first to Rome for a couple days, then I went up to Florence. I'd get up early in the morning, and I'd wander around, and I sat on a park bench, sort of between seven in the morning and noon, then in the afternoon I'd go visit museums and whatever. And I had a notebook, an Italian notebook, and I wrote myself long essays on what was going on and what I was worried about. And it helped me get my mind organized. Then in the afternoons I wouldn't do anything. Then at the end of the third quarter I went through the organizational changes. Just recently I pulled out my original memo and it was amazing, the degree to which I had my mind around it, the overlap must have been 80 to

90 percent [between what he wrote in Florence and what eventually was implemented].

Both "letters" were spontaneous and unpremeditated, although the issues they dealt with had been fermenting in Reed's mind for many months. Then it took several more months, after his return to headquarters, to sort out the good ideas from the bad, partly through discussions with friends and colleagues. And then several more months had to pass before ways were found to implement them. But without the "letter from the beach" and the "letter from the bench" it is doubtful that Reed could have found such a fresh perspective on the issues confronting his company.

This example still raises the question of how much the beach and the bench actually mattered. Certainly the creative solutions to Citicorp's problems would never have come about if anyone else had been sitting on them. The question is, would Reed have come up with the problem and the solution if he had stayed in his Manhattan office? While this question is unanswerable, the evidence does suggest that unusual and beautiful surroundings—stimulating, serene, majestic views imbued with natural and historical suggestions—may in fact help us see situations more holistically and from novel viewpoints.

How one spends time in a beautiful natural setting seems to matter as well. Just sitting and watching is fine, but taking a leisurely walk seems to be even better. The Greek philosophers had settled on the peripatetic method—they preferred to discuss ideas while walking up and down in the courtyards of the academy. Freeman Dyson's education at Cambridge, England, owed much less to what he heard in the classroom or read in the library than to the informal and wide-ranging conversations he had with his tutor while strolling the paths around the college. And later, in Ithaca, New York, it was through similar walks that he absorbed the revolutionary ideas of the physicist Richard Feynman: "Again, I never went to a class that Feynman taught. I never had any official connection with him at all, in fact. But we went for walks. Most of the time that I spent with him was actually walking, like the old style of philosophers who used to walk around under the cloisters." Will the new generation of physicists, crouched in front of their computer screens, have equally interesting ideas?

When ordinary people are signaled with an electronic pager at random times of the day and asked to rate how creative they feel, they tend to report the highest levels of creativity when walking, driving, or swimming; in other words, when involved in a semiautomatic activity that takes up a certain amount of attention, while leaving some of it free to make connections among ideas below the threshold of conscious intentionality. Devoting full attention to a problem is not the best recipe for having creative thoughts.

When we think intentionally, thoughts are forced to follow a linear, logical—hence predictable—direction. But when attention is focused on the view during a walk, part of the brain is left free to pursue associations that normally are not made. This mental activity takes place backstage, so to speak; we become aware of it only occasionally. Because these thoughts are not in the center of attention, they are left to develop on their own. There is no need to direct them, to criticize them prematurely, to make them do hard work. And of course it is just this freedom and playfulness that makes it possible for leisurely thinking to come up with original formulations and solutions. For as soon as we get a connection that feels right, it will jump into our awareness. The compelling combination may appear as we are lying in bed half asleep, or while shaving in the bathroom, or during a walk in the woods. At that moment the novel idea seems like a voice from heaven, the key to our problems. Later on, as we try to fit it into "reality," that original thought may turn out to have been trivial and naive. Much hard work of evaluation and elaboration is necessary before brilliant flashes of insight can be accepted and applied. But without them, creativity would not be what it is.

So the reason Martha's Vineyard, the Grand Tetons, or the Big Sur may stimulate creativity is that they present such novel and complex sensory experiences—mainly visual ones, but also birdsong, water sounds, the taste and feel of the air—that one's attention is jolted out of its customary grooves and seduced to follow the novel and attractive patterns. However, the sensory menu does not require a full investment of attention; enough psychic energy is left free to pursue, subconsciously, the problematic content that requires a creative formulation.

It is true that inspiration does not come only in locations sanctioned by the board of tourism. György Faludy wrote some of his best poems while facing daily death in various concentration camps, and Eva Zeisel collected a lifetime of ideas while imprisoned in the most notorious of Stalin's prisons, the dreaded Ljublianka. As Samuel Johnson said, nothing focuses the mind as sharply as the news that one will be executed in a few days. Life-threatening conditions, like the beauties of nature, push the mind to think about what is essential. Other things being equal, however, it would seem that a serene landscape is a preferable source of inspiration.

CREATING CREATIVE ENVIRONMENTS

While novel and beautiful surroundings might catalyze the moment of insight, the other phases of the creative process—such as preparation and evaluation—seem to benefit more from familiar, comfortable settings, even if these are often no better than garrets. Johann Sebastian Bach did not travel far from his native Thuringia, and Beethoven composed most of his pieces in rather dismal quarters. Marcel Proust wrote his masterpiece in a dark cork-lined study. Albert Einstein needed only a kitchen table in his modest lodgings in Berne to set down the theory of relativity. Of course, we do not know whether Bach, Beethoven, Proust, and Einstein may not have been inspired at some time in their lives by a sublime sight and spent the rest of their lives elaborating on the inspiration thus obtained. Occasionally a single experience of awe provides the fuel for a lifetime of creative work.

While a complex, stimulating environment is useful for providing new insights, a more humdrum setting may be indicated for pursuing the bulk of the creative endeavor—the much longer periods of preparation that must precede the flash of insight, and the equally long periods of evaluation and elaboration that follow. Do surroundings matter during these stages of the creative process?

Here it may be useful to make a distinction between the macroenvironment, the social, cultural, and institutional context in which a person lives, and the microenvironment, the immediate setting in which a person works. In terms of the broader context, it goes without saying that a certain amount of surplus wealth never hurts. The centers of creativity—Athens in its heyday; the Arab cities of the tenth century; Florence in the Renaissance; Venice in the fifteenth century; Paris, London, and Vienna in the nineteenth; New York in the twentieth—were affluent and cosmopolitan. They tended to be at the crossroads of cultures, where information from different traditions was exchanged and synthesized. They were also loci of social change, often riven by conflicts between ethnic, economic, or social groups.

Not only states but also institutions can foster the development of creative ideas. The Bronx High School of Science and the Bell Research Laboratories have become legendary because of their ability to nurture important new ideas. Every university or think tank hopes to be the place that attracts future stars. Successful environments of this type provide freedom of action and stimulation of ideas, coupled with a respectful and nurturant attitude toward potential geniuses, who have notoriously fragile egos and need lots of tender, loving care.

Most of us cannot do a great deal about the macroenvironment. There is not that much we can do about the wealth of the society we live in, or even about the institutions in which we work. We can, however, gain control over the immediate environment and transform it so that it enhances personal creativity. On this score, there is much to learn from creative individuals, who generally take great pains to ensure that they can work in easy and uninterrupted concentration. How this is done varies greatly depending on the person's temperament and style of work. The important thing, however, is to have a special space tailor-made to one's own needs, where one feels comfortable and in control. Kenneth Boulding preferred to think and work in a cabin overlooking the Colorado Rockies, and he also used to get into the hot tub intermittently to gather his thoughts. Jonas Salk liked to work in a studio where, in addition to the material he needed for writing on biology, there was a piano and an easel for painting. Hazel Henderson, who lives in a rather isolated community in north Florida to avoid the constant distractions of the urban centers, describes her daily routine:

> I like to run for about two miles every morning, and I have a special place to run to, which is a very beautiful spot, just about a mile from here, where there's a beautiful salt marsh, it's looking over the city. And if you look to the left, it's just absolutely wild and beautiful. And there are my favorite blue herons and curlews

and there's fish jumping and you can feel this teeming, living activity. And then if I look this way, to the right, there's this beautiful little city with its little spires, it's very harmonious. And, you know, there is a kind of balance between the natural system and the human system.

Robertson Davies crafts his intricate fiction in a house he built fifty miles north of Toronto, on a prehistoric seashore rich in fossils, "in a very nice position looking down, down the valley toward Toronto so that we can see the lights and look toward it and be glad that we're not there." The sociologist Elise Boulding has worked out almost monastic routines to help the rhythm of her creative thinking:

> An early morning walk, and reflection. In that year, 1974, I spent a lot of time on my knees; I have a little prayer plot that's at the back of the hermitage. I am not sure I would do that now. In 1991 I am a different person than I was in '74. But, do a certain amount of reading, you know, like the saints and those who have been through spiritual journeys, and simply reflection. A lot of reflection, meditation. I have spent time in Catholic monasteries, and I value very much the hours of office and so on. In '74 I followed the hours of office and sang them. Again, I am not sure I would do that now. But just lots of quiet, a lot of time spent just looking out of the window at the mountains, and meditating.

In Finland many people know Pekka, an elderly Lapp whose official job is to supervise the social services in the northernmost part of the country. But Pekka also travels widely: He spends his vacations visiting Tibet to learn the beliefs and lifestyles of the monasteries, or Alaska, in search of the vanishing Inuit culture. When he is in Helsinki on government business, he is known for never sitting down until he feels that the office where the meeting is held feels right. If it does not, he will take the elevator down to the street, walk around until he finds some branches, or stones, or flowers he likes. He will bring these objects back to the office, place them here and there on the desk or file cabinets, and when he feels that the environment looks serene and harmonious, he is ready to start business. Those who have to deal with Pekka generally feel that his impromptu interior decoration also helps them to have a better meeting and come to more satisfying decisions.

Elisabeth Noelle-Neumann, an innovative and successful German scientist and businesswoman (a few years ago, in a list of the one hundred most influential women in Germany published by a business magazine, she was ranked number two), has mastered the art of personalizing her environment. Her office, in a remodeled fifteenth-century farmhouse, is furnished with graceful antiques; her home on the shores of Lake Constance is filled with books and rare objects that reflect her personality. Because she spends so much time traveling from one place to another (about fifty thousand miles every year just by car), her Mercedes 500 is another important working space. While the chauffeur drives, Noelle-Neumann reads and writes surrounded by favorite audiotapes, bottles of mineral water, sheafs of notepaper, and bundles of ballpoint pens of various colors. Wherever she goes, she takes a familiar microenvironment with her.

To a certain extent everyone tries to accomplish something similar to what Elisabeth and Pekka do. We usually do it with our homes by filling them with objects that reflect and confirm our uniqueness. Such objects transform a house into a home. When we moved into a summer home in Montana, all it took to make the alien environment familiar was for my wife to place on the mantelpiece two colorful wooden ducks we had had for some time. With the ducks safely nesting along the wall, the empty space became immediately cozy and comfortable.

We need a supportive symbolic ecology in the home so that we can feel safe, drop our defenses, and go on with the tasks of life. And to the extent that the symbols of the home represent essential traits and values of the self, they help us be more unique, more creative. A home devoid of personal touches, lacking objects that point to the past or direct toward the future, tends to be sterile. Homes rich in meaningful symbols make it easier for their owners to know who they are and therefore what they should do.

In one of my studies we interviewed two women, both in their eighties, who lived on different floors of the same high-rise apartment house. When asked what objects were special to her in her apartment, the first woman looked vaguely around her living room, which could have passed for a showroom in a reasonably pricey furniture store, and said that she couldn't think of anything. She gave the same response in the other rooms—nothing special, nothing personal, nothing meaningful anywhere. The second woman's living room was full of pictures of friends and family, porcelain and silver inherited from aunts and uncles, books she loved or that she intended to read. The hallway was hung with framed drawings of her children and grandchildren. In the bathroom the shaving tools of her deceased husband were arranged like a tiny shrine. And the life of the two women mirrored their homes: the first followed an affectless routine, the second a varied, exciting schedule.

Of course, furnishing one's house in a certain way does not miraculously make one's life more creative. The causal connections are, as usual, more complicated. The person who creates a more unique home environment is likely to be more original to begin with. Yet having a home that reinforces one's individuality cannot but help increase the chances that one will act out one's uniqueness.

It used to be said that a man's home is his castle, in deference to the fact that at home one feels more secure and in control than anywhere else. But increasingly in our culture it could be said that a man's—and especially a woman's—car is the place where freedom, security, and control are most deeply experienced. Many people claim that their car is a "thinking machine," because only when driving do they feel relaxed enough to reflect on their problems and to place them in perspective. One person we interviewed said that about once a month, when worries become too pressing, he gets into his car after work and drives for half the night from Chicago to the Mississippi. He parks and looks at the river for about half an hour, then drives back and reaches Chicago as the dawn lights up the lake. The long drive acts as therapy, helping him sort out emotional problems.

Cars can be personalized by a variety of means: The make we buy, the color, the accessories, and the music system all contribute to an at-home feeling in a vehicle that affords both privacy and mobility. In addition to cars, offices and gardens are spaces that can be arranged to provide environments that reflect a personal sense of how the universe ought to be. It is not that there is one perfect pattern by which to order our surroundings. What helps to preserve and develop individuality, and hence enhance creativity, is an environment that we have built to reflect ourselves, where it is easy to forget the outside world and concentrate completely on the task at hand.

PATTERNING ACTIVITIES

It is not only through personalizing the material environment that we are able to enhance creative thought. Another very important way to do so is by ordering the patterns of action we engage in. Manfred Eigen, the Nobel Prize winner in chemistry, plays Mozart at the piano almost every day to take his mind off the linear track. So does the writer Madeleine L'Engle. Mark Strand walks his dog and works in the garden. Hazel Henderson, who struggles daily with the problems of the various environmental groups she helps organize, gardens and takes walks to refresh her thinking. Some ride bikes and some read novels; some cook and others swim. Again, there is no best way to structure our actions; however, it is important not to let either chance or external routine automatically dictate what we will do.

Elisabeth Noelle-Neumann rarely eats at the times other people usually eat but has her own strict schedule that fits her own needs. Richard Stern has

a sort of rhythm. I've imposed on time a rhythm which has enabled me to function. Function as a writer, function as a father, a husband—not always the best one—as a university professor, colleague, friend.

He goes on to specify in more concrete terms what he means by "rhythm":

My guess is that though it resembles other people's rhythms, that is, anybody who does work either has a routine or imposes on his life certain periods in which he can be alone or in which he collaborates. At any rate, he works out a sort of schedule for himself and this is not simply an external, exoskeletal phenomenon. It seems to me it has much to do with the relationship of your own physiological, hormonal, organic self and its relationship to the world outside. Components can be as ordinary as reading the newspaper in the morning. I used to do that years ago, and I stopped for years and years, which altered the rhythm of my day. One drinks a glass of wine in the evenings at certain times, when the blood sugar's low, and one looks forward to it. And then of course those hours in which one works.

Most creative individuals find out early what their best rhythms are for sleeping, eating, and working, and abide by them even when it is tempting to do otherwise. They wear clothes that are comfortable, they interact only with people they find congenial, they do only things they think are important. Of course, such idiosyncrasies are not endearing to those they have to deal with, and it is not surprising that creative people are generally considered strange and difficult to get along with. But personalizing patterns of action helps to free the mind from the expectations that make demands on attention and allows intense concentration on matters that count.

A similar control extends to the structuring of time. Some creative people have extremely tight schedules and can tell you in advance what they will be doing between three and four in the afternoon on a Thursday two months from today. Others are much more relaxed and in fact pride themselves on not even knowing what they will be doing later on today. Again, what matters is not whether one keeps to a strict or to a flexible schedule; what counts is to be master of one's own time.

Longer stretches of time show the same variable structure. Freeman Dyson and Barry Commoner believe that one should make a major career change every ten years or so to avoid becoming stale. Others seem perfectly satisfied delving deeper and deeper into a narrow corner of their domain throughout their lives. But what none of the persons we interviewed ever said was that he or she did this or that because it was the socially expected thing to do at that particular time.

So it seems that surroundings can influence creativity in different ways, in part depending on the stage of the process in which a person is involved. During preparation, when one is gathering the elements out of which the problem is going to emerge, an ordered, familiar environment is indicated, where one can concentrate on interesting issues without the distractions of "real" life. For the scientist it is the laboratory, for the businessperson the office, for the artist the studio. At the next stage, when thoughts about the problem incubate below the level of awareness, a different environment may be more helpful. The distraction of novel stimuli, of magnificent views, of alien cultures, allows the subconscious mental processes to make connections that are unlikely when the problem is pursued by the linear logic learned from experience. And after the unexpected connection results in an insight, the familiar environment is again more conducive for completing the process; evaluation and elaboration proceed more efficiently in the sober atmosphere where the logic of the domain prevails.

However, at any point in time, what matters most is that we shape the immediate surroundings, activities, and schedules so as to feel in harmony with the small segment of the universe where we happen to be located. It is nice if this location is as fetching as a villa on Lake Como; it is a far greater challenge when fate throws you into a Siberian gulag. At either extreme, what counts is for consciousness to find ways to adapt its rhythms to what is outside and, to a certain extent, to transform what it encounters outside to its own rhythms. Being in tune with place and time, we experience the reality of our unique existence and its relationship to the cosmos. And from this knowledge original thoughts and original actions follow with greater ease.

The implications for everyday life are simple: Make sure that where you work and live reflects your needs and your tastes. There should be room for immersion in concentrated activity and for stimulating novelty. The objects around you should help you become what you intend to be. Think about how you use time and consider whether your schedule reflects the rhythms that work best for you. If in doubt, experiment until you discover the best timing for work and rest, for thought and action, for being alone and for being with people.

Creating a harmonious, meaningful environment in space and time helps you to become personally creative. It may help you achieve a life that reflects your individuality, a life that is rarely boring and rarely out of control; a life that makes others realize the possibilities for uniqueness and growth inherent in the human condition. But creating such a life does not guarantee that you will be recognized as a genius, as a historically significant creative figure. To achieve historical creativity many other conditions must be met. For instance, you must be lucky, for to excel in some domains you might need the right genes, you might have to be born in the right family, at the right historical moment. Without access to the domain, potential is fruitless: How many Congolese would make great skiers? Are there really no Papuans who could contribute to nuclear physics? And finally, without the support of a field, even the most promising talent will not be recognized. But if creativity with a capital *C* is largely beyond our control, living a creative personal life is not. And in terms of ultimate fulfillment, the latter may be the most important accomplishment.

PART II

THE LIVES

SEVEN

THE EARLY YEARS

There is a certain amount of voyeurism involved in reading—and writing—about eminently creative people. It is a little like watching celebrity shows like *Lifestyles of the Rich and Famous,* where one is allowed to peek behind the facade into the living rooms and bedrooms of people whom we envy from afar. But there is also a perfectly legitimate reason for reflecting on what happens to exceptional individuals from early childhood to old age. Their lives suggest possibilities for being that are in many ways richer and more exciting than most of us experience. By reading about them, it is possible to envision ways of breaking out from the routine, from the constraints of genetic and social conditioning, to a fuller existence. It is true that the accomplishments of these creative persons are to a great extent influenced by sheer luck—the good fortune of having been born with exceptional genes, or of having had a supportive environment, or happening to be at the right place at the right time. But many people with similar luck aren't creative. So beyond these external factors where luck holds sway, what allows certain individuals to make memorable contributions to the culture is a personal resolution to shape their lives to suit their own goals instead of letting external forces rule their destiny. Indeed, it could be said that the most obvious achievement of these people is that they created their own lives. And how they achieved this is something worth knowing, because it can be applied to all our lives, whether or not we are going to make a creative contribution. Hence what follows is not intended as light entertainment but as an exploration of how human potential can be expanded.

CHILDHOOD AND YOUTH

In our culture—perhaps in all cultures—some of the most cherished stories relate the childhoods of heroes. If a man or woman is held in high esteem, the popular imagination wants to find a sign of greatness as soon as possible in that person's life, to justify and explain the success that followed. Here is one such story.

As the fog slowly lifts, one after the other the bare hilltops burst from the shadows and blaze in the sunlight. A shepherd boy reaches into the pocket of his cape for an old crust and chews on it uneasily. His dog has been looking for some time toward the valley where the old mill stands, as if something is afoot down in the darkness. And now the ewes begin to stir. A yearling, scared by the tension in the air, starts to bleat as if lost.

Then the shepherd boy hears the dry clip-clop of hooves coming up the rocky path and almost immediately sees the outline of a rider emerge from the shadows below. Who could this stranger be? He has only a slender sword at his side, so he is not a warrior; he wears none of the sacred symbols of the clergy; he seems to lack the caution of a traveling merchant. Yet he is certainly no peasant, richly dressed as he is in blue velvet hose and a golden mantle. What other sort of man can there be, who can ride so easily through the lonely hills of Tuscany in the Year of Our Lord 1271?

The rider smiles down at the boy, shifting in the saddle. His eyes slowly circle the horizon.

"Well, I think I am good and lost. I was trying to find the shortest road from Florence to Lucca, but after a full night's traveling, I seem to have left all human dwellings behind. Where are we, actually?" he asks, turning toward the boy. "And what name do they call you?"

The shepherd gestures in the direction opposite to where the sun was rising. "If you followed the creek down there for two leagues, you'd be in the Valley of the Mugello. To the left is the road to Florence, and to the right the one to Lucca. And my name is Angiolo, son of Bondone."

At this the rider nods, then yawns. He looks around at the ridges rising and falling like the waves of a tawny sea, and then, as if shaking himself awake from sleep, he slides off the saddle.

"Sweet Mother of God, but I am tired. I hope, Angiolotto, that you have some fresh ewe's milk, because I haven't stopped to eat since this past noon. Don't worry, I will pay you well for it," he says, jingling coins in the fancy red leather purse that hangs from his belt.

Angiolo uncovers a piece of cheese and the jar of milk he has kept behind a slab of granite. He apologizes to the rider for having no bread to offer, but the gentleman takes out a fresh chestnut pie from his saddlebag, which they share.

After they have eaten in silence for a while, the boy cannot resist asking "May I query, my Lord, what takes you to Lucca? I would wager you are not from these parts."

"And the wager would be yours. I was born up in Lombardy, on the banks of the river Po. My master is Teboldo of the Visconti who earlier this year was crowned pope as His Holiness Gregory, the tenth of that name. I rode out two moons ago from the Flaminian gate of Rome on a mission from him."

Angiolo is not sure what all this meant, but being a curious boy, he keeps questioning further. "And what kind of a mission would that be?"

The rider smiles. "His Holiness wants the best craftsmen to come to Rome and make the Eternal City as beautiful as it deserves to be. I am supposed to find master builders, sculptors, and painters, and convince them to enter the service of His Holiness."

The boy thinks about this for a while. "How do you find out who the best craftsmen are?"

"Oh, one asks questions, listens to stories. One looks at the work in churches, in palaces." Here a shade of smugness passes over the features of the rider. "But I have also my own special test. I ask any man who is supposed to be good to draw a perfect circle, a cubit across, freehand. If he is really good, he will draw something that looks quite round. But few do come close without a compass or a string held at the center."

Angiolo rummages among the ashes of last night's fire and comes up with a stick of charcoal. "What?" he asks. "You mean like this?" And with one smooth movement, he draws a perfect circle on the slab of stone from which they were eating.

The pope's envoy scratches his head. He looks at the boy, looks at the circle on the stone. He looks away at the hills, now almost melting in sunshine.

"Not bad, not bad at all. How about drawing natural things? Have you ever drawn people, or, say, animals?"

Now it is Angiolo's turn to smile. He glances at the fat ram, sunning itself at his feet, the leader of the flock, and with a few quick strokes he has sketched it so vividly that all it lacks is the Lord's breath for it to start bleating. The rider from Rome becomes very thoughtful.

This is a version of the story of how the great painter Giotto was discovered, a story that all schoolchildren in Italy have heard or read at some time or other, probably many times through their lives. There are illustrations in schoolbooks and texts of Angiolotto drawing his circle on the stone with the startled envoy looking on, or of his drawing the sheep while the rider holds his head in amazement. The story goes on to tell how the envoy took the boy to the workshop of the famous Cimabue, there to learn the fine points of painting. Giotto—as the boy was soon called—began to paint one astonishing picture after another. His fame soon surpassed that of his master, and he became known as the greatest artist in Italy, perhaps in all of Christendom. It is a story that most educated people in Europe know and cherish.

Unfortunately, like many good stories, this one reflects more our psychological needs than reality. When I recently searched for material on Giotto's childhood at a leading university library, I found 102 volumes on the painter. None of them claimed to have any information about Giotto's childhood or, for that matter, about the first thirty years of his life. A typical biography starts as follows: "According to old documents, Giotto was born in 1266, at Vespignano di Mugello or in Florence, *but nothing is known about his youth, there are only legends.* The only facts are those of his artistic beginnings in Assisi, but even these are unclear and difficult to establish" (italics added).

All the volumes agree that Giotto's style was extraordinarily novel, that he resurrected the dead art of painting and prepared the way for the renaissance of the arts that was to come a century later. But the precocity of his genius is the stuff of myth, and the legends that sprang up around his life are an indication of how much we need events to be predictable, to make sense. If someone becomes outstanding, we want to believe that unmistakable signs of greatness were there early for all to see. Whether it is the Buddha, Jesus,

Mozart, Edison, or Einstein, genius must have revealed itself in the earliest years of life.

In fact, it is impossible to tell whether a child will be creative or not by basing one's judgment on his or her early talents. Some children do show signs of extraordinary precocity in some domain or other: Mozart was an accomplished pianist and composer at a very early age, Picasso drew quite nice pictures when he was a boy, and many great scientists skipped grades in school and astonished their elders with the nimbleness of their minds. But so did many other children whose early promise fizzled out without leaving any trace in the history books.

Children can show tremendous talent, but they cannot be creative because creativity involves changing a way of doing things, or a way of thinking, and that in turn requires having mastered the old ways of doing or thinking. No matter how precocious a child is, this he or she cannot do. Mozart in his teens might have been as accomplished as any musician alive, but he could not have changed the way people played music until his way of making music was taken seriously, and for this to happen he had to spend at least a decade mastering the domain of musical composition and then produce a number of convincing works. But if the real childhood accomplishments of creative individuals are no different from those of many others who never attain any distinction, the mind will do its best to weave appealing stories to compensate for reality's lack of imagination.

We all know the mechanism that generates such stories, because we have used it to make our own lives, or those of our children, more interesting and more sensible. For example, little Jennifer has a poem published in the junior high literary magazine; soon her parents tell their friends about the clever things she used to say as a toddler, and how she liked to listen to nursery rhymes, and how early she was able to recognize written words, and so on. If Jennifer then goes on to be a real writer, the stories of her childhood are likely to become ever more clearly focused on her precocity. Not because anyone is consciously trying to alter the truth, but because as one tells a tale over and over, the tendency is to highlight what in hindsight we feel are the important parts and to eliminate details that contradict the point of the story. Our sense of inner consistency demands it, and the audience will also appreciate the story more.

With each telling, Jennifer's childhood becomes more remarkable. Thus are myths born.

Prodigious Curiosity

Children cannot be creative, but all creative adults were once children. Thus it makes sense to ask what creative individuals were like when they were children, or what sorts of events shaped the early lives of those persons who later accomplished something creative. But when we look at what is known about the childhoods of eminent creative persons, it is difficult to find any consistent pattern.

Some children who later astonished the world were quite remarkable right out of their cradles. But many of them showed no spark of unusual talent. Young Einstein was no prodigy. Winston Churchill's gifts as a statesman were not obvious until middle age. Tolstoy, Kafka, and Proust did not impress their elders as future geniuses.

The same pattern holds for the interviews we conducted. Some of our respondents, like the physicist Manfred Eigen, or the composer and musician Ravi Shankar, displayed unusual gifts in their respective domains before their teens. Others, such as the chemist Linus Pauling, or the novelist Robertson Davies, blossomed in their twenties. John Reed, CEO of Citicorp, made a decisive impact on the banking industry in his forties; Enrico Randone, president of the giant Assicurazioni Generali insurance conglomerate of Italy, left his mark on the company he led while in his late seventies. John Gardner discovered he had a gift for politics in his midfifties, when President Johnson asked him to be the first secretary of Health, Education and Welfare, and Barry Commoner decided to break away from academic science and start his environmental movement at about the same age. In all these instances of late blooming, the earlier years provide at best only glimpses of extraordinary ability in the domain they eventually turned to.

If being a prodigy is not a requirement for later creativity, a more than usually keen curiosity about one's surroundings appears to be. Practically every individual who has made a novel contribution to a domain remembers feeling awe about the mysteries of life and has rich anecdotes to tell about efforts to solve them.

A good example of the intense interest and curiosity attributed to creative persons is the following story told about Charles Darwin's youth. One day as he was walking in the woods near his home he noticed a large beetle scurrying to hide under the bark of a tree. Young Charles collected beetles, and this was one he didn't have in his collection. So he ran to the tree, peeled off the bark, and grabbed the insect. But as he did so he saw that there were two more specimens hiding there. The bugs were so large that he couldn't hold more than one in each hand, so he popped the third in his mouth and ran all the way home with the three beetles, one of which was trying to escape down his throat.

Vera Rubin looked out of her bedroom window and saw the starry skies for the first time when she was seven years old, after her family had moved to the edge of the city. The experience was overwhelming. From that moment on, she says, she could not imagine not spending her life studying the stars. The physicist Hans Bethe remembers that from age five on, the best times he had were when playing with numbers. When he was eight years old he was making long tables of the powers of two and of the other integers. It's not that he was especially brilliant at this, but he enjoyed doing it more than anything else. John Bardeen, the only person to be awarded two Nobel Prizes in physics, was good in school—he skipped from third to seventh grade—but did not get interested in math until he was ten. After that, however, math became his favorite pastime; whenever he could, he solved math problems. Linus Pauling, also a double Nobel Prize winner, fell in love with chemistry before he even entered school, while helping his father mix drugs in his pharmacy. The physicist John Wheeler remembers: "I must have been three or four years old in the bathtub and my mother bathing me, and I was asking her how far does the universe go . . . and the world go . . . and beyond that. Of course she got stuck as much as I have always been stuck since."

Robertson Davies wrote continuously in school and won prizes for his essays. As a young child, Elisabeth Noelle-Neumann, doyenne of public opinion research in Europe, built imaginary communities: "My favorite toy when I was a child was not dolls but wooden pieces to build up a village—trees, houses, fences, animals, and very different houses, for example, a town hall. And I would spend two or three days at a time when I was ten, twelve years old thinking up stories about the lives of the people in the village." Jacob Rabinow, one of the most prolific inventors in terms of the number and variety of patents registered, became fascinated with his father's shoe-making machine as a small child in Siberia, and since then he has explored and tried to understand every machine he has encountered. The neuropsychologist Brenda Milner describes herself as follows:

> The thing that has driven me my whole life, and I have always maintained this, is curiosity. I am incredibly curious about things, little things I see around me. My mother used to think that I was just very inquisitive about other people's business. But it was not just people, it is things around me. I am a noticer.

The sociologist David Riesman says: "If you ask what drives me, I would say it's curiosity." Yet none of these individuals—not Darwin, not Riesman—were prodigies or even gifted children as we now define them. But they had a tremendous interest, a burning curiosity, concerning at least one aspect of their environment. Whether sounds or numbers, people or stars, machines or insects—the fascination was there, and generally it lasted all through the person's life.

It is true that these memories of childhood may be even more open to retrospective distortion, to the kind of romancing that has led us to mistrust the accounts of prodigious early abilities, like the one about Giotto. Perhaps these stories are also largely post hoc fabrications. I am reasonably sure, however, that they are not. When people in the eighth or ninth decade of their life describe their first fascination, they do so with a concreteness that seems genuine. Sometimes the material evidence is also present: an old telescope built in childhood, a battered book that served as inspiration many years ago, a juvenile poem or sketch. So while these people may not have been precocious in their achievements, they seem to have become committed early to the exploration and discovery of some part of their world.

But where does such intense interest come from? That, of course, is the really important question. Unfortunately, here too a definite answer has to wait until we know a lot more about creativity than we do now. Perhaps the best general answer we can give at this point is that each child becomes interested in pursuing whatever activity gives him or her an edge in the competition for resources—the attention and admiration of significant adults being the most important resource involved. Whereas later in life creative individuals learn to love what they do for its own sake, at first this interest is often motivated by competitive advantage. A child who gets recognized for her ability to jump and tumble is likely to become interested in gymnastics. A boy whose drawings get more favorable comments than those of his friends will become interested in art.

It is not necessarily the sheer amount of talent that matters but the competitive advantage one has in a particular milieu. A girl with very modest musical gifts may become intensely interested in music if everyone else around her is even less musical. On the other hand, a boy who is very good with numbers is unlikely to get involved in mathematics if his brother is already known as gifted in math, because as the younger sibling he would have to grow up in the older one's shadow. He may choose to develop his second best suit and become interested in something else instead.

In some cases, the competitive edge is the result of the child's heredity—what's bred in the bone, so to speak. Especially among musically and mathematically gifted children, superior performance shows itself with such force that the audience has no choice but to recognize it (provided, of course, that the audience knows enough about music or math). In such cases, children will usually accept the gift of their ancestors and become more and more interested in developing it. In other cases—probably the majority—the initial curiosity is sparked by some feature of the social environment. Subrahmanyan Chandrasekhar, who won the Nobel Prize in physics in 1983, was the nephew of the first scientist in India to earn the same prize in 1930. As a boy, everyone in the family expected him to emulate the eminent uncle. Chandrasekhar knew that if he wanted to be accepted and admired by his relatives he had better become interested in science.

However, not every creative scientist was interested in science as a child, nor was every creative writer committed to writing at an early age. A good example of the kind of career shifts that are common is the case of young Jonas Salk, who eventually discovered the polio vaccine named after him:

> Well, as a child, I wanted to study law, so as to be elected to Congress and make just laws. This was when I was eight years old or thereabouts, ten years old. And then I decided to study medicine for reasons that had to do with my mother feeling that I wouldn't make a good lawyer because I could never win an argument with her.

Hilde Domin, the eminent German poet, wrote her first poem when she reached middle age, after her mother's death; and she did not start publishing her poetry until even later. Jane Kramer, who became a pioneering TV producer and later dean of the Columbia School of Journalism, was not aware of her vocation until she was in her twenties. György Faludy switched to poetry only after he discovered that he could not draw. Another poet, Anthony Hecht, said:

> When I was very young, I felt that the greatest aptitude I had was for music, not poetry. And I think this was an inhibiting factor for me in trying to be a poet. I was thinking too much in musical terms, I was trying too hard to achieve musical effects, I was thinking and wishing that I could convert poetry into abstract music. And one of the things I had to learn was to stop thinking that way. It took enormous concentration and determination.

But even though these individuals may not have known what specific form their curiosity would take, they were open to the world around them and interested in finding out about it, in living life as fully as possible.

Few paths were as convoluted as that of the chemist Ilya Prigogine, who received the Nobel Prize in 1977. The son of a Russian emigré aristocrat, as a young man in Belgium he was mainly interested in philosophy, art, and music. His family, however, insisted that he study a respectable profession, and so he enrolled in law at the university. As he read criminial law, he became interested in the psychology of the criminal mind. Dissatisfied with superficial knowledge, he decided to understand better the underlying brain mechanisms that might explain deviant behavior, and this led him to the study of neurochemistry. Enrolled in the chemistry department of the university, he realized that his initial interest was perhaps too amibitious, and started basic research in the chemistry of self-organizing systems.

But Prigogine continued to be inspired by his initial curiosity; he gradually realized that the statistical unpredictability in the behavior of simple molecules might shed light on some of the basic problems of philosophy, such as the question of choice, of responsibility, of freedom. Whereas the physical laws of Newton and Einstein were deterministic and expressed certitudes that applied equally to the past and to the future, Prigogine found in the unstable chemical systems he studied processes that could not be predicted with certainty, and that could not be reversed once they happened.

If you can say that the universe is deterministic, a kind of automaton, then how can we hold to the idea of responsibility? All of Western philosophy was dominated by this problem. It seemed to me that we had to choose between a scientific view which was negating humanistic tradition, or a humanistic tradition which was trying to destroy what we learned from science . . . I was very sensitive to this conflict because I came to science, to hard science, from the human sciences. . . . But what I learned from thermodynamics confirmed my philosophical point of view. And gave me the energy to continue to look on a deeper interpretation of time and of the laws of nature. So, I would say, it is a kind of feedback between the humanistic and the scientific point of view.

The synergy between the humanistic and the scientific quest has served Prigogine well. In addition to illuminating basic thermodynamic processes, his ideas have inspired a great variety of scholars in the natural and the social sciences. Concepts he familiarized such as "dissipative structures" and "self-organizing systems," have found their way into discussions of urban planning and personality development. But like the molecular systems he studies, Prigogine's career could not have been predicted from a knowledge of his initial interests alone. It took the subtle interaction between his curiosity, the desires of his parents, the opportunities offered by the intellectual environment in which he lived, and the results of his experiments to give shape to that conceptual system we now associate with his name.

The Influence of Parents

In most cases it is the parents who are responsible for stimulating and directing the child's interest. Sometimes the only contribution of the parents to their child's intellectual development is treating him or her like a fellow adult. Donald Campbell, whose many novel methodological and theoretical contributions have enriched contemporary psychology, is one of the many respondents who feel "blessed" that his parents never talked down to them, and listened to their opinions about all sorts of adult issues. What the novelist Robertson Davies says is typical of many other respondents:

> My parents were like all parents. A hundred different things, it's very hard to describe what they all were. But one of the things they were, which I very, very greatly appreciate: They were very generous. They never denied their children anything that would help them. And they were very generous to me because I showed an aptitude for education, and so they helped me get a lot of education. And also they helped me to get a kind of grounding in music and literature by their example and their advice, and just by sending me where that was to be found. And so I have great cause to be grateful to them. And though often we had strong differences of opinion, I always feel that they were very kind and generous to me.

In other cases the entire family is mobilized to help shape the child's interest. Elisabeth Noelle-Neumann and each of her sisters had an aunt or uncle deputized to take them to museums and concerts at least twice a month. It was important, she says, that each sister had her own exclusive area of expertise—the one who was always taken to the ballet was not taken to the art museum and vice versa. This way sibling rivalry was minimized and personal interest reinforced.

A fairly typical childhood is the one recalled by Isabella Karle, one of the leading crystallographers in the world, a pioneer in new methods of electron diffraction analysis and X-ray analysis. Her parents were Polish immigrants with minimal formal education and limited means. Yet even during the worst years of the Great Depression Isabella's mother saved from her housekeeping money so that the family could take two-week vacations to explore the East Coast. The parents took their children to the library, to museums, and to concerts. Before starting first grade in the Detroit public schools, Karle had been taught by her parents to read and write in Polish. "They were very good at introducing us to the world," says Isabella, "even though their resources were limited." She remembers being an excellent student, receiving her Ph.D. degree in chemistry fifteen years after entering the first grade. Her early interest centered on historical novels before she was introduced to chemistry. Yet she never took a science course until her junior year in high school, when a counselor advised that taking one would make it easier for her to get into a good college. So from a list of courses in biology, chemistry, and physics, she pointed at random to the one in the middle. "And the chemistry," she says, "fascinated me absolutely." So even though a child need not develop an early interest in a domain in order to become creative in it later, it does help a great deal to become exposed early to the wealth and variety of life.

Strong parental influence is especially necessary for children who have to struggle hard against a poor or socially marginal background. Lacking other advantages, such as good schools and access to mentors, it is almost impossible to succeed without parental support and guidance. Oscar Peterson, the renowned jazz pianist, remembers that when he was a child his father, who was a porter on the Canadian railroads, used to set him the task of learning to play a piece of music every time he left on a trip from Montreal to Vancouver. As soon as he came back, his father made sure that Oscar had done his homework. If not, he would get "his bum kicked." But the most important influence of the family was building Oscar's sense of strong personal standards and self-confidence, and encouraging his love for music:

> They didn't try to tether me and keep me in line. They would see me doing something and they'd say, "I think you know better than that. I think if you look in the mirror and take a good hard look you know you don't really mean that. That's not you." So they let me know that they had great expectations, more so than I was living up to at that moment.

> My family gave me, first of all, the love of music. They helped me appreciate some of the music that I was hearing, and that of course catapulted me into the medium. But they also gave me a set of personal rules to live by that kept me from getting into some of the troubles that musicians were getting into at the time. And they gave me a certain amount of self-esteem by feeling that if I wanted to, I would do well.

The sense of self-respect and discipline Oscar Peterson absorbed at home stood him in good stead later on, when the temptations of the jazz world became acute. While many of his peers succumbed to the easy enticements of sex, drugs, and liquor, respect for his parents and their values kept Peterson on a steady course:

> They let me know they would never tolerate or accept that [taking drugs]. I won't call any names, but a very famous musician once offered me cocaine—I guess it was cocaine—no, heroin, excuse me. As he called it, "a hit with heroin." And I told him quite frankly, "I would never be able to go home if I did this." And that's the thing that terrified me more than anything else. I couldn't figure out what I would tell my mother—far less my father—if I came home with a habit. There would be no reason for it. It wasn't a fear of what he would do to me, it was a fear of— maybe destroying him altogether. I didn't know how I could ever explain this to him.

John Hope Franklin, the African-American historian, remembers that his father, a lawyer, read all the time, so the son grew up thinking that reading was what adults did night and day; and he remembers his mother as always supportive and encouraging. Franklin credits both parents for providing the intellectual and moral foundations of his life:

> I come from an educated parentage, you know. My mother was a schoolteacher and a graduate of the teachers' training program at Roger Williams University in Tennessee, and my father also attended that school. That is where they met. Then he went on to Morehouse College in Atlanta and graduated from there. Then he studied law. He read law in the office of a lawyer. That's what was frequently done around 1900. And he passed the bar with the second highest scores. He graduated from the University of Michigan and took the bar in the Indian Territory, not yet a state. Oklahoma was not yet a state. And so my parents had an enormous influence on my intellectual as well as my social development. I learned from them the value of studying and reading, that sort of thing. I learned from them too certain elements of honesty and integrity. I didn't have to wonder later whether I should or should not do certain things. It was part of my being because of their influence.

Manfred Eigen, the Nobel Prize–winning chemist, learned from his father the music he still plays and the high standards of performance his father expected. The historian William McNeill's father was also a historian, whose synthetic view of the past influenced his son's professional development. Freeman Dyson also remembers his parents fondly:

> Well, I was extremely lucky, of course, in having the parents that I had. They were both of them remarkable people. My father was this unusual combination of a composer and administrator, so that was a great inspiration to me, the feeling that you can do lots of things and do them all well. And my mother was equally unusual in a way because she was a lawyer and had read very widely and in fact was more of a companion to me than my father. They were both of them such strong characters. And yet still they left me complete freedom to do my stuff, which was science. And neither of them was a scientist, but they understood what it was about.

Parental influence is not always positive. Sometimes it is perceived as having been fraught with tension and ambivalence. Hazel Henderson modeled herself on her loving mother but resented the fact that she was so submissive to her patriarchal husband. Speaking of her father, she said:

> He would tend to be authoritarian because that's the way men were supposed to be. And so Mother never won an argument. I didn't want to be like him, although I realized that power was useful. And I did want to be like him in terms of, uh, well, I want to be effective, and I don't want to be trashed I don't want to be a doormat. And so that was a tremendous tension in my childhood, what the hell to do with this. And so I think that, although I never verbalized it or thought about it at the time, I ended up deciding that I was really going to unite love and power.

Often, especially in the case of artists, parents are horrified by the direction their child's interest is taking. Mark Strand, a U.S. poet laureate, started out with an interest in the arts. His parents "were not pleased when I announced my intention to be a painter. Because they were worried about how I would earn a living. And it was even worse when I expressed my intention to become a poet. They thought all poets starved, or were suicides or alcoholics." György

Faludy had to take many university courses in various subjects to please his father, before turning to poetry. The generation of women represented in our sample were discouraged by their families to consider science as a possible career. What chance did they have to become physicists or chemists? Better stick with the plan of trying to become high school teachers.

As the above quotes suggest, parents were not simply a source of knowledge or intellectual discipline. Their role was not limited to introducing their children to career opportunities and facilitating access to the field. Perhaps the most important contribution was in shaping character. Many respondents mentioned how important a father or mother had been in teaching them certain values. Probably the most important of these was honesty. An astonishing number said that one of the main reasons they had became successful was because they were truthful or honest, and these were virtues they had acquired from a mother's or a father's example. Robertson Davies says this about his parents, both of whom were writers:

> They were very sincere about what they wrote, and I was brought up—I would not say strictly, because there was nothing harsh about it—but my parents brought me up in a kind of religious atmosphere so that I had a very profound respect for truth, and I was perpetually being reminded, because my parents were very great Bible-quoters, that God is not mocked.

The German physicist Heinz Maier-Leibnitz, who trained two of his students to the Nobel Prize, believes that the responsibility of a scientific mentor is not only to be honest himself but also to make sure of the honesty of his coworkers:

> I don't know whether the word *honesty* is the best word. It's the search for truth in your work. You must criticize yourself, you must consider everything that may contradict what you think, and you must never hide an error. And the whole atmosphere should be so that everybody is like that. And later, when you are head of a lab or an institute, you must make a great effort to help those who are honest, those who don't work only for their careers and try to diminish the work of others. This is the most important task that a professor has. It's absolutely fundamental.

Why is honesty considered so important? The reasons given share a common core, even though they vary depending on the respondent's domain of activity. The physical scientists said that unless they were truthful to their observations of empirical facts, they could not do science, let alone be creative. The social scientists stressed that unless their colleagues respected their truthfulness, the credibility of their ideas would be compromised. What the artists and writers meant by honesty was truthfulness to their own feelings and intuitions. And businesspersons, politicians, and social reformers saw the importance of honesty in their relationship with other people, with the institutions they led or belonged to. In none of these fields could you be ultimately successful if you were not truthful, if you distorted the evidence, either consciously or unconsciously, for your own advantage. Most of the respondents felt fortunate to have acquired this quality from the example of parents.

Only in a few cases does parental influence appear as a thoroughly negative force, an example of what the child wants to avoid in the future. Parents who are always quarreling, who are materialistic, who are unhappy with their lives, show their children ways not to be. But by and large it seems that parents are still the main source of the curiosity and involvement with life that is so characteristic of these creative individuals. This is true even when the parent is no longer alive.

Missing Fathers

A notable contradiction to the importance of parental help is the fact that so many creative people lost their fathers early in life. This pattern is especially true for creative men. About three out of ten men and two out of ten women in our sample were orphaned before they reached their teens.

George Klein, one of the founders of the new domain of tumor biology, is one of them. In a book of essays he describes at length the effect on his life of his father's death. He attributes both his sense of almost arrogant autonomy and the feeling of responsibility that drives him to the fact that he didn't have a living father to fear and to depend on. A young boy deprived of a father may feel a great sense of liberation, a freedom to be and do anything he wants to; at the same time, he may feel the tremendous burden of having to live up to the expectations he himself has attributed to the absent father.

A fatherless boy has the opportunity to invent who he is. He will not have to stand in front of a powerful, critical father and justify himself. On the other hand, he will not have the opportunity to grow up and become a friend and peer to his father. The relationship remains frozen in time, and the psyche of the child always carries the demanding memory of the all-powerful parent. It is possible that the complex and often tortured personality of creative individuals is in part shaped by this ambivalence. George Klein ends his essay entitled "The Fatherless" with the following lines:

Father, little brother, my son, my creator, you who will never allow me to know you, come, oppress me, crush me, mold me into whatever you want—into someone I never was, never will be, if only I could tell you that . . . What would I really want to tell you? Perhaps only this: It is wonderful to live—thank you for making that possible for me. I probably would have killed you if you had lived, but I was never truly able to live while you were dead.

Although few mention their loss with such insight and pathos, in most cases the father's early death seems to leave a drastic mark on the son's psyche. Wayne Booth was raised in a Mormon family, where fathers are looked up to as God's representatives, almost godlike themselves. So when his father died, young Wayne felt a double blow: first the natural grief over losing his father and then the shock to his most basic beliefs: If Father was so powerful, how could he die? But again, with the grievous loss came an unusual gain: In the highly hierarchical Mormon family, at a very young age, he replaced his absent father, gaining the respect and high expectations vested in the oldest male. Wayne Booth's approach to his vocation reflects this ambivalence of his early years. On the one hand, his approach to teaching, to literature, to criticism is informed by a deep respect for order and tradition; on the other hand, he has kept questioning accepted truths, maintaining into his seventies the open curiosity usually associated with youth.

Sometimes the father, though alive, is virtually inaccessible to the son. Such was the case with the Indian composer and musician Ravi Shankar:

I have to talk about my father a little. See, he was a seeker. In the sense that he was always seeking for knowledge. And he was such a learned person. In every subject. Starting from Sanskrit, to music. He was a lawyer by profession, he was in the Privy Council in London, he was with the League of Nations when it started in Geneva. He did his political science in French, almost nearing his fiftieth year. And toward five, six years from the end of his life, he gave up everything and started giving talks on Indian philosophy, even at Columbia University and foundations in New York. He earned a lot on different occasions, and he was offered fantastic jobs, you know, paying a lot of money, but he never saved. He didn't really look after us that way. My mother was separated from him at a very early stage. And he married an English lady, whom I haven't seen but I have heard about. So from my childhood on I saw my mother very unhappy and very lonely. But you know, she was such a great lady. She spent all her energy, time, and everything for the sake of us children. With very little money that we had, she really struggled to give education to my brothers.

My father, as I said, was a very lonely person himself. And he lived away from his family, always. I have hardly seen him. If I add them up—two days, three days, maybe a week. [The] longest was once two weeks in Geneva we spent when he was with the League of Nations. I haven't seen him for more than maybe two or two and a half months altogether. So I had nothing to do with my father, unfortunately, though I respected him and liked him very much. But I grew up very lonely myself, because I was the youngest. My mother was my best friend.

The mere fact of not having a father is not what affects the later life of such children; what counts is the meaning they extract from the event. The death of a father is as likely to destroy a son's curiosity and ambition as it is to enhance it. What makes the difference is whether there is enough emotional and cognitive support for the bereaved child to interpret the loss as a sign that he must take on adult responsibilities and try harder to live up to expectations. And here the mother becomes crucial, because in most cases it is to protect and comfort a loving mother that the child tries to work hard and succeed.

The effects of a parent's death are often quite complicated. Brenda Milner's father, whom she adored, died of tuberculosis when she was eight. He had been a pianist and music critic for the *Manchester Guardian*. Because his job allowed him to spend many mornings at home, he took Brenda's education in hand and taught her the arithmetic tables and had her read Shakespeare to him. His death was "the worst emotional experience" in her life. After this event, Milner was drawn to science in part to avoid being overly influenced by her artistic mother, for whom she had a great affection when separated, but with whom she quarreled if they spent more than a quarter of an hour together in the same room. "I wanted to show that I was doing my own thing and not my mother's thing," she says of her decision to pursue science. "It was selfish, perhaps, but it is very claustrophobic living with someone who has so much emotional investment in you when you are a child."

In some cases, the support to the orphaned child may come from the larger community. When Linus Pauling's father died, the nine-year-old boy was taken on as the responsibility of the other pharmacists in Portland. Each day after school he would go to a different drugstore, and help his father's colleagues prepare medications, thus developing the initial interest in the mysteries of chemistry that he had first acquired by helping in his father's store. Being orphaned certainly did not dampen Pauling's interest in the world around him:

I don't think that I ever sat down and asked myself, now what am I going to do in life? I just went ahead doing what I liked to do. When I was eleven years old, well, first I liked to read. And I read many books. My father is on record as having said, a few months before his death when I was just turning nine, that I was very interested in reading and had already read the Bible and Darwin's *Origin of Species*. And he said I seemed to enjoy history. I can remember when I was twelve and had a course in ancient history in high school—first year—I enjoyed reading this textbook so that by the first few weeks of the year I had read through the whole textbook and was looking around for other material about the ancient world. When I was eleven, I began collecting insects and reading books in entomology. When I was twelve, I made an effort to collect minerals. I found some agates—that was about all I could find and recognize in the Willamette Valley—but I read books on mineralogy and copied tables of properties, hardness and color and streak and other properties, of the minerals out of the books. And then when I was thirteen I became interested in chemistry. I was very excited when I realized that chemists could convert certain substances into other substances with quite different properties. And this was essentially the basis of chemistry. The difference in their properties interested me. Hydrogen and oxygen gases forming water. Or sodium and chlorine forming sodium chloride. Quite different substances from the elements that combined to form the compounds. So ever since then, I have spent much of my time trying better to understand chemistry. And this means really to understand the world, the nature of the universe.

While creative adults often overcome the blow of being orphaned, Jean-Paul Sartre's aphorism that the greatest gift a father can give his son is to die early is an exaggeration. There are just too many examples of a warm and stimulating family context to conclude that hardship or conflict is necessary to unleash the creative urge. In fact, creative individuals seem to have had either exceptionally supportive childhoods or very deprived and challenging ones. What appears to be missing is the vast middle ground.

Another aspect of the family background that shows the same pattern is the social class of parents. Many creative individuals came from quite poor origins and many from professional or upper-class ones; very few hailed from the great middle class. About 30 percent of the parents were farmers, poor immigrants, or blue-collar workers. However, they didn't identify with their lower-class position and had high aspirations for their children's academic advancement. The psychologist Bernice Neugarten's father was a recent immigrant from Europe with little schooling who struggled to make ends meet during the depression. When she came home to Nebraska on a break from college, her father asked, "How do you like it?" Bernice explained that she was beginning to develop an inferiority complex because at the University of Chicago she was surrounded by so many Ph.D. students. "What is that?" her father asked. She answered, "If you just go to college and get a bachelor's degree, Dad, that is not as far as you can go. People can go on and take something you call a master's degree and something you call a doctor's degree." At which point her father waved a finger at her and said, "Then you should do that!"

Only about 10 percent of the families were middle-class. A majority of about 34 percent had fathers who held an intellectual occupation such as professor, writer, orchestra conductor, or research scientist. The remaining quarter were lawyers, physicians, or wealthy businessmen. These proportions are quite different from what one would expect from the frequency of such jobs in society as a whole. Clearly it helps to be born in a family where intellectual behavior is practiced, or in a family that values education as an avenue of mobility—but not in a family that is comfortably middle-class.

The Mirror of Retrospection

In looking back at childhood, it is inevitable that what we see is colored by what happened in the years in between, by present circumstances, and by future goals. A person who is relatively happy and content may remember more sunshine than there actually was, and someone wounded by life may project more misery into the past. We do know that adults who feel positively about themselves describe their childhoods in more favorable terms. What remains unclear is which is the cause and which is the effect? Do these adults have a positive self-concept because they had happier childhoods, or do they remember their childhoods as happier because their adult self-concept is positive?

In some of the interviews with fine artists that I conducted on and off for over twenty years, I noticed an intriguing pattern. An extremely successful young artist in 1963 described his childhood as perfectly normal, even idyllic. He went out of his way to assure me that none of the conflicts and tensions one reads about in biographies of artists had been present in his case. Ten years later, the same artist was having trouble professionally: His paintings were no longer fashionable, critics and collectors seemed to avoid him, his sales had plummeted. Now he began to mention events in his childhood that were definitely less rosy. His father had been aloof and punishing, his mother pushy and possessive. Instead of talking about the lovely summer days spent in the orchard, as he had ten years earlier, now he dwelt on the fact that he had often wet his bed and on the resulting consternation this caused his parents.

Ten years later still, the artistic career of this no longer young man was pretty much washed up. His work was definitely in disfavor, and he had gone through two messy divorces, a severe drug habit, and

was trying to control his alcoholism. Now his description of childhood included alcoholic fathers and uncles, physical abuse, and emotional tyranny. No wonder the child had failed as an adult. Which
version of his early years was closer to the truth? Did the therapy he
underwent when things began to unravel help him see more clearly a
past he had repressed? Or did the helpful therapist provide him with
a script that explained and excused why he had failed? There is no
way to choose among these alternatives with any assurance. It is possible that the early success had been a fluke, and the later failure was
ordained by a miserable childhood. Or it may have been that the
artist failed through no fault of his own, punished by the fickle
changes in taste and market. In any case, there is a powerful pressure
to make the past consistent with the present. Yielding to this pressure
provides a sense of subjective truth whether or not it conforms with
objective events in the past.

So it is possible that the reason our successful creative adults
remember their childhoods as basically warm is that they *are* successful. In order to be consistent with the present, their memory privileges positive past events. Biographers convinced that the early childhood of creative individuals must include suffering may indeed find
much evidence of grief that was not mentioned in our interviews.
Similarly, if biographers assume that a creative person must have had
a happy childhood, they will presumably find quite a bit of evidence
for that, too. The issue does not seem to be what were the objective
facts involved. What matters more is what the children make of these
facts, how they interpret them, what meaning and strength they
extract from them—and how they make sense of their memories in
terms of the events they encounter later in life.

On to School

It is quite strange how little effect school—even high school—seems
to have had on the lives of creative people. Often one senses that, if
anything, school threatened to extinguish the interest and curiosity
that the child had discovered outside its walls. How much did schools
contribute to the accomplishments of Einstein, or Picasso, or T. S.
Eliot? The record is rather grim, especially considering how much
effort, how many resources, and how many hopes go into our formal
educational system.

But if the school itself rarely gets mentioned as a source of inspiration, individual teachers often awaken, sustain, or direct a child's interest. The physicist Eugene Wigner credits László Rátz, a math teacher in the Lutheran high school in Budapest, with having refined and challenged his own interest in mathematics ("no one else could evoke a subject like Rátz"), as well as that of his schoolmates the mathematician John von Neumann, and physicists Leo Szilard and Edward Teller. Clearly, the teacher must have been doing something right.

What made these teachers influential? Two main factors stand out. First, the teachers noticed the student, believed in his or her abilities, and *cared*. Second, the teacher showed care by giving the child extra work to do, greater challenges than the rest of the class received. Wigner describes Rátz as a friendly man who loaned his science books to interested students and gave them tutorials and special tests to challenge their superior abilities. Rosalyn Yalow, who earned a Nobel Prize in medicine although trained as a physicist, remembers her interest in mathematics being awakened in tenth grade, when she was only twelve years old, by a teacher named Mr. Lippy. This is what she says about him, and the other teachers who had been influential:

> I was a good student, and they always gave me lots of extra work to do. I took geometry from Mr. Lippy. He soon brought me into his office. He'd give me math puzzles and math beyond what was formally given in the class, and the same thing happened in chemistry.

John Bardeen became interested in math about the same age, influenced by a teacher who noticed his abilities, encouraged him, and suggested problems he could work on. As a result of this extra attention, when he took high school algebra at age ten he won the end-of-the-year prize in a competitive math exam. The first teacher who took a close interest in the young Linus Pauling was a high school chemistry teacher by the name of William V. Greene:

> He gave me a second year of chemistry so that I got credit for two years of high school chemistry. I was the only student in the second-year chemistry class. He asked me a number of times to stay for an hour at the end of classes and help him operate the bomb calorimeter.

To keep up interest in a subject, a teenager has to enjoy working in it. If the teacher makes the task of learning excessively difficult, the student will feel too frustrated and anxious to really get into it and enjoy it for its own sake. If the teacher makes learning too easy, the student will get bored and lose interest. The teacher has the difficult task of finding the right balance between the challenges he or she gives and the students' skills, so that enjoyment and the desire to learn more result.

But given how famous the students in our sample became a few decades later, it is surprising how many of them have no memory of a special relationship with a teacher. This is especially true of those outside the sciences. Perhaps because a precocious math ability is easier to detect, teachers seem more willing to encourage future scientists than students gifted in the arts or the humanities. In fact, teachers are sometimes tarred uniformly with a black brush. George Klein found all but one of his teachers mediocre and felt that as a teenager he learned more about philosophy and literature from debates with some of his schoolmates than from any of his classes. Brenda Milner, the neuropsychologist, remembers how frustrated she was in school because she could not draw, sing, or do any of the things that were considered "creative" by her teachers. Because she was fiercely competitive, yet inept in the skills prized by her school, she turned into a workaholic in the subjects she *was* good at:

> I used to go home and undo the sewing that I had done so badly during the day, crying. I was also crying when I tried to draw a map of the Great Lakes and I could not get them to connect. There was nothing I liked to do better than algebra equations at night. I mean, it was just a pleasure. But I was not good at these handy-crafty things. And they liked to give prizes for artwork and all of the things I was bad at. You never got any recognition for Latin and algebra, and so on.

Some of these exceptional students remember extracurricular activities more favorably than school subjects. Robertson Davies began to think of himself as a writer when he won most of the literary prizes offered by his school. John Bardeen knew he was good at math when he outperformed his older classmates in a prize competition. Elisabeth Noelle-Neumann could get away with much in school because she wrote poems the teachers thought were beautiful. The future Nobel Prize physicists at the Lutheran school of Budapest were excited by the monthly competition that Rátz made up for his students. Every month a new set of problems was published in the intramural math journal, and the students discussed and debated them at length in their free time. Whoever solved the problems most elegantly by the end of the month won a great deal of recognition from his peers as well as from the teacher.

The Awkward Years

The teenage years are not an easy time for anyone. No matter how much care parents devote to their children during this phase of life, no matter how well suited the culture is to avoiding conflict between adults and adolescents, inevitable tensions emerge when children are between the ages of twelve and twenty. The necessity to adapt to physical changes, to regulate sexual urges, and to establish independence and autonomy while maintaining ties with family and peers are tasks that confront adolescents very suddenly and generally cause quite a lot of misery all around.

Talented teenagers not only are not immune but have some special obstacles to surmount. For instance, they must devote time to the development of their interests and talents, which usually means that they are alone more often than other teens—practicing their music, writing their essays, or solving their math problems. They are on the whole less happy and cheerful as a result (though when alone they are significantly less miserable than their peers are).

Youths with special talents also tend to be less sexually aware and less independent from their families than the norm. This is an important factor in their development, because it means that they spend relatively more time in the protected, playful stages of life in which experimentation and learning are easier to achieve. Sexually active adolescents meld quickly into the program of the genes, and if they achieve autonomy too early they become burdened by social responsibilities like getting a job, keeping house, and rearing children. Thus they have less freedom to try out the new ideas and behaviors that are essential to the development of creativity. At the same time, a youth who is not too interested in sex and depends on his parents is likely to be unpopular, a typical nerd.

Another reason for the lack of popularity is that the intense curiosity and focused interest seem odd to their peers. Original ways of thinking and expression also make them somewhat suspect. Unfortunately, one cannot be exceptional and normal at the same time. Parents often fret and plot to make their talented children more popular without realizing the inherent contradiction. Popularity, or even the strong ties to friends so common in adolescence, tends to make a young person conform to the peer culture. If the peer group itself is intellectual, as in the case of George Klein and a few others, then the conformity supports the development of talent. But in most cases it is not. Then loneliness, however painful, helps protect the interests of the adolescent from being diluted by the typical concerns of that stage of life.

None of the creative people we interviewed remembers being popular in adolescence. Some of them seem to have had a reasonably untroubled time, and others think back on those years with barely disguised horror; however, nostalgia for the teenage years is almost entirely absent. Marginality—the feeling of being on the outside, of being different, of observing with detachment the strange rituals of one's peers was a common theme. Of course, a feeling of marginality is typical in adolescence, but in the case of creative people there are concrete reasons for it.

Some, like the sociologist David Riesman, recognize the necessity—in fact, the positive contribution—of this outsider role: "I had the advantage of my marginality—marginal to the upper class, marginal to my school friends, and so on, but also marginal because of my views, and at times, insulated." Others experienced long periods of illness, which required separation from school and peers. The physicist Heinz Maier-Leibnitz spent three months in bed and the rest of one school year recovering from a lung ailment in the Swiss mountains. Brenda Milner and Donald Campbell complained of poor coordination in youth, which made playing sports or dancing rather difficult. These people did not persevere in their creative careers *because* they were more lonely than other children. However, when they found themselves on the outside, they were able to profit from it instead of lamenting their loneliness.

Those who were somewhat precocious intellectually—such as John Bardeen, Manfred Eigen, Enrico Randone, and Rosalyn Yalow—experienced another sort of marginality. They were promoted into higher grades and therefore grew up surrounded by older teenagers with whom they did not form close friendships. John Gardner remembers: "I moved very rapidly through school. This was a period when you were allowed to move as fast as you wanted to—provided you were able. So I finished the first eight grades in five years, and the result was I was with children older and bigger than myself."

Performance in school matters more in some domains than in others. In mathematics and the sciences, the exposure one gets in high school is necessary for further advancement. Doing well in advanced courses is not sufficient, but it is a necessary condition for being accepted to a good college and then to a good graduate department, which in turn is a necessary step to a later career. But performance in high school is a poor indicator of future creativity in the arts and the humanities.

Young artists, especially visual artists, are notoriously uninterested in academic subjects, and their scholastic records usually reflect this. It is probably for this reason that the French—who reckon mental ability in rather rigid rational terms—say *bête comme un artiste* (dumb as an artist) when they want to put down someone's intellect. Certainly Eva Zeisel, an accomplished artist whose ceramic creations are exhibited in many museums, including the Museum of Modern Art in New York, felt that she was "not considered the bright child in the family" (it is true that she was being compared to uncles Michael and Karl Polanyi and cousin Leo Szilard). She tells how when she was seventeen she overheard a couple talking about her a few rows back at a concert: "Her grandmother is such a clever, bright, intellectual person. Her mother is such a beauty. And now look at her . . ."

Michael Snow, the versatile Canadian artist-musician-filmmaker, admits that he wasn't a very good student in high school and was surprised to be awarded the art prize in his senior year. Ravi Shankar started touring with a musical troupe at age ten, and after that his education was conducted by his guru, an elder musician.

THREADS OF CONTINUITY

In some cases, the continuity of interest from childhood to later life is direct; in others it is strangely convoluted. Linus Pauling's interest in the material composition of the universe started when he worked in his father's drugstore. Elisabeth Noelle-Neumann's interest in her countrymen's opinions and values can be traced to her games with the imaginary inhabitants of the toy villages she built. Frank Offner remembers an important early event in his life:

> I know that I always wanted to play and make things like mechanical sets. . . . When I was six or seven years old, we were in New York and I remember at the Museum of Natural History there was a seismograph which had a stylus working across the smoked drum, and there were a couple of heavy weights, and I asked my father how it worked and he said, "I don't know." And that was the first time . . . you know, like all kids do, I thought my father knew everything. But so I was interested in how that worked, and I figured it out.

What makes this memory so interesting is that all through his life, some of Offner's most important inventions involved a stylus moving across a drum. For instance, he invented a crystal-operated pen recorder, "which made the cardiograph a hundred times better than anything anyone had done before," and he perfected the first EEG machines. Yet Offner saw nothing especially meaningful in this continuity, and when it was pointed out to him, he shrugged it off.

There are also cases in which the individual's adult theme harks back to the interests of an earlier generation. C. Vann Woodward, who revolutionized the way we understand the history of the American South, traces his interest in his vocation far back:

> That interest was born out of a personal experience of growing up there and feeling very strongly about it, one way or the other. I have always told my students: "If you are not really interested in this subject and do not feel strongly about it, don't go into it." And of course much of my writing was concerned with those controversies and struggles that were going on at the time, and what their background and their origins and their history were.
>
> The place I grew up was important. The environment and the time following the Civil War and Reconstruction. There was talk about that from my earliest recollections. It is the defeated who really talk and think about a war, not the victors. And I grew up in a family that came of slave-owning stock and were planters, and then in the small towns where we lived my father was a superintendent of the public schools.

The artist Ellen Lanyon's maternal grandfather came to the United States from Yorkshire, England, to paint murals for the World's Columbian Exposition of 1893. Because she was his oldest grandchild, Ellen had the feeling that she was destined to inherit her grandfather's calling and his creative spirit.

And when I was about twelve years old, my grandfather died. My father and mother put together his equipment that was left plus new tubes of paint, et cetera, and it was presented to me on my twelfth birthday as a sort of, you know, a gesture. Passing the torch or something. And so I started painting, and I painted a self-portrait, the first thing I tried. I can absolutely remember the place, the room, you know, and everything. I don't know what happened to the painting. It's somewhere. I think that my mother has it. But in any case, I think that's the kind of beginning that sets a pattern for a person.

Nowhere was intergenerational continuity more clearly evident than in the case of the physicist Heinz Maier-Leibnitz. He is a descendant of Gottfried Wilhem Leibniz (1646–1716). At a distance of more than two and a half centuries, the parallels in their lives are quite astonishing. G. W. Leibniz is identified in the *Encyclopædia Britannica* as "philosopher, mathematician, political advisor." Maier-Leibnitz is an experimental nuclear physicist and has been a scientific adviser to the German government. The elder one was one of the founders of the German Academy of Sciences in 1700; the younger was one of its recent presidents. G. W. Leibniz was elected a foreign member of the French Academy of Sciences because of his attempts to renew German and French intellectual cooperation after a war between the two countries; about 250 years later Maier-Leibnitz received the same honor for the same reasons. G. W. Leibniz developed an "algebra of thought" according to which all reasoning was supposed to be reducible to an ordered combination of basic elements. His descendant has been working on a procedure by which the truth value of television and newspaper stories could be evaluated by breaking them down into basic propositions.

It should be added, however, that for each creative person whose life seems like a seamless unfolding from childhood into old age, or whose interests seem preordained even before birth, there is another whose later career seems to be the product of chance or of an interest that appears seemingly out of nowhere long after the early years are past.

WHAT SHAPES CREATIVE LIVES?

We are used to thinking about the way a life unfolds in a deterministic fashion. Even before modern psychoanalysis, it was believed that adulthood is molded by the events experienced in infancy and childhood: "As the twig is bent, so the tree grows." "The child is father to the man." Certainly after Freud it has become even more of a commonplace to assume that whatever ails us psychically is the result of some unresolved childhood complex. And by extension, we seek the causes of the present in the past. To a large extent, of course, such assumptions are true.

But reflecting on the lives of these creative individuals highlights a different set of possibilities. If the future is indeed determined by the past, we should be able to see clearer patterns in these accounts. Yet what is astonishing is the great variety of paths that led to eminence. Some of our respondents were precocious—almost prodigious—and others had a normal childhood. Some had difficult early years, lost a parent, or experienced various forms of hardship; others had happy family lives. A few even had normal childhoods. Some encountered supportive teachers; others were ignored and had bad experiences with mentors. There were some who knew early in life what career they would pursue, while others changed their direction as they matured. Recognition came early to some and late to others.

This kind of pattern—or rather, the lack of it—suggests an explanation of development that is different from the usual deterministic one. It seems that the men and women we studied were not shaped, once and for all, either by their genes or by the events of early life. Rather, as they moved along in time, being bombarded by external events, encountering good people and bad, good breaks and bad, they had to make do with whatever came to hand. Instead of being shaped by events, they shaped events to suit their purposes.

Presumably many children who started out with talents equal or superior to those of the ones we met in this group fell by the wayside either because they lacked resolve or because the conditions they encountered were too harsh. They never had an understanding teacher, a lucky break that led them to a scholarship, a mentor, a job that would keep them on track. So the Paulings and the Salks are the survivors, the gifted few who also were fortunate enough to make use of the opportunities that came their way.

According to this view, a creative life is still determined, but what determines it is a will moving across time—the fierce determination to succeed, to make sense of the world, to use whatever means to unravel some of the mysteries of the universe. If the parents are loving and stimulating, great, that is just what a son or daughter needs to build the future. If the parents die, this is terrible, but what can a young child do? Lick the wounds and make the best of it.

Of course, this still leaves the question, So where does this fierce determination, this unquenchable curiosity come from? Perhaps that question is too reductionistic to be useful. Many causes could be at the root of curiosity: genetically programmed sensitivity, stimulating early experiences, and, if Freud was right, a repressed sexual interest. It may not be so important to know precisely where the seeds come from. What *is* important is to recognize the interest when it shows itself, nurture it, and provide the opportunities for it to grow into a creative life.

EIGHT

THE LATER YEARS

Until very recently, creative persons tended to learn their craft by apprenticing to a master, or by teaching themselves the elements of a domain through trial and error. Higher education was open to very few, and until two centuries or so ago, it was mostly reserved for scholars and clergymen. Copernicus was a church canon who taught himself mathematics and astronomy, Gregor Mendel was a monk, and Galileo was trained as a physician. But nowadays it is almost unthinkable for a person to change a domain without first having learned it in college. Even poets and painters are expected to get advanced degrees.

COLLEGE AND PROFESSION

For many of our respondents, the years in college and graduate school were a high point—if not *the* high point—of life. This is the period when they found their voice, when the vocation became clear. Often they had come from small provincial settings where they felt odd and disoriented. College provided soulmates and teachers who were able to appreciate their uniqueness.

For some individuals it was also in college that they could first assert their independence: David Riesman chose the law instead of the medical career favored by his father; others, like Jonas Salk, switched in the opposite direction. Isabella Karle, like most other women who went into science, had to convince her parents that this was a better choice than becoming a teacher. John Gardner, who wanted to become a writer, decided to go into psychology instead. Anthony Hecht, who loved music and mathematics as a teenager, was seduced by literature into becoming a poet.

But these were not necessarily easy years either. Linus Pauling, despite his brilliance, had to work through college at a schedule that few undergraduates would now consider possible. After enrolling at the Oregon Agricultural College on the advice of a friend's parents:

> I made a little money by odd jobs, working for the college, killing dandelions on the lawn by dipping a stick in a bucket containing sodium arsenate solution and then stabbing the stick into the dandelion plant. Every day I chopped wood, a quarter of a cord perhaps, into lengths—they were already sawed—into a size that would go into the wood-burning stoves in the girls' dormitory. Twice a week I cut up a quarter of a beef into steaks or roasts, and every day I mopped the big kitchen, the very large kitchen area. Then at the end of my sophomore year, I got a job as a paving engineer, laying blacktop pavement in the mountains of southern Oregon.

Even in college, the performance of the future creator is rarely off the scale. When Brenda Milner was taking her college exams in Cambridge with twelve other students in her cohort, she was overwhelmed by the brilliance of a fellow student whose theoretical ideas, she felt, were way beyond hers. She was sure that he would set the standard on the exam and she would not get a "first," thereby forfeiting her chances for a fellowship. "But in the end it was so funny—he never took the exams. He was brilliant, but not focused. I think he was found in a little backroom in London with some rats in a bath, or something. But I did very well on the exams because I had this man to pace me." In a similar vein, Rosalyn Yalow remembers:

There was another girl in college with me and we took a number of courses together. When we took physical chemistry, she got ninety, I got sixty, and everybody else got thirty. She actually took a master's degree with Hans Bethe at Cornell but then dropped out for a number of years when her husband came back from the army. She eventually finished her Ph.D. but never really made anything with it. Inherently, she was probably smarter than I was, but she didn't have the same drive.

Milner calls it *focus,* Yalow calls it *drive*—this advantage they had over more brilliant fellow students. After curiosity, this quality of concentrated attention is what creative individuals mention most often as having set them apart in college from their peers. Without this quality, they could not have sustained the hard work, the "perspiration." Curiosity and drive are in many ways the yin and the yang that need to be combined in order to achieve something new. The first requires openness to outside stimuli, the second inner focus. The first is playful, the second serious; the first deals with objects and ideas for their own sake, the second is competitive and achievement oriented. Both are required for creativity to become actualized.

If teachers help or hinder the development of creative individuals in high school, they do so even more in college. College teachers are important in two ways. First, they can ignite a person's dormant interest in a subject and provide the right intellectual challenge that leads to a lifelong vocation. Second, they often exert themselves in various ways to make sure that the student is noticed by other important members of the field. A college graduate in the sciences is unlikely to be admitted to a good laboratory without her college teacher writing enthusiastic letters to the lab director; a student in literature or the arts is helped enormously in placing his first poems or paintings if his teacher is willing to put in extra effort and pull a few strings. A B.A. degree (or even a Ph.D., for that matter) is just not worth much in terms of a career without the active support of one's teachers, a support that is needed to attract the attention of the gatekeepers at the next higher levels.

Isabella Karle met one such teacher early in her college career:

The man who was my first professor at Wayne State University took a personal interest in me. He said: "Well, you're going on to graduate school, of course?" And I said: "What's that?" And he told me about it, and I said it sounded like a good idea, so he and I kept up a correspondence after I went to the University of Michigan for a number of years. He advised me on the courses to take, the kind of things that may interest me, so that was very nice of him.

Anthony Hecht heard about John Crowe Ransom while he was in the army, and as soon as he was demobilized he enrolled at Kenyon College to study with the older poet. Not only did Ransom publish Hecht's first verse in the *Kenyon Review*, which he had been editing ("that was the beginning of my publishing career"), but when a member of the English Department became ill he hired Hecht to teach a freshman English course ("that was the beginning of my teaching career").

Entering a career requires a great deal of determination and a good dose of luck. In fact, the majority of the people we interviewed mentioned luck most frequently as the reason they had been successful. Being in the right place at the right time and meeting the right people are almost necessary to take off within a field. And unless one becomes visible in a field, it is very difficult to make a creative contribution to it. This is true even of those individuals who seem most isolated, most alienated from their culture. It is difficult to imagine Martin Luther's ideas spreading very far if they had not been voiced in what was then the center of German intellectual life, or of Kafka's work making a great impact if he had written in Urdu, or if he had not been noticed by critics in nearby Vienna, which at the time was the center of modernist experimentation.

Almost all the women scientists of the generation we interviewed mentioned that without World War II it would probably have been impossible for them to get graduate training, fellowships, postdoctoral positions, and faculty appointments. But because so many men were fighting in the war, and professors needed graduate student assistants, these women were grudgingly admitted into higher education. When Rosalyn Yalow was accepted to Illinois as a graduate student in physics in 1941, she was the second female—the previous woman having matriculated in 1917. "They had to make a war so that I could go into graduate school," she said. This is almost exactly the story told by Brenda Milner, Isabella Karle, and Margaret Butler.

It is very possible that if these women had been born just a decade earlier, they would have been prevented from making a creative contribution to their respective domains.

SUPPORTIVE PARTNERS

The individuals in our sample had, as a rule, stable and satisfying marital relationships. Some of those in the arts started out having a vigorous and varied sex life, but most of them married early and stayed married to their spouses for thirty, forty, or more than fifty years.

One of the exceptions was octogenarian Bradley Smith, the photographer who answered our question about what accomplishment in his life he was most proud of with the terse words: "Making love, probably." He claims that he became sexually active at age six and never looked back. To the question about what fuels the inspired mental associations that lead to his art, he said: "Well, I think probably sex and songs. If I was asked to reduce it to what keeps me going, I think that the creative instinct is fed by sex and music. Without them I think that you would wither, pretty much." The sculptor and cinematographer Michael Snow concurs: "Well . . . an important aspect of creativity is sex or sexual desire . . . if I can put it in the colloquial, I'm still horny, but I was much more horny then [referring to thirty years earlier]." A musician's wife, after the interview, turned to us and said, in front of her husband: "What he didn't tell you is that all through life what inspired him was girls." The writers described fiery romantic lives in their youth, but they all eventually settled down to domestic bliss.

But the majority conformed to a more sedate sexual pattern. Recent studies suggest that the amount of dalliance, marital infidelity, and sexual experimentation is much less than earlier estimates had suggested. When asked which of their accomplishments they were most proud of, a great many of our respondents—and almost as many men as women—mentioned their family and children. When explaining what enabled them to accomplish what they had achieved, several pointed to the indispensable help of their spouses. And these answers did not ring perfunctory.

Hans Bethe, one of the leading physicists earlier this century and teacher of many of the later ones, volunteered: "My wife has very much influenced my life and made me happy. Before I was married, I was never very happy. I had happy times, moments, weeks, but since I am married I am more or less continually happy. We talk a lot over meals, we are very fond of walking in the mountains." Not a bad endorsment, after fifty-four years of marriage. Anthony Hecht expresses himself in almost identical terms:

> I felt somehow as though I were floundering as a human being in many ways, making many errors and wasting my time and not being happy. Not that I was not happy before, but those periods of happiness were brief. But since my marriage to Helen and the birth of our son there has been an almost beatific tranquillity and serenity that has made everything seem worthwhile.

Robertson Davies has also been married for fifty-four years, having met his wife when they were both trying for a stage career at the Old Vic. She was a prompter and knew every word of the classic repertoire from start to finish.

> And Shakespeare has played an extraordinary role in our marriage as a source of quotations and jokes and references, which are fathomless. I feel that I am uncommonly lucky because we've had such a terribly good time together. It's always been an adventure and we haven't come to the end yet. We haven't finished talking, and I swear that conversation is more important to marriage than sex. It has been enormously helpful in my work because my wife sort of clears the way so that I can get down to business and work without interruptions.

Hecht agrees: "The only thing you need for poetry that seems to me essential is quiet—and time. And if you have a spouse who is understanding, he or she will see to it that you are not interfered with and that time and quiet are available to you." This theme of the spouse as a protective buffer against the intrusions of the world was repeated again and again by practically all the stably married individuals. Linus Pauling, who had been married to his college sweetheart for fifty-eight years before being widowed, gave this very politically incorrect advice to a hypothetical young scholar:

You ought to go up to Corvallis, Oregon [where the University of Oregon is located], and look around for some young woman who's majoring in home economics. This is of course what happened to me. I was fortunate, I believe, that my wife felt her duty in life and her pleasure in life would come from her family, her husband and her children. And that the way that she could best contribute would be to see to it that I was not bothered by the problems that are involved in the household; that she would settle all of these problems in such a way that I could devote all of my time to my work. So I was really fortunate in that way.

John Gardner, whose political career involved a great deal of stress, believes that he was able to maintain his sanity primarily because of a harmonious family life:

> We've been married fifty-seven years now, fifty-seven years yesterday, and I have a very, very strong family orientation. My two daughters, who are now in midlife, and their children—four grandchildren. We're a very close unit and that's very important to me. I think it's an important counterbalance, particularly to an active life, particularly to a life that's very abrasive—fighting, leading in the public arena, and so forth.

Inevitably there were also badly strained marriages. Achieving a creative result in any field is stressful enough for one person to bear; it is much harder on one's partner. In fact, it is surprising what a strong sense of responsibility these individuals generally felt for keeping their relationships stable. John Reed divorced after twenty-seven years of marriage; during a period of one year when his wife was hospitalized, he took time off from his rapidly ascending career to take care of their four children, aged two through twelve. "I spent the year playing Daddy with them, which turned out in retrospect to be the best investment decision I ever made. Raising kids is a far more rewarding thing than earning money for a company, in terms of a sense of satisfaction." Jacob Rabinow's wife, who has been married to the inventor for almost sixty years, summarizes the situation philosophically: "Living with an inventor is like being a golf widow, but it's not for Sundays only!"

Ravi Shankar separated from his first wife in 1967. Several years later he met his present wife, whom he married twelve years after their meeting:

> And believe me, I feel very much happier now. I feel at peace, and it is something which I have missed. I have been running around so much, never giving time to my family. And I have to blame myself, you know, not being able to be a family person. But for the first time I am now going through this wonderful experience. My wife is not a performing musician, but she's a musician, she's also a dancer. She's very sympathetic and very helpful to me. And I love her and feel very much at peace now.

The Women's View

The married women in our sample also felt that their husbands had freed them to concentrate on their work. The sculptor Nina Holton answered the question about what she was most proud of in her life as follows: "It's the combination of having been so lucky to have had a very good family life, a husband whom I love and who has been most marvelous, plus my own interest in so many things, particularly sculpture, which made it a life which was so complete, and in a way stunning." The marriage of historian and scriptwriter Natalie Zemon Davis, who teaches at Princeton, has survived much separation from her husband, who teaches at Toronto. They call each other every day and spend most weekends together.

In addition, husbands often served as mentors to their wives and helped them to get started on their careers. Margaret Butler says that she was able to overcome her employer's skepticism about women scientists in great part because "I had an awful good backing in my husband. He's the one." In 1945 Elisabeth Noelle-Neumann founded the opinion polling institute she now directs with the help of her more experienced husband. Developmental psychologist Bernice Neugarten's advice about balancing family and work life for professional women is:

> A laid-back approach is the way you can do it, better than getting all uptight, if you can afford to be that, if you can manage that. My husband was very sympathetic about this. He said: "Do as you like, anything I can do to help, as long as the kids are cared

for, I have no worries about that, use your time as you want." And
that was very important. I had other women friends who were not
so supported by their husbands in those years. We are talking about
the forties and fifties, not the nineties.

But the unequal gender roles also inject strong ambiguities into
the married life of creative women. Elise Boulding, who had played
the cello and studied music in college, married a year after gradua-
tion. Her husband, Kenneth, had already achieved an international
reputation as an economist. He introduced her to the literature of
the social sciences and to new perspectives for understanding the task
of achieving world peace, which was one of her chief concerns. She
took an M.A. degree in sociology and was ready to launch herself
into a career in the social sciences. Then the Bouldings had five children, spaced two years apart. The children were very welcome, but
the ten years with diapers left her very far behind her husband professionally. It was not easy afterward to be always in his shadow, and it
took long years for Elise to find her own scholarly identity and self-
assurance.

The poet Hilde Domin was married to an eminent classical
scholar. Although their marriage was strong and happy, Hilde felt
that her husband was jealous of her attempts to write verse. When
she first showed one of her poems to him, he said acidly: "Well, look
at what the cat dragged in." It was not until after he died that she
began to devote herself wholly to writing, and not so long after she
became one of the most widely read poets in Germany.

Because of this tension between two usually strong individuals, the
relationship sometimes cannot take the strain. Yet most of the time
the divorce is amicable, and the former spouses keep seeing each
other on friendly terms. Hazel Henderson (who since remarried)
says:

I was divorced ten years ago. I'm still very friendly with my ex-
husband, but I had to come to terms with the fact that I couldn't
be a wife the way this culture defines what a wife is. And he had
every right to try to find a wife. But we didn't get divorced until
our daughter was eighteen, and so I think that we fulfilled our
obligations pretty well, and we have a good relation with each
other and with her.

Brenda Milner, who says her husband had been "enormously helpful" in her career, later divorced but insists: "he is my best friend, probably. I mean, there is no bitterness between us at all, quite the contrary. To this day we influence each other a lot. We talk a great deal."

These accounts of the relationships of creative individuals are so diverse that they cannot prove any one point. But they can *disprove* a generally held notion that people who achieve creative eminence are unusually promiscuous and fickle in their human ties. In fact, the opposite seems closer to the truth: These individuals are aware that a lasting, exclusive relationship is the best safeguard of that peace of mind they need in order to focus on their creative pursuit. And if they are lucky, they find a partner who fills that need.

THE MAKING OF CAREERS

Creativity is rarely the product of a single moment; perhaps more often it is the result of a lifetime, like Darwin's slow accumulation of facts and hypotheses that resulted in his epoch-making description of the evolutionary process. It is true that in mathematics and the sciences generally, a few short papers—such as Einstein's 1905 articles on special relativity—may make enough of a difference to change an entire domain of learning. The physicist Freeman Dyson believes that his scientific stature was established by the two papers he published in 1948 in the *Physical Review,* which took him six months to puzzle about, a few hours to see the solution of, and another six months to write. However, even in such exceptional cases, if we add in the years that preceded these great events, the years of training and thinking, then the creative process shows its real magnitude: much longer than it appears when one pays attention to just the single crucial episode.

Most creative achievements are part of a long-term commitment to a domain of interest that starts somewhere in childhood, proceeds through schools, and continues in a university, a research laboratory, an artist's studio, a writer's garret, or a business corporation. As this list suggests, occupational paths vary enormously depending on the domain in which a person is active. The career of a poet is very different from that of a high-energy physicist or the CEO of a banking conglomerate. Moreover, the career lines of men and women can vary a great deal even within the same sub-field. Are there in fact *any* commonalities we can talk about in such a diverse group?

There is one sense in which the careers of all creative individuals are similar: They are not careers in the ordinary sense of the term. Most of us join an organization at an entry level, perform a pre-scribed role for a number of years, and leave at a higher level. What we do during this period is more or less known in advance, and oth-ers could do the same job if we didn't. A worker may start as a tool-maker and leave as a foreman; a teacher may teach for thirty years and become a principal; a soldier may become a sargeant; a young lawyer may end up as a partner of the firm, and so forth. These roles are relatively fixed, and we fit into them. It is true that in the postin-dustrial economy we are now entering this pattern may become less rigid, but I would still be very surprised if most people do not con-tinue to follow career lines that are laid out for them.

In contrast, creative individuals usually are forced to invent the jobs they will be doing all through their lives. One could not have been a psychoanalyst before Freud, an aeronautical engineer before the Wright brothers, an electrician before Galvani, Volta, and Edi-son, or a radiologist before Roentgen. These individuals not only discovered new ways of thinking and of doing things but also became the first practitioners in the domains they discovered and made it possible for others to have jobs and careers in them. So creative indi-viduals don't *have* careers; they *create* them. In addition, these pio-neers must create a field that will follow their ideas, or their discov-ery will soon vanish from the culture. Freud had to attract physicians and neurologists to his camp, the Wright brothers had to convince other mechanics that aeronautics was going to be a feasible career. Because careers can take place only within fields, if a person wants to have a career in a field that does not exist, he or she must invent it. And that is what people who create new domains do.

But what about writers, musicians, and artists? These are some of the oldest professions. So it must be wrong to claim that a creative poet creates the role of a poet. Yet there is a very real sense in which this actually is true. Each poet, musician, or artist who leaves a mark must find a way to write, compose, or paint like no one has done before. So while the role of artists is an old one, the substance of what they do is unprecedented. Two examples, one from the sciences and one from the arts, may illustrate what is involved in creating creative careers.

Rosalyn Yalow's parents had no education, but they read to their children and expected them to go to college. For whatever reason—and Rosalyn tends to believe it has to do with genetic inheritance—she always felt sure that, somehow or other, she would make it in the world. She still keeps a picture of herself as a three-year-old, wearing boxing gloves, standing above her elder brother lying on the ground (the brother went on to work at the post office). In school she found herself enjoying math:

> I was good in math and I was a good student in general. And I worked hard and I was responsive when they wanted to give me extra things to do.
>
> Q: Did you do these things because they asked you to do it? Because you saw it as the way to succeed?
>
> A: No, I did not see it as the way to succeed. I did it because I liked doing it. You know, Otis, the physics teacher, would use demonstrations for the principles of physics. Well, you had to do work to get the demonstrations to work. So he would give me the job of trying it. And it was interesting. I liked doing it. I was willing to spend the extra time to do these things.

During high school and college, Yalow was fortunate in getting a string of science and math teachers who recognized her ability and motivation and who kept challenging her with increasingly difficult tasks. During this period she also read Marie Curie's biography, which made a great impression on her and from then on served as a distant role model. In college, in the 1930s, she formed the opinion (shared by most scientists of her generation) that "physics was the most exciting field in the world." She was particularly attracted to artificial radioactivity because she sensed that it was a tool that could open up many areas of science and could become important in chemistry and biology as well.

Because of the great breakthroughs in physics during this period, her college teachers advised Yalow to go on to graduate school and become a physicist. At this time there were very few jobs in pure physics anywhere. Even such future greats as Eugene Wigner or Leo Szilard were pressured by their parents to specialize in engineering so that they could fall back on recognizable careers if necessary. Yalow loved physics, but to be on the safe side she took up stenography so she could have a secretarial job if all else failed.

But she was lucky again. In part because World War II had left so many openings in graduate school, she was accepted at the University of Illinois and was given assistantships and research experience. The other fortunate conjunction was that a whole generation of new technology was coming on line: the cyclotron, the betatron, all the new machines that made it possible to study the isotopes whose characteristics she felt might lead to important scientific applications.

She was hired in 1947 by the Bronx Veterans Administration Hospital to work in the radiotherapy department. Everyone else was an M.D., while Yalow had never taken a biology course in her life. But by working closely with physicians, she began to learn how her knowledge of the physics of radiation could help solve puzzles about human physiology and disease. In 1950 she joined forces with a physician, Solomon A. Berson, and a few years later they formed a department of radioisotope service, which then became a department of nuclear medicine. There had been no such departments before; Yalow was one of those who "invented" nuclear medicine. Now people can have routine careers in that field, but half a century ago, it did not exist.

It was while working in the nuclear medicine lab that Yalow became involved in a series of experiments that eventually led to her most important breakthroughs. In the course of trying to figure out why some people suffered from diabetes, her lab succeeded in using radium H for measuring not only insulin but also peptide hormones and the antigens that the body produced. This resulted in the development of the radioimmunoassay method (RIA), which Yalow and Solomon Berson first used in 1959 to study insulin concentration in the blood of diabetics but which soon was successfully applied to hundreds of other diagnostic tasks. As a result, Yalow received some of the most coveted prizes in the field of medical research. In 1976 she was the first woman to be awarded the Albert Lasker Prize for basic medical research, and in 1977 she received the Nobel Prize in physiology and medicine.

Nothing about Yalow's career was routine. Only her basic physics training had been conventional. But after that, she specialized in the still young domain of radiation physics. Later, she was among the first scientists to apply radiation physics to biological problems. And she was the first person to discover a way to use radioisotopes to measure what goes on inside the human body. There was no blueprint she could follow in her career. There was no job, no role for doing the kind of things she ended up doing. Of course, many favorable circumstances had to converge: the development of theory in physics; the availability of large machines for producing and measuring radiation, left over from the war effort; World War II itself, which allowed Yalow to get the education she needed; supportive parents and all the encouraging teachers in her childhood; and finally, the recognition of an already established field (in her case, medicine) that would legitimate her attempts to develop a new one. Without this rare convergence it is unlikely that Yalow could have achieved what she did. But she had to put together all these pieces by herself without a manual. How did she do it?

Yalow explains her success very simply: "I was always interested in learning and I was always interested in using what I learned." Basically, she spent her life talking to physicians, finding out what problems they encountered, trying to think of a way to solve the problems by experiment, then running the experiments. Experimental results are rarely conclusive; so she reflected on what she found, talked it over with colleagues, ran some new tests, and repeated the cycle several times before—if she was lucky—something interesting turned up. For instance, this is her account of her major discovery:

> *Something comes up, and you recognize that it has happened.* I mean, just like the way in which radioimmunoassay developed. We were testing a hypothesis that diabetics destroyed insulin quickly, and this is why adult diabetics did not have enough insulin. So we gave labeled insulin [that is, insulin marked chemically so that its site and rate of absorption can be measured] and we saw that it did not disappear quickly; it disappeared more slowly! So now we had to examine why it disappeared more slowly. And we discovered the antibody. So we attempted to quantify the amount of antibody, and when we did that we realized that we could measure the insulin reciprocally. We did not set out to develop a radioimmunoassay; it fell out from an unrelated question. Now, then, when we had the radioimmunoassay, we said, "Ah! We can use it to measure all kinds of things."

Of course, this makes it sound all so easy. It compresses into a few sentences years of exciting but exhausting work. Nevertheless, the general outline is the same whether the breakthrough occurs in art or physics, poetry or business: A new way of doing things is discovered because the person is always open to new learning and has the drive to carry through the new idea that emerges from that learning. It may be interesting to compare Yalow's career with that of an artist.

When I interviewed Michael Snow in 1994, the streetlights of the city of Toronto were festooned with colorful banners featuring the most famous image Snow had created: *Walking Woman,* the outline of a strangely dynamic and seductive female figure. The banners were announcing three separate retrospective shows of his work: one taking up most of the temporary exhibit space in the huge Ontario Gallery of Art and two in fashionable venues in other parts of town. At the same time, concerts of his music were being held, and some of his experimental films were being shown. The day of the interview he answered long-distance calls from Lisbon concerning a show of his work next fall; from the Centre Pompidou in Paris, where some of his sculpture had been vandalized; and a request to borrow a few of his paintings to complement an exhibition of the works of the Belgian artist Rene Magritte. Michael Snow's career has certainly reached an apogee few artists ever reach. And like the careers of other creative artists, it was not one that could be traced to any existing pattern.

Snow says that his protean interests "started with confusion, which I tried to dispel by concentrating on one or the other of the mediums I was interested in." One of the media was music:

My mother was a very fine classical pianist—she still is, at ninety; she is not a professional and never wanted to be, but she really can play very well. She wanted me to take piano lessons and I refused. She tried in many ways to convince me that I should, and I just wouldn't do it. I guess it was in my second year in high school, I happened to hear some jazz things on the radio which really more than impressed me. I never heard anything like it, and it just knocked me out. And I started to become interested in jazz, and in a really kind of zealous way I listened to everything, and I met other people who were interested in it. I wanted to play that way so I started to try to teach myself how to play.

We had two pianos, one upstairs and one in the basement. I used to play in the basement. And once my mother came down and listened to me for a while before she made herself evident. And we talked. You know, the first thing she said is, "You're playing the piano. How can you be playing the piano?" So we had this little talk and I said, "Well, I just became interested in playing it." And she said, "Well, you should take lessons." And I said, "No, I'm doing OK."

So Snow went on to join experimental jazz groups, spent some years in New York learning from the local music scene, founded his own group, did some recordings, and ended up having quite an influence on the development of contemporary Canadian music. He had the same unorthodox approach to the other forms of art he set his hand to. In high school, the one subject he did well in was drawing. So he decided to go to art school, where he met an influential teacher—that is, a teacher who responded to his work, commented on it, suggested books to read and artists to look at. He also suggested that Snow submit a couple of paintings to a group show of the Ontario Society of Artists. "And they were accepted, which turned out to be kind of sensational because no student work had ever been accepted before." As he finished college, his abstract paintings were beginning to attract attention. But Snow, like many of his contemporaries, was impatient with the limitations of two-dimensional surfaces and suspicious of the use that paintings were put to by those who bought them, usually as part of a "decorating scheme." So he moved into sculpture, photography, holography, and filmmaking, exploring the possibilities of various media and materials. All through this time he still felt confused and unsure of himself. The first time he realized that he was becoming a real artist was in the mid-1960s:

Yeah, it's almost embarrassing. In a way it depends on recognition. Certainly it does. I guess there shouldn't be anything embarrassing about that. The film *Wavelength* won a prize in a film festival. It got a lot of publicity, and I won five thousand dollars, the grand prize. I didn't think of it going on down the ages at all. I thought, I'm going to make this thing, and I hope it's good. And then it won the prize, and it got me a lot of publicity, and I was asked to do a tour of Europe with my films, and I was in some collections, and I did have a career, yeah.

Although all creative persons, in breaking new ground, must create careers for themselves, this is especially true for artists, musicians, and writers. They are often left to their own devices, exposed to the vagaries of market forces and changing tastes, without being able to rely on the protection of institutions. It is not surprising that so many promising artists give up and take refuge in teaching, rehabbing old houses, and designing for industry rather than flounder forever in the uncharted seas of so vague a profession. Those who persevere and succeed must be creative not only in their manipulation of symbols but perhaps even more in shaping a future for themselves, a career that will enable them to survive while continuing to explore the strange universe in which they live.

THE TASK OF GENERATIVITY

According to the developmental psychologist Erik Erikson, the defining task of a person's middle years is to achieve generativity. This involves being able to pass on both one's genes and one's memes. The first refers to leaving children, the second to leaving one's ideas, values, knowledge, and skills to the next generation. It is much easier to come to terms with one's mortality when one knows that parts of oneself will continue to live on after one's death.

There is often a presumption that these two ways of being generative—the physical and the cultural—are at odds with each other. The Romans had a saying: *libri aut liberi* (books or children), referring to how difficult it was to have it both ways. In fact, in many cultures it has been the case that those who wrote the books—the monks in early Christendom, the Tibetan lamas, or Buddhist monks—were not supposed to have children, at least officially. Yet there are of course many notable exceptions, and the people in this book in general are among them. Most respondents had children whom they appreciated greatly ("my children" was probably the most common answer to the question about their proudest accomplishment), and they had the opportunity to see their ideas carried on by students or followers.

Here is the historian John Hope Franklin:

I would say that one of the major sources of pride is the cadre of Ph.D.'s that I trained at the University of Chicago and who are now many of them distinguished historians. They range from being in government service of one kind or another all the way to department chairman at various institutions, and they have produced a very considerable body of writing, largely on the nineteenth century, which is my own specialty. So that aside from my own personal creativity, I would say my projection in them is itself a great source of satisfaction. That is, the fact that they have taken what I have taught them and what they have learned in the process of associating with me and they have gone on to replicate my career in some ways.

Ravi Shankar expresses a similar idea but focuses on a different aspect of the master-student relationship, the effects of the younger on the elder partner:

I feel more creative when I am amidst musicians, namely, my advanced students. When they are around me, even one of them, when I'm teaching him I become much more, you know, animated, and the music just gushes out like a fountain. All that I have learned and all that I have thought. And by doing that you go on growing, you know? And when I teach, that's what I said earlier, that you learn at the same time. Because you're doing new things, without trying to.

The physicist Heinz Maier-Leibnitz also answered the question about his greatest source of pride in terms of his relationship with students:

But then I came to Munich and having all those students, and being able to do more than I could do by myself, and having them becoming independent, this was really quite something which I shall never forget. When you teach, you know, it's not like learning from a book. What you do is present yourself, whether you like it or not. The hope is that the students will learn by looking, by feeling what the teacher feels.

Brenda Milner, who decided against having children because she did not think herself cut out for the role of a mother, nevertheless is very explicit about the importance of being generative:

I think that your only chance to achieve, well, not immortality, because there is no such thing, but your only way of continuing really to have an influence, is through students. I mean, [Donald O.] Hebb is active through his students. I am only one of them. He has had a variety of students who have gone into different fields, but you see his influence, the influence of his thinking. Even if you look at Peter, my ex-husband, who was greatly influenced by Hebb. I feel that this is very important. It keeps you part of the ongoing stream, even as you get older.

TAKING A STAND

Although one of the most obvious traits of creative individuals is utter absorption in their projects, this single-mindedness does not prevent them from becoming deeply involved with historical and social issues. Sometimes the involvement comes after the person has already achieved renown in a particular field, but it can also be part of the warp and woof of a person's entire adult life. The number of individuals in our sample who have run risks in defense of their beliefs is rather astonishing. The two causes that generated the greatest concern were environmental deterioration—including here the nuclear arms race—and the Vietnam War. In the second half of our century these two issues appear to have mobilized creative people the most.

After winning a Nobel Prize in chemistry in 1954 and being listed at least by one publication as one of the twenty greatest scientists of all time, Linus Pauling turned his energies to warning his colleagues and the population at large about the dangers of nuclear war. He organized conferences and demonstrations during which he occasionally was detained by the police. He was accused of being a Communist, and his passport was revoked, even though in 1962 he had been awarded the Nobel Peace Prize. Physicist Viktor Weisskopf devoted much of his energies to fighting the arms race as a board member of the Union of Concerned Scientists.

Benjamin Spock, the author of the baby book that has supposedly sold more copies than any book in the world except for the Bible, also became a vigorous protester against the nuclear arms race and later against the Vietnam War. He too was detained by the police several times and finally tried to organize a third party and ran for the U.S. presidency in an attempt to implement his beliefs. A similar course was taken by Barry Commoner, who abandoned a blossoming scientific career in order to organize a movement for environmental responsibility. He also ran unsuccessfully for the U.S. presidency. And so did Eugene McCarthy, although in his case, as a U.S. senator, the presidential attempt was not a career change.

The actor Edward Asner became heavily involved in union and antiwar activities, and the photographer Bradley Smith spent time in Southern jails as a result of trying to organize workers in the cotton fields of Louisiana and Mississippi. The artist Lee Nading has been arrested by several sheriffs in the Southwest for defacing public property, because he used to paint giant hex signs on roads leading to nuclear installations. Natalie Davis exiled herself to Canada in protest against the Vietnam War. John Gardner left his position of power in Washington to organize grassroots movements such as Common Cause. György Faludy spent many years in concentration camps, first under the Nazis because he was a Jew, then under the Communists because he wrote poems critical of Stalin and the system. Eva Zeisel was put in solitary confinement in Ljublianka prison for more than a year because she had insisted on making beautiful dinnerware in the factory she ran for the Soviets instead of just making it as cheap as possible.

The saga of Naguib Mahfouz is a good example of the troubles that an artist with integrity can run into. Mahfouz is a shy, retiring man who loves the relaxed, dreamlike rhythms by which the Cairo leisure classes live: "After graduating from the university, I wanted to have a job and a new lifestyle: to work until afternoon, walk around in the evenings, go to a club, go to a cafe." But when he described realistically in his novels what his countrymen did and thought, and the profound changes in values that have washed over Egypt in the past several generations, he incurred the displeasure of the government and was kept under house arrest for years. And then, ironically, his objective descriptions of the way people lived also alienated the fundamentalist Islamic factions that thought Mahfouz did not respect the absolute authority of religion and was offensive toward it. At one point the writer signed a statement denouncing "cultural terrorism" and was quoted as saying, "The censor in Egypt is no longer the state; it's the gun of the fundamentalists." Recently the police discovered a death list that included Mahfouz near the top; the government then offered him armed bodyguards. But unlike other threatened intellectuals, Mahfouz refused protection. Then one October evening in 1994, as the eighty-two-year-old novelist was walking to his favorite coffeehouse to relax in the company of other writers, a Mercedes pulled up behind him, and a man jumped out and stabbed Mahfouz in the back.

Again, these trends certainly don't suggest that creative individuals are inevitably interested and involved in the world around them and that they are willing to pay a heavy price for their beliefs. But these accounts do disprove the often-voiced opposite conclusion, that exceptional artists and scientists are too selfish, too wrapped up in their work, to care much for what is happening in the rest of the world. If anything, it seems that the curiosity and commitment that drive these people to break new ground in their respective fields also direct them to confront the social and political problems that the rest of us are all too content to leave alone.

Beyond Careers

As creative individuals begin to be known and successful, they inevitably take on responsibilities beyond the ones that made them famous, even if these do not involve radical activism. There are two main reasons why this is so, one internal, the other external.

The internal reasons come into play when the creative person runs out of steam or runs out of challenges. For example, it is possible that a particular branch of science or style of art will reach a ceiling or become obsolete. Certainly the great intellectual excitement that blew through physics during the 1920s and 1930s has abated considerably, while other branches of science attract the interest of bright young investigators. Jazz is no longer what it was fifty years ago, the novel is said to be dead, and painting is retro. Those who have dedicated their lives to these endeavors are tempted to look for greener pastures. Or it may be that the domain is still exciting but the person himself has run out of ideas or feels boxed in by the limitations of his specialty or by the shortcomings of his lab and his tools. When this happens, the university scientist may look for a deanship, the inventor turns into a consultant, and the artist looks in earnest for a teaching job.

The external pressure to diversify comes from the demands the environment places on the individual. There are many administrative positions in which a respected name is a great asset. Government agencies and private foundations like their executives to have a reputation for creativity, and there are innumerable ad hoc jobs that are attractive. Generally it is not money, or even power, that tempts the creative person to accept such offers, but the feeling that there is something important that needs to be done and that he or she is the one who can do it.

Most of the women scientists in our sample—Margaret Butler, Rosalyn Yalow, Vera Rubin, Isabella Karle—devote a great deal of their time to traveling around the country and lecturing high school girls about the importance of taking math courses before it is too late, before they realize, in college, that they would like to major in science but can't because they don't know enough math. The lives of many bright women are blighted, they feel, because of this lack of foresight. All four are also involved in various scientific associations, especially those catering to women scientists. Butler is active in local politics, and Yalow lectures extensively about radiation safety.

Creative scientists are sooner or later drawn into the politics and the administration of science, and if they are any good at it, they will have a second or third career "doing God's work" rather than their own. Manfred Eigen still runs his huge laboratory at the Max Planck Institute in Göttingen, where he hopes to demonstrate the processes of selection in inorganic molecules—thereby showing how evolution proceeded even before life appeared on our planet. But he spends more and more time on such activities as awarding grants and fellowships on behalf of the German Science Foundation and traveling to official conferences—as well as playing the piano.

Another German scientist, Heinz Maier-Leibnitz, had a long and distinguished teaching and research career before he switched, in the 1950s, to building and directing the first European nuclear research reactor in Grenoble. He retired from that to accept the presidency of the *Vorschungsgemeinshaft*, the equivalent of our National Science Foundation. In this job he lobbied government officials and politicians on behalf of research programs, supervised the administration of grants and fellowships, and struggled with the media to preserve a positive image of science. When he retired again, he started writing best-selling cookbooks, while continuing informally his role as a wise old man of science, contributing articles and attending conferences.

It would be easy to believe that at least artists, musicians, and writers may be left alone to follow their inspiration and to work in the solitude of their studio. But such is not the case. Robertson Davies describes his current activities, showing both the internal and external forces that distract him from writing:

At the moment I am rather busy because I just completed a novel and it is in the stage where it goes to publication, and that means a lot of discussion with the publishers and correction of their edited version. That sort of thing. And that is quite timetaking. Also I have a number of public speeches lined up which I must prepare and give. Because I take a lot of pains with public speeches and I don't like to say shallow silly things.

And then I am going to have to do quite a bit of traveling in connection with the new book because, you know, nowadays a writer is not permitted simply to write a book, he has to be sort of a traveling showman and go around and read passages from it and talk to people.

And I am involved in getting my papers together and preparing them to go to the National Archives in Ottawa, and that is far more trouble than I thought. And another thing which I find quite demanding is that for the past several years a biographer has been writing a book about me and I have to find ridiculous photographs of myself as a baby and that sort of nonsense, and it is very difficult to say, "No, I won't do it," because biographers are determined people, and if you don't do as they wish, they will find it by themselves and God knows what they will turn up with. So you have to be tactful.

Davies's account highlights another task that creative individuals begin to turn to after they become successful: to preserve the record of their lives. Letters have to be sorted and labeled for the archives, papers collected and annotated, paintings collected in museums, memories recorded in biographies. When poets, musicians, and artists become well known they are increasingly asked to sit on award and fellowship committees. Their opinion is asked in the matter of grants, and journalists call to find out what their thoughts are on religion, sex, and politics. As Davies says:

> One of the problems about being a writer today is that you are expected to be a kind of public show and public figure and people want your opinions about politics and world affairs and so forth, about which you don't know any more than anybody else, but you have to go along or you'll get a reputation of being an impossible person, and spiteful things would be said about you.

Of course, this kind of expectation of universal knowledge, which ends up diluting and cheapening the person's unique vision and genuine expertise, does not afflict writers only. The same idea is expressed by the physicist Eugene Wigner:

> By 1946, scientists routinely acted as public servants as scientists, publicly addressing social and human problems from a scientific viewpoint. Most of us enjoyed that, vanity is a very human property. . . . We had the right and perhaps even the duty to speak out on vital political issues. But on most political questions, physicists had little more information than the man on the street.

The Question of Succession

For those who have built an institution during their lifetime, one of the consuming concerns becomes the issue of succession. Who will lead the company? Who will direct the laboratory after the present chief retires? Will the institution survive the departure of the person who devoted his or her life to it? These questions become extremely important in later life. Few of these individuals would subscribe to the resigned quip of the Marquise de Pompadour: "After us, the flood."

Robert Galvin spent most of his last three years as CEO of Motorola making sure that the "right" person would be in line to succeed when it was time for him to retire. A wrong choice would have meant jeopardizing the future of a dynamic, prosperous company employing tens of thousands of workers, which he had spent his life energies strengthening.

Elisabeth Noelle-Neumann started her public opinion polling institute in 1945, right after the end of World War II. She expanded it from a husband-and-wife operation into one of the largest and most respected firms of its kind, employing several hundred full-time and thousands of part-time workers. Much of the institute's success is due to her personal contacts among German social and political leaders, to her breakthroughs in sampling methodology, and to her drive. Understandably, now that she is in her seventies, she is worried about the future of her creation. Who, among those who work for her, is most likely to preserve the company's prestige and success?

George Klein, who has built up a large laboratory for tumor cell research at the Karolinska Institute in Stockholm, is still far from retirement, but he too spends increasing time debating which of the five dozen or so scientists working for him should be groomed to take over the lab. A person has to be intellectually brilliant, fiscally astute, and reasonably unselfish in order to head a lab successfully. If for instance Klein promotes a successor who is too concerned with his or her own career to the point of exploiting the ideas of the rest of the staff, he or she is likely to alienate the best researchers, who will then leave and go work somewhere else. Institutions are fragile things. And when they are built around a creative person, their survival is more threatened than usual.

The Matter of Time
One thing such people don't have too much of is time on their hands. It is difficult to imagine any of them being bored, or spending even a few minutes doing something they don't believe is worthwhile. Eva Zeisel says: "When some people at my age ask me what to do, I say, 'You must have an obsession.' You must always have too little time instead of too much." Bradley Smith is convinced that one is forced to become creative in order to avoid repetition and boredom. "You do not have time. The input is coming in all of the time. You do not have time to get bored."

Now in their seventies, eighties, and nineties, they may lack the fiery ambition of earlier years, but they are just as focused, efficient, and committed as before. "Come Friday," says John Hope Franklin with a chuckle, "I also say 'Thank God it's Friday,' because then I look forward to two uninterrupted days of work at home." In one fashion or another, their work—the focused application of all of their skills to a worthy, self-chosen goal—continues until they die or are incapacitated. But then why call what they are doing work? It may just as easily be called play.

The majority of people in every culture invest their lives in projects that are defined by their society. They pay attention to what others pay attention to, they experience what others experience. They go to school and learn what should be learned; they work at whatever job is available; they marry and have children according to the local customs. It is difficult to see how it could be otherwise. Would it be possible to have a stable, predictable life if most people were not conformists? If we couldn't count on plumbers doing their jobs, teachers teaching, and doctors abiding by the rules of the medical profession? At the same time, a culture can evolve only if there are a few souls who do not play by the usual rules. The men and women we studied made up their rules as they went along, combining luck with the singleness of their purpose, until they were able to fashion a "life theme" that expressed their unique vision while also allowing them to make a living.

THE SLINGS AND ARROWS OF FATE

As is obvious by now, creative people are certainly not immune to the disappointments and tragedies that cast shadows on the lives of everyone else. They are fortunate, however, to have a calling that makes it possible for them to dwell as little as possible on what might have been and go on with their lives.

Occasionally one of the interviewees would break down in tears when talking about the death of a parent or spouse. In a few cases, it was evident that deep emotional scars were left by the worst blow an adult can suffer—the death of a child. These and many lesser tragedies—wars, imprisonment, failures, financial troubles—were amply present in the histories of these people. But the hurt did not turn into an emotional swamp in which they foundered; instead, it helped to strengthen their resolve.

Some of the most permanent wounds were inflicted by professional mentors. Subrahmanyan Chandrasekhar still remembers the humiliation he felt sixty years ago when the great astrophysicist Sir Arthur Eddington made light of Chandrasekhar's scientific prospects. Frank Offner still smarts from the petty jealousy of one of his graduate school supervisors who discouraged him from taking advantage of early career opportunities and blackened his reputation behind his back.

The ability of these people to minimize obstacles is well illustrated in how the women responded to our persistent queries about the difficulties they encountered, as women, in their careers. Most of them denied that sex bias or the burden of role conflict produced by dual expectations had any great negative effect on their lives. The general attitude seemed to be "So what else is new?" and "Let's get on with what needs to be done." Not that these women are unaware of the difficulties women face in many careers. In fact, they could be very passionate in decrying the special burdens of women. But they just didn't see that the issues were relevant to their own case. Vera Rubin's answer is typical:

> I think I was terribly naive all along and when I came upon obstacles I don't think I took them very seriously. I just felt that the people who presented obstacles really did not understand that I really wanted to be an astronomer. And I tended to ignore them or dismiss them, so I don't think the obstacles have been severe. In general, I think they were just a lack of support. I always met teachers who told me—in college, in graduate school—to go and find something else to study . . . they didn't need astronomers . . . I wouldn't get a job . . . I shouldn't be doing this. And I really just dismissed all that. I just never took it seriously. I wanted to be an astronomer and I didn't care whether they thought I should or should not. So, somehow or other I just had the self-confidence to ignore all those bits of advice.
>
> It didn't seem to matter. I mean, the problem with a question like that is that I survived. There must be lots of people—lots of women especially—who would have liked to have been astronomers, and all of this did matter and therefore they didn't survive.

This kind of "naïveté," generated by confidence and a merging of self-interest in a larger project—such as astronomy, in Rubin's case— acts as a buffer between creative individuals and the forces of entropy that frustrate their personal goals.

Yet entropy cannot be kept at bay forever. Sooner or later death stops the journey of discovery. Even worse, physical deterioration may set in and spoil the last years of life. At seventy-three, the historian William McNeill still chops logs in his rural retreat and leads an otherwise vigorous life. But at the end of the interview he muses:

> Well, the other thing that you haven't touched upon that certainly seems to be important is good health. You know, being able to assume that your body does what it should, without paying any special attention. Now this is absolutely essential to getting things done. And I've wondered, if you were really sickly, what would happen? If something really twisted your whole experience of the world—some severe pain or something else. It would be just a different world, that's all. There certainly have been individuals who've had miserable, persistent pain, persistent difficulties of a physiological kind. I've never been in that position, so I don't know what it would be like. But it seems to me—well, it would be very hard to get up with a ringing headache and do anything.

As McNeill notes, the reason pain is so dreadful is that it forces us to pay attention to it, and so it interferes with concentration on anything else. So chronic pain could end all serious work. Of course, as he also mentions, some individuals are able to overcome even this obstacle. Michel de Montaigne, one of the most creative minds of the sixteenth century, suffered all through his life from kidney stones and a variety of other diseases. Yet he continued traveling, engaging in politics, and writing his famous essays. Stephen Hawking, immobilized in his wheelchair by Lou Gehrig's disease, unable to control even the vocal chords in his body, continues to develop his cosmological theories and travel around the world. But in this respect also our group was fortunate. Their health held up to the end, and they did not have to test themselves to see how their creativity could survive chronic pain.

NINE

CREATIVE AGING

There is still quite a bit of controversy among scholars about the relationship between age and creativity. When the topic was first studied, the findings suggested that creativity peaked in the third decade of life, and less than 10 percent of all great contributions were made by persons over sixty. Opinions differ, however, about what qualifies as a great contribution. When we look instead at total output, the picture changes. In the humanities the number of contributions appears to hold steady between thirty and seventy years of age; the trend is similar in the sciences, and only in the arts is there a sharp decline after sixty. In our sample productivity did not decline either; if anything, it increased in the later years. Linus Pauling at ninety-one claimed that he had published twice as many papers between the ages of seventy and ninety than in any preceding twenty-year period.

Recent studies suggest that not only quantity but quality is retained with age, and some of the most memorable work in a person's career is done in the later years. Giuseppe Verdi wrote *Falstaff* when he was eighty, and that opera is in many ways one of his best— certainly very different in style from anything ever written before.

Benjamin Franklin invented the bifocal lens when he was seventy-eight; Frank Lloyd Wright completed the Guggenheim Museum, one of his masterpieces, when he was ninety-one years of age; and Michelangelo was painting the striking frescoes in the Pauline chapel of the Vatican at eighty-nine. So although performance in many areas of life may peak in the twenties, the ability to change a symbolic domain and thus contribute to the culture may actually increase in the later years.

WHAT CHANGES WITH AGE?

One question in the interview asked about the major changes the person had experienced in the past two or three decades of life, especially with regard to his or her work. The answers are illustrative of how these creative individuals perceive the process of aging.

In general, the respondents did not see much change between their fifties and seventies, or sixties and eighties. They felt that their ability to do work was unimpaired, their goals were substantially the same as they had always been, and the quality and quantity of their accomplishments differed little from what they had been in the past. Generalized complaints about health or physical well-being were almost entirely absent. Not a single person, even among those well above eighty, had anything but a positive attitude toward how they were doing physically, even though they were realistically aware of specific decrements and limitations.

Surprisingly, when all the answers are taken into account, the number of positive changes reported is almost twice the number of negative ones. Part of this rosy picture is probably due to the tendency to put one's best foot forward in an interview situation. But given the general frankness of the responses, I am left with the belief that we are dealing with something deeper than impression management. After all, it should not be surprising that if these people have carved out unique lives for themselves, they should also approach the end of life creatively.

The answers to the question about what has changed in the last twenty to thirty years fall naturally into four basic categories. They deal with changes in physical and cognitive capacities, in habits and personal traits, in relationships with the field, or in relationships with domains. In addition, changes in each of these four categories tend to have either a positive or a negative valence—thus generating eight possible kinds of outcome.

Physical and Cognitive Capacities

As we would expect, the most frequent changes mentioned had to do with the person's abilities to perform physically or mentally. About a third of the responses fell into this category. But we didn't expect that the number of negative changes reported would be balanced by an equal number of positive ones. How could this be true, given the generally dismal opinion we have of old age?

Psychologists have long made a distinction between two broad types of mental abilities. The first is what they call fluid intelligence, or the ability to respond rapidly, to have quick reaction times, to compute fast and accurately. This ability is measured by tests asking a person to remember strings of numbers or letters, recognize patterns embedded in more complex figures, or draw inferences from logical or visual relationships. This type of intelligence is supposedly innate and little affected by learning. Its various components peak early—on some tests it is teens who perform best, on some others it is twenty- or thirty-year-olds. Each later decade shows some decrease in these skills, and after age seventy the decline is usually quite severe even among otherwise healthy individuals.

The second type of mental ability is known as crystallized intelligence. It is more dependent on learning than on innate skills. It involves making sensible judgments, recognizing similarities across different categories, using induction and logical reasoning. These abilities depend more on reflection than quick reaction, and they usually increase with time, at least until sixty years of age. In our sample of creative individuals, it is this kind of mental ability that is supposed to be improving, or at least staying stable, even in the ninth decade of life.

When we look at what the interviews say, we find that the most common complaint is a decline in energy, or a slowing down in one's activity. This is a problem especially for performers: Ravi Shankar recalls nostalgically that even ten years ago he was like a tornado, cutting records in England, flying to India to do the soundtrack of a movie, jetting to California for a concert, all without missing a beat; whereas now, at seventy-four, he prefers to stay home, take his time, and focus on a few students and select performances.

A few scientists also mention that they are getting slower and more cautious. Physicist Hans Bethe says that he makes more mistakes in calculations at eighty-eight years of age—although he is also more alert at catching mistakes than he used to be. Heinz Maier-Leibnitz, another physicist in his eighties, feels that while his appetite for doing things has increased, his energy no longer keeps up with his desire. Sociologist James Coleman recalls that twenty years ago he used to travel to a different city, check into a hotel incognito, and work four days and nights without interruptions with just a few hours thrown in for sleep—a regimen that he would not follow now.

But an almost equal number of people said that in the last decades their mental abilities have remained the same, or have improved, a claim made most often by respondents in their sixties or seventies. This positive claim is based on the contention that because of greater experience and better understanding they can now accomplish things faster and better than before. For instance, Robert Galvin, who was seventy when he was interviewed, reports that his business decisions have become sharper and more effective because after intense study he now understands better the forces involved in international trade:

> We understood as we traveled around the world that there were some markets that were open and some that were not. Europe was fairly open, Japan was very closed. And we instinctively knew that that was not tolerable. We didn't know what to say about it, we couldn't write a fancy memo about it. We could only say things in an elementary way. So we went back to school, to learn from scholars. Scholars had this important concept called the principle of sanctuary that was as applicable in business as in war. And all of a sudden what we instinctively knew became clearer to us. We now could think sharper and faster on issues of international trade.

Barry Commoner feels that now he is much smarter and knows a lot more than he did a few decades ago. Isabella Karle believes that experience provides her with a knowledge that is more complex than it was earlier. Several agree with the poet Anthony Hecht that time has honed their skills. All of these positive developments are examples of crystallized intelligence, the ability to use information available in the culture for one's own ends. As far as it was possible to determine, men and women gave exactly the same 1:1 ratios of positive to negative cognitive outcomes.

Habits and Personal Traits

The second category of changes people reported involved issues of discipline and attitude. These were mentioned about a quarter of the time, and here positive outcomes outnumbered the negative ones two to one. Negative changes almost always involved too much pressure and too little time, with the person taking the blame for not learning to avoid overcommitment. Other trait-related problems included increasing impatience and guilt over not keeping physically fit.

The positive outcomes featured diminished anxiety over performance, being less driven, and exhibiting more courage, confidence, and risk taking. Several respondents echoed Anthony Hecht's words:

> I probably am a little more trustful in unconscious instincts than I was before. I'm not as rigid as I was. And I can feel this in the quality and texture of the poems themselves. They are freer metrically, they're freer in general design. The earliest poems that I wrote were almost rigid in their eagerness not to make any errors. I'm less worried about that now.

Several respondents mentioned having learned from past mistakes or criticism of their work. This kind of learning could be quite painful. John Reed, the CEO of Citicorp, believes that after being "bloodied" in the market, when the stock of his company took a severe plunge that he blames himself for not having foreseen, his whole way of exerting leadership had to be modified:

> My approach to business has been much changed over the past ten years. I don't think I've lost any of my spark, or creativity, but I'm not quite as free. I don't have that absolute enthusiasm. It's been tempered by the realization that you can be wrong. I know some of my shortcomings, in spades, and I'm quite sensitive to them. And what I'm doing now, I'm doing quite well, but it's all discipline, it's not natural. In other words, I have disciplined myself to do these things and get them done, and I am working at it very hard. But it's not fun, and up till now, most things I have done have been fun.

C. Vann Woodward has the historian's privilege of correcting his own shortcomings more easily, by bringing out a new edition of his work:

Well, I have learned more and I have changed my mind and the reasons and conclusions about what I have written. For example, that book on Jim Crow. I have done four editions of it and I am thinking about doing a fifth, and each time it changes. And they come largely from criticisms that I have received and those criticisms come largely from a younger generation. I think the worst mistake you could make as a historian is to be indifferent to or contemptuous of what's new. You learn that there is nothing permanent in history. It is always changing. So, as one who writes about it, I am one of those who change, but I hope not for the worse.

The negative impact of time pressure was turned around by several respondents who felt good about having become masters of their own time. Again we see that the same event, in this case excessive demands on one's time and psychic energy, can have either a positive or a negative valence, depending on what the person does with it.

But even when a person copes successfully with mushrooming demands, it is often impossible to master time completely. Elisabeth Noelle-Neumann describes how her methods of work have changed:

They have become more orderly, more systematic. I developed many techniques during the last twenty years to cope with this terrible lack of time—it has become worse and worse. I thought it couldn't be, but still time got to be shorter.

The astronomer Vera Rubin is very graphic in her description of the demands on her time:

The biggest challenge is to try to get enough time to do science. There are professional meetings, there are all kinds of organizations, there are committees. I am available at any hour of the day or night for any woman astronomer who has a problem, and that is certainly well known. So I may spend an hour a day involved in that kind of thing. It is just very hard to keep the time to do science, and I still really, really want to do it.

And I am more privileged than most because I don't teach. But I think our expectations of what we can accomplish have gotten so high. I mean, there is the telephone and the fax and the computer. On bad days I have seventeen or twenty-four E-mail messages. Most days I really can barely handle my mail. I get lots of preprints and reprints and letters, and I don't have a secretary, which would help at some level. But if I read all the reprints and preprints and letters I could spend the whole day just dealing with what comes in that day.

While men and women mentioned equal proportions of negative outcomes in terms of habits and traits, women reported more than twice as many positive outcomes as men did. Apparently creative women have an easier time adapting psychologically to the later years. Compared to men, they were especially likely to mention greater serenity and fewer internal pressures. Here again is Vera Rubin:

Thirty years ago it was totally different. I would have questioned whether I would ever really be an astronomer. I mean, I had enormous doubts early on in my career. It was just nothing but one large doubt whether this would really work. It wasn't that I was able to persevere. I was unable to stop! I just couldn't give it up, it was just too important. It just never entered the realm of possibility. But I never was sure, really sure, that it was going to work and I would ever really be an astronomer.

Relationships with the Field

Another fourth of the responses dealt with changes in the relationship with colleagues, students, and institutions. Again, the number of positive and negative outcomes were about equal, but with one intriguing difference: All the negative outcomes were mentioned by men, whereas the positive ones were equally divided between the genders. Men apparently miss more the lack of formal institutional membership that age usually entails; they suffer more from retirement with its decrease in prestige and power. Eugene McCarthy left the U.S. Senate long ago; the sociologist David Riesman misses the scholarly conferences he no longer attends because he doesn't like to travel; the physicist Viktor Weisskopf, like many of his colleagues in the sciences, is no longer involved in active research.

But with age it is also possible to acquire a greater centrality in the field, or to develop new forms of association, especially with students. George Stigler spends more time on the prestigious journal he is editing; Ravi Shankar is planning the new center for the teaching of traditional music that the Indian government is about to build for him. The anthropologist Robert LeVine has decreased his trips to visit fieldwork sites in Africa, but he spends more time training third-world students. Manfred Eigen leads a giant laboratory in Göttingen, works closely with his twelve Ph.D. students, and is active in various scientific societies and government agencies.

Relationships with Domains

The last category of answers that respondents gave to the question of what has changed in their life during the past decades has to do with the acquisition of knowledge. Contrary to the previous cases, where positive and negative outcomes were roughly in balance, the 17 percent of the responses that fell into this category were uniformly positive. It seems that the promise of more and different knowledge never lets us down. We can lose physical energy and cognitive skills, we can lose the power and prestige of social position, but symbolic domains remain always accessible and their rewards remain fresh till the end of life.

Some individuals discovered a broader set of possibilities within the domain they had been pursuing; one example is Nina Holton, who is fascinated with what she has been learning about sculpting in bronze. Some branched into new enterprises related to their past work: Freeman Dyson is now writing about science for the general public as enthusiastically as he used to do active science, and currently has a dual career as a mathematical physicist and a writer. Others discover an entirely new interest: Heinz Maier-Leibnitz writes cookbooks after having been president of the German Science Foundation.

Still others simply look forward to being able to read more widely and to explore hitherto neglected realms of knowledge. Or they claim that in the past years they have learned to enjoy life more fully.

Often the changes are not so much a matter of aging, or of the person deciding to change, but are dictated by the interaction with the medium, by the logic of the domain itself. The painter Ellen Lanyon describes the evolution of her style in the past decades:

For a lot of my early work, I was labeled a sophisticated primitive because I was doing Chicago street scenes, but they were influenced by Sienese egg tempera painters of the fourteenth century. And consequently there was a certain kind of naive approach to perspective which was also premeditated. I was not naive. I was using a certain style. And in the late forties, that was quite appropriate. It was part of what was going on also in American imagery and especially regional imagery. Then because I moved through a period of time where I wanted to work on a larger scale, I worked with oil paint. And then in the very early sixties, by chance, I started to work from photographs. I worked from old family photographs. I worked from newspapers, sports photographs. I worked from old rotogravures that I found in Italy. And it was all figure painting. It was all nostalgia. You know, at that time to work from photographs was a taboo. I was actually working through the photograph and translating a sort of space or a pattern on the canvas that in its way resembled and was a view, a photographic view, of a particular situation that had occurred. It froze time. It stabilized a situation. Some of those photographs of the family were of deceased people. And a secondary reason was to take my own personal history, document it, establish it in time, so to speak, and therefore it was out and finished. I could set it aside, and I could go on. And that was a very important thing for me. And so therefore the work changed because imagery changed and moved.

Next I went into the use of acrylics, which by that time were pretty well improved, and one could work with them. And I spent about five years training myself in the use of acrylics. So that now most people don't even know they're not oil paintings. In the process of doing that I resolved that I would also change the content. So I made another sort of cerebral decision, and I chose to work with the object, not the still life, but the object. And I went through a whole series of things, and it is at that point that the work became much more, I would say, metaphysical. The objects began to take on their own life. And it worked through a whole series that had to do with stage magic, early experiments with physics and chemistry. That started in about 1968, and the work is still involved in that general area. Then animals, birds, insects came in through the stage production. I mean, it all sort of proliferated and moved along.

This quote illustrates well how inexhaustible domains can be. In this case the different media of paint—egg tempera, oils, photographs, acrylics—different art-historical influences, changing emotional priorities, and maturing reflections on experience all interact and provide an endless series of developments that Lanyon can explore throughout her life. It is for this reason that changes in the domain are seen as being always positive; they allow a person to keep being creative even when the body fails and when societal opportunities become restricted.

ALWAYS ONE PEAK MORE

It is easy to see why these individuals see age in a more positive light than we may have expected. Every one of them is still deeply involved in tasks that are exciting and rewarding, even if they are ultimately unattainable. Like the climber who reaches the top of the mountain and, after looking around in wonder at the magnificent view, rejoices at the sight of an even taller neighboring peak, these people never run out of exciting goals. The actor Edward Asner expresses the sentiments of the whole group when he says that what absorbs his attention now is

demonstrating that my acting ability is better than it's ever been, doing it across the board, doing it however and whichever way I can. In as many ways as I can. Radio, commercials, voice-overs, narrations for documentaries, on-stage, TV, films. It doesn't matter. I thirst to . . . burst at the seams, eager for the chase.

We asked respondents to tell us what their current challenges were, what goals absorbed their energies more than anything else. All the answers were enthusiastic, describing in great detail the person's current involvement. It was clear that, like Asner, everyone was still "eager for the chase." The lone exception only confirmed this conclusion. Freeman Dyson, the one respondent who had nothing particular to work on at the moment, said that this was therefore a very creative period for him, because idleness was a necessary precursor of a productive burst: "I'm fooling around not doing anything, which probably means that this is a creative period, although of course you don't know until afterwards. I think that it is very important to be idle. So I am not ashamed of being idle."

Some individuals, like the columnist Jack Anderson, let the challenge be determined by outside events; he was sure that interesting and important issues would keep coming up and present him with opportunities for involvement:

> I always try to make the most important task the one that I am working on. I try to keep motivated by assigning a high priority to whatever it is that I am working on. I do not want to live in the past. I have had a few achievements in the past, but that is done and that is over with and I am glad that I did well. But that does not mean anything today. It is what I do today and what I do tomorrow that is important.

This kind of future orientation was typical. There was very little reminiscing and dwelling on past success in this group; everyone's energies were focused on tasks still to be accomplished.

The most frequent challenge was working on a book and writing of four or five articles during the next year. Some had outstanding research agendas to complete. A good example is the answer of Isabella Karle, whose esoteric technical jargon cannot entirely disguise the excitement bubbling under the surface of her quest:

> Well, right now, I'm studying a peptide system that makes channels in cell membranes, and it transports potassium ions from one side of the cell to the other. I am collaborating in this work with a man in India. He has been able to isolate and purify—I say this because many natural products come in many slightly different versions, and unless you can separate out these various different versions, you can't grow a crystal because it won't repeat properly, and a crystal has to have the molecules repeat in a certain fashion. He has prepared the materials and he has grown the crystals. In fact, the same material grows somewhat different crystals from different

solvents. And I'm now looking at the third crystal form. Each one of them shows how a channel is formed. There is a helical peptide. The peptide has a big bend in it, and two peptides come together in an hourglass fashion like so. [She gestures.] They are hydrophobic on one side. That means that they are compatible with the kind of materials that make up cell walls. On the inside, they're hydrophilic, that is, they attract water or polar substances. So this channel in the crystal, in all the crystal forms, is filled with water, but it is interrupted in the middle by a hydrogen bond between two moieties, so that if you had a water molecule, it would not go through the middle of it, through the midportion. These materials are used as antibiotics, and that's how they perform their work. Well, it's very important in biochemistry, biophysics, to try to figure out how ions are transported because our bodies do that in all kinds of ways for all the foods that we eat, the minerals that we need.

Another answer that suggests the multifaceted nature of these people's commitments is the schedule Rosalyn Yalow describes for her recent past, a schedule in which scientific research, policy making, family times, and public service are all intertwined:

Well, let's see, on the 24th of February I lectured at Memorial Hospital here in New York City, and then at four and at six I met with women in science at Mount Sinai Hospital; then I went to the Eastern students' research meetings in Miami, and then I gave endocrine gland rounds the next day. Then I came home [to New York City] and I went to Auburn University where I lectured and interacted for three days; then I went to the Pittsburgh conference, which is on spectroscopy, and then to an analytical chemistry meeting in New Orleans. I came back and then I went to the Stewart Country Day School and I spoke to their seventh to twelfth graders; then I went to Albany—all of this in the same week.

Yesterday I spoke at New York Academy of Science to their high school gang. Next week I am going out to Lawrence Livermore Laboratories in California. I am on the advisory committee for Lawrence Livermore and Los Alamos. But I am giving a radiation lecture and I am speaking to the women's group.

Then I go to see my daughter and her husband and my grand-child, and I come back on the 29th. On the 31st I leave to go to Nashville, where I am speaking at Vanderbilt for two days and then two days at the University of the South at Sewanee. Then I go back to California. The American Chemical Society is having a three-day symposium for which they borrowed my title "Radiation Society." I get back from there and then I am going out to Las Vegas for the American College of Nuclear Physicians, for some sort of meeting on radiation.

For those like Barry Commoner, George Klein, Elisabeth Noelle-Neumann, and Enrico Randone, who had been responsible for institutions—a business, a research lab—the main challenge is to continue helping the institutions survive. Here is what Robert Galvin says about his continued involvement with Motorola:

> Having given up the direct and operating leadership of our corporation, I wish to remain fully active and influential in our institution. I am putting an incrementally greater amount of attention on those factors that I think will have leverage impact on the performance of the institution in the decades ahead, not just the weeks and months ahead. I think there are some significant fundamentals that show promise of allowing a commercial institution to elevate its performing capabilities. One of them that is on the Class A list for me is the vocational skill of creativity, the potential for changing the quality of leadership, which relates to the functions of anticipation and commitment.

Some people have found unanticipated challenges thrust upon them, as it were. Jonas Salk was planning to devote himself to science policy and philanthropy when the AIDS epidemic intruded on the world. Salk found the challenge too compelling to resist; he went back into the lab to try to find some immunological means to prevent the disease, just as he did with the polio vaccine many decades earlier. A dozen years after retiring from a faculty position at the University of Chicago, Bernice Neugarten felt so distressed by reading statistics about the plight of poor children that she returned to doing policy research full-time.

The dedication may be interpreted by some as workaholism, an obsessive inability to enjoy any other aspect of life except achievement. But this would be missing the point. For most of them work is not a way to avoid a full life, but rather is what makes a life full. The television producer Robert Trachinger shows the multifaceted nature of this process:

I really want to enjoy life now. I've kicked back. I've always been a very hard worker, A-type personality. I used to have high blood pressure problems and take pills. Now I don't have to take pills anymore. I do some yoga, I do some tai chi. Teaching remains my great love because I get so much love and response from students, so much caring. That's important to me because the lonely ghetto kid is still very much a part of me. I have enough money now to live comfortably without working, or without teaching for that matter. I enjoy going to Europe and teaching young people in Europe, and consulting with the schools that I'm beginning to set up, departments of television and filmmaking. And I caution them about buying into our form of television because it's kind of cultural imperialism. Most of what they watch on television in Europe is American television, and it erodes their cultures. So we talk about these values. How do you develop responsible filmmakers and television makers who are not out simply to titillate audiences and make bucks?

I'm going to school. I attend great books courses, and I'm fascinated by reading. If you came upstairs, I have easily fifteen hundred to two thousand books, many of which I've not read, but I hope as I get older I'll read, and I'll have more and more time to read. And I counsel young people. I am not a sage by any means, but I've lived sixty-seven years, and there are some things I do sense and do know. Caring is a good feeling, and we've lost our appetite for it.

THE SOURCES OF MEANING

According to Erik Erikson, the last psychological stage that people confront in their lives is what he called the task of achieving integrity. What he meant by this is that if we live long enough and if we resolve all the earlier tasks of adulthood—such as developing a viable *identity*, a close and satisfying *intimacy*, and if we succeed in passing on our genes and our values through *generativity*—then there is a last remaining task that is essential for our full development as a human being. This consists in bringing together into a meaningful story our past and present, and in reconciling ourselves with the approaching end of life. If in the later years we look back with puzzlement and regret, unable to accept the choices we have made and wishing for another chance, despair is the likely outcome. In Erikson's words: "A meaningful old age . . . serves the need for that integrated heritage which gives indispensable perspective on the life cycle. Strength here takes the form of that detached yet active concern with life bounded with death, which we call wisdom . . ."

The notion of integrity connotes the ability to tie together, to relate to others outside oneself. Erikson thought that the perspective of an older person is based on a new definition of identity, which could be summarized in the sentence "I am what survives me." If toward the end of life I conclude that nothing of myself is likely to survive, despair is likely to take over. But if I have identified with some more enduring entities, my survival will provide a sense of connection, of continuity, that keeps despair at bay. If I love my grandchildren, or the work I have accomplished, or the causes I have championed, then I am bound to feel a part of the future even after personal death. Jonas Salk calls this attitude "being a good ancestor."

In our study, we did not pursue directly the question whether and to what extent our respondents had achieved a sense of integrity about their lives. But answers to one question shed some light on the issue of what entities serve as the kernel for the identity of this sample, the kernel around which a sense of integrity is likely to develop. The question was "Of the things you have done in life, of what are you most proud?" This question is certainly not ideal for studying integrity, because several respondents felt put off by the word *proud* and others were not too happy about singling out a particular accomplishment as the source of their greatest pride. Nevertheless, people by and large answered the question in a way that suggested they were thinking of the most meaningful, important thing they had done in life and therefore of their main link to the future.

Categorizing the answers was very simple, since all of them were basically of one type. As we would expect in a group of such successful people, what they had achieved in their professional life was the first kind of answer. About 70 percent of the accomplishments mentioned as sources of pride had to do with one's work. Surprisingly, however, 40 percent of the women and 25 percent of the men (a good 30 percent of the whole group) mentioned the family first as what they were most proud of. These are some of the reasons two men give for their answers:

Well, I get a great deal of satisfaction and pride out of my children. And my grandchildren are absolutely delightful, because I do not have to worry about when they misbehave. I turn them over to their mothers. Grandchildren are created, I suspect, for grandfathers to play with. My grandchildren are all beautiful and a great source of enjoyment. Now, my children, some of them have had problems. But basically they have turned out pretty good. I have been very pleased with what they have accomplished and I am concerned about those who have problems. I am much more concerned about them than any of my problems. And I am more pleased with their accomplishments than I think I am pleased with my own. I get a good deal of pride from my family.

Loyal, hard work in keeping the family at its best; my wife and children. The fact that my marriage has been stable—better than stable, it's been really very fortunate—has been essential, I think, for my particular character. I think if I had not had a stable sense that that part of my life was OK, better than OK, if I'd had children that were failures, or if I'd had a divorce, I'm sure it would have affected the tone of what I write and also the teaching; I don't think I could have done the same kind of aggressive, bouncy teaching.

As the second excerpt suggests, pride in family is often combined with pride in work. One could even conclude that although the family was mentioned first in this answer, its importance is subsidiary to that of writing and teaching. It sounds almost as if the family matters primarily because it enables the writer to concentrate all his energies on his task. In fact, family and work are usually so inextricably related that it is difficult to say from a single answer whether fame and accomplishments are valued because they enhance the family's well-being, or the other way around. It is striking, however, that no other themes intruded on this simple duality. In the last years of the twentieth century, among sophisticated people of supremely high achievement, one may have expected a greater variety and more esoteric topics on which to build a life's narrative. It certainly appears to vindicate Freud's deceptively simple answer to an inquiry about the secret for a happy life: "Love and work," he said, and with those two words he may have run out of all the options.

In looking more closely at the answers, another interesting pattern appears. Some of the respondents—about 70 percent of the 70 percent who mentioned work first as source of pride—speak primarily about extrinsic reasons for feeling proud, such as the great contributions they have made, the recognition and prizes they received, their renown among colleagues. The remaining 30 percent emphasize intrinsic reasons—the cultural advance made possible by the accomplishment or the personal rewards of a difficult job well done.

The physicist John Bardeen, although mentioning extrinsic reasons, emphasizes more the intrinsic importance of what he had been working on:

> I think the theory of superconductivity. The two things I'm most noted for are being coinventor of the transistor and the theory of superconductivity. The transistors, of course, had much more worldwide impact than superconductivity, but superconductivity was more of a challenge. And the theory had much greater impact on other fields of physics. As for theoretical contribution, it opened up some new ways of thinking about the structure of the nucleus and particles of high energy. So it contributed more to a deeper understanding of what the universe is all about, I think.

The more extrinsic responses tended to dwell on the number of copies the person's books have sold, on the directorships of large research organizations he or she held, on the canvases displayed in important exhibitions—in other words, on the highlights of a job résumé. The economist George Stigler has a refreshingly direct answer to the question:

> I guess I have to say the things in which I succeeded in impressing other people with what I have done. And those would be things like the two areas of work in which I received the Nobel Prize, and things like that. So those and certain other work that my profession has liked would be the things, as far as my professional life goes, of which I'm most proud.

Every other Nobel Prize winner, however, gave reasons that were intrinsic. Perhaps among these individuals who are so close to the top rungs of achievement in their fields, only those who have reached the highest stages can afford the luxury of playing down the importance of worldly success. But a closer look at the answers suggests that it is unwise to place too much weight on the answers to this single question. The reason becomes evident when we realize that almost 40 percent of the men gave responses that were coded as intrinsic, whereas none of the women did—every single one of them talked about the pride they felt in the extrinsic aspects of their contributions.

Yet these women enjoy their work, are in awe of its importance, and are much less interested in the fame and power it might bring them. It would be difficult to find a male scientist as much in love with his work as Vera Rubin, an artist as committed as Eva Zeisel, a historian as excited by his craft as Natalie Davis. So why did the women tend to emphasize external success in their answers to the question about pride? Probably because women have a much more difficult time gaining recognition than men, and therefore when they get it, it means more to them. In any case, this contradiction illustrates the important point that trying to interpret a single answer out of the total context of what we know about a person can be deceptive. Isabella Karle, who at first gives a rather extrinsic response, in a different part of the interview, sounds definitely more interested in the intrinsic aspects of her work:

> I've been successful in the sense that I've had all sorts of scientific awards and have been elected to memberships in what are considered the "elite" societies. And I get invited to speak at all sorts of universities all over the world, and that's all very nice. But I think that the biggest satisfaction is just doing, finding out something about nature that hasn't been known before. There's the satisfaction of seeing what some of these molecules look like. There's a satisfaction in seeing how they may react. I suppose it's very personal—why other people find satisfaction in playing a Beethoven sonata faultlessly, or painting a picture.

Following the ins and outs of our respondents' answers to the question about pride, we conclude that, like the rest of the world, they also stress in their personal narratives the twin themes of work and love. These are the sources from which they build a meaningful story about their pasts and a bridge to the future. The fact that they were lucky in having achieved a greater renown than most people get for their efforts does not seem to make much difference. There is no evidence that being awarded one or two Nobel Prizes gives a person a greater claim on wisdom or a surer defense against despair than having lived a full life as an honest plumber and parent.

FACING THE INFINITE

At the time of the interviews, all of our respondents were still actively involved in family and work projects that reflected the main themes of their lives. But often their interest had broadened to include larger issues: politics, human welfare, the environment, and occasionally transcendent concerns with the future of the universe. Interestingly, in entering the last decades of life, none of them appears to have embraced an orthodox religious faith. Fear of death did not loom large, certainly not enough to send them to seek solace in a faith that had been alien at an earlier age. Those few who, like sociologist Elise Boulding, had strong religious foundations to start with continued in their beliefs. Yet even when a ritualized faith was missing, a broader faith seemed to be much in evidence: a faith in an ultimately meaningful universe, which imposes requirements of awe and respect—and curiosity—on men and women.

Perhaps the closest anyone has come to a conversion experience is the pediatrician Benjamin Spock, who has been taken to task by religious fundamentalists and preachers such as Norman Vincent Peale for having introduced permissiveness into American child rearing and thus corrupted the national character. Now in his nineties, Spock is writing a book on spirituality. But his understanding of spirituality is a far cry from that of institutionalized religions:

Spirituality, unfortunately, is not a stylish word. It's not a word that gets used. That's because we're such an unspiritual country that we think of it as somewhat corny to talk about spirituality. "What is *that?*" people say. Spirituality, to me, means the nonmaterial things. I don't want to give the idea that it's something mystical; I want it to apply to ordinary people's ordinary lives: things like love, and helpfulness, and tolerance, and enjoyment of the arts or even creativity in the arts. I think that creativity in the arts is very special. It takes a high degree and a high type of spirituality to want to express things in terms of literature or poetry, plays, architecture, gardens, creating beauty any way. And if you can't create beauty, at least it's good to appreciate beauty and get some enjoyment and inspiration out of it. So it's just things that aren't totally materialistic. And that would include religion.

All through her adventurous life the ceramist Eva Zeisel has tried to help the disadvantaged and has used her artistic gifts in part to advance left-wing causes that she sincerely believed would make the world a better place to live in. Now in her eighties, she looks at her past with objective eyes and, while not regretting anything, is no longer sure that her motives were the wisest. Although, without regrets or despair, she finds that the only thing she can absolutely rely on is the work she has produced, and perhaps the old goal of "doing good"—although she is less sanguine about that:

I was thinking how to convey my accumulated wisdom to my granddaughter. And one of the things that I thought to tell her is that one tries to do good and one tries to produce something. I find that my craft helped me very much to make life meaningful, because once you make a pot and it is outside of you, it makes your life kind of justified and not flimsy. After all you go through, at the end you die, and it makes life much more . . . well, satisfying. It justifies your existence. . . .

Then the question of doing good for society. Don't forget that all our contemporaries and ourselves had some big ideology to live for. Everybody thought he had to either fight in Spain or die for something else, and most of us had to be in prison for one reason or another. And then at the end it turns out that none of these great ideologies was worth your sacrificing anything for. Even doing personal good is very difficult to be absolutely sure about. It's very difficult to know exactly whether to live for an ideology or even to live for doing good. But there cannot be anything wrong in making a pot, I'll tell you. When making a pot you can't bring any evil into the world.

For Mark Strand, the poet's responsibility to be a witness, a recorder of experience, is part of the broader responsibility we all have for keeping the universe ordered through our consciousness:

> Yeah, I think that it grows out of a sense of mortality. I mean, we're only here for a short while. And I think it's such a lucky accident, having been born, that we're almost obliged to pay attention. In some ways, this is getting far afield. I mean, we are—as far as we know—the only part of the universe that's self-conscious. We could even be the universe's *form* of consciousness. We might have come along so that the universe could look at itself. I don't know that, but we're made of the same stuff that stars are made of, or that floats around in space. But we're combined in such a way that we can describe what it's like to be alive, to be witnesses. Most of our experience is that of being a witness. We see and hear and smell other things. I think being alive is responding.

The physicist John Wheeler is puzzling out something that sounds very similar to Strand's position, a quest after what has been called the anthropic principle, the idea that the world exists because *we* exist, an idea that usually occupies theologians and philosophers. According to them, we know about the existence of the universe because we are conscious of it. Perhaps the universe is there *only because* we are conscious of it. But Wheeler, always a scientist, would like to formulate this vague and ambitious proposition so that it actually could be tested:

> Right now I am animated by an idea which may be totally wrong but I can't tell until I go further. That this great show that is going on around us, the world, that somehow we play a vital part in bringing it about. Thomas Mann expresses this somewhere in a vivid passage. And how can that idea be expressed in a way that is so clear that it is testable. Well, here I go, reading the works of the German philosopher Heidegger, and talk with everybody who has a prospect of being able to contribute to my point of view on this issue. It would be nice to bring it all to a particular focus on a particular issue that could be decided yes or no, but at this point the whole thing is so big, so unformed, that it is better to nurture it.

Jonas Salk, in addition to his immunological research, the concern for his institute, and membership on the board of philanthropic foundations, for many years considered the broader implications of evolutionary theory as it affects the evolution of culture and consciousness:

> I have continued to be interested in some larger questions, more fundamental questions, about creativity itself. This institution [the Salk Institute for Biological Studies] was established with the idea in mind that there would be a crucible for creativity, a center for the study of creativity to explore with individuals who have exhibited that quality in the course of their lives. I see us human beings as a product of the process of evolution—I would say creative evolution. We have now become the process itself, or part of the process itself. And from that perspective, I have become interested in what I call universal evolution, the phenomenon of evolution in itself as manifest in what I call prebiological evolution, evolution of the physical, chemical world, then biological evolution, then what I call metabiological evolution, evolution of the mind by itself, the brain-mind. And now I'm beginning to write about teleological evolution, which is evolution with a purpose. So my purpose now is to try to understand evolution, creativity, in a purposeful way.

Younger scientists often look at their elders with a certain discomfort. They scoff at the Wheelers, the Spocks, and the Salks, implying that they might be going slightly soft in the head, because in old age they seem to throw all caution to the wind, break out of disciplinary boundaries, and start concerning themselves with the big problems of existence. While occasionally there might be grounds for dismissing these attempts as the vaporings of senility, the examples in our sample point in a different direction. Older scientists and artists who have spent decades within a narrow segment of their domain often feel a sense of liberation when, after they have left their mark on the discipline, they begin to explore the world outside the artistic studio or the scientific laboratory. As they do this, the problems they address are almost certainly going to be more intractable than the ones they faced earlier in life. Should they therefore desist, or be ridiculed? Or should we feel sorry instead for their critics, who judge human endeavor only by the strict and often sterile rules of a single discipline?

Of course these questions are rhetorical, because as far as I am concerned the quest these men and women are embarking on is exactly what makes their stage of life so exciting and worthwhile. It is difficult to see how it could be otherwise. Wisdom and integrity cannot be found in any single domain. A broader viewpoint that breaks across disciplinary boundaries is needed, a way of understanding that combines knowing and sensing, feeling and judging. In facing this task one cannot expect to succeed in the public eye, as one can when a field of culture recognizes one's contributions to art, business, or science. But by this time a person aspiring to wisdom knows that the bottom line of a well-lived life is not so much success but the certainty we reach, in the most private fibers of our being, that our existence is linked in a meaningful way with the rest of the universe.

DOMAINS
OF CREATIVITY

TEN

THE DOMAIN OF THE WORD

N ow in this third part of the volume we look more closely at specific domains of creativity. For as chapter 4 showed, even though there are important common features to the creative process, in order to really see what happens in its concreteness we have to consider each domain separately. At a very abstract level, creativity in physics and poetry shares common traits; but such a level of abstraction misses many of the most interesting and vital aspects of the process. Therefore, this chapter and the next two present a number of cases from the same domain, in order to get a more detailed understanding of what is involved in producing a cultural change.

We start with a brief analysis of the goals and working methods of five writers—three poets and two novelists. Starting with writers makes sense because of all the cultural domains literature may nowadays be the most accessible. It is not easy to describe how theoretical physicists work in a way that is understandable to laypersons (among whose ranks I count myself). But we all read stories, we all write to a certain extent, so the craft of professional writers is not abstruse. However, even within the somewhat homogeneous domain of literature there are large differences. Not only the obvious difference between poet and novelist, but within each of these subdomains there are innumerable variations in terms of which part of the long tradition of poetry, for instance, the writer draws on; whether the writer works in a classical mode or as an experimentalist; what genre he or she prefers, and so on. Despite the fact that in the last analysis each writer is unique, the five sketches that follow give a flavor of what is involved in literary creation. But before getting down to cases, it may be useful to consider the more general question: Why are we interested in literature?

Among the oldest symbolic systems in the world are those organized around the content and the rules of language. The first narrative stories telling of real or imaginary events, the myths and campfire tales of our ancestors, extended dramatically the range of human experience through imagination. The rhyme and meter of poetry created patterns of order that must have seemed miraculous to people who had yet scarcely learned to improve on the precarious order of nature. And when the discovery of writing made it possible to preserve memory outside the fragile brain, the domain of the word became one of the most effective tools and greatest sources of pride for humankind. Perhaps only art, dance, and music are more ancient; the beginnings of technology and arithmetic probably contemporaneous.

What makes words so powerful is that they enrich life by expanding the range of individual experience. Without stories and books, we would be limited to knowing only what has happened to us or to those whom we have met. With books we can join Herodotus during his travels to Egypt, or be with Lewis and Clark on their epic journey to the Pacific, or imagine what it might be like to travel beyond our galaxy a few hundred years hence.

But more important, the written word allows us to understand better what is happening within ourselves. In recording real or imaginary events, the writer arrests the evanescent stream of experience by naming its aspects and making them enduring in language. Then by reading and repeating a verse or passage of prose, we can savor the images and their meanings and thus understand more accurately how *we* feel and what *we* think. Fragile thoughts and feelings are transformed by words into concrete thoughts and emotions. In this sense, poetry and literature allow the creation of experiences that we would otherwise not have access to; they take our lives to higher levels of complexity.

Poetry and literature do not achieve their effect by simply presenting information. Their effectiveness rests on formal properties—on the music of the verse, the vividness of the imagery. When asked about the relative importance of intuition and intellect in their work, scientists tend to say something like "It is most effective when intuition and intellect are both involved." A writer, Madeleine L'Engle, answered the question as follows: "Your intuition and your intellect should be working together . . . making love. That's how it works best." The two statements have the same content, but which one is more effective? The image of intellect and intuition making love is more likely to arrest our attention, and get us to think about what is involved in the dialectical process of thought. It is also a more precise description, because it brings attention to the fact that the involvement is *between* the intellect and the intuition; it is not just a dry functional connection but one that actually resembles the relationship of love. So the choice of words, the construction of images and stories, is as important for the writer as the content of the message.

It has been said that all the stories have already been told, that there is nothing left to say. At best, a writer's job is to pour new wine in old bottles, to retell in a new way the same emotional predicaments that humans have felt since the beginnings of time. Yet many authors find this a worthwhile challenge; they think of themselves as gardeners whose task is to cultivate perennial ideas generation after generation. The same flowers will bloom each spring, but if the gardener slacks off, weeds will take over.

The writers whose works are described in this chapter share an obvious love for their craft. They have an almost religious respect for their domain and believe with the Gospel of St. John that "In the beginning was the Word." At the same time, they know that the power of words depends on how they are used; so they enjoy playing with them, stretching their meanings, stringing them in novel combinations, and polishing them until they shine. Playful as they are with words, all of them are also deadly in earnest. They are all involved in creating imaginary worlds that are as necessary for them as the physical world they inhabit. Without the symbolic refuges they create, the "real" world would not be very interesting. All of them feel that it is writing that gives them their identity; that if they could not write, their life would lose much of its meaning. At the same time, the goals and approaches of the five writers are

quite different. Some feel that they have a central message they want
to convey, others tend to react more to experience; some emphasize
tradition, others spontaneity.

TO BE A WITNESS

Mark Strand was living in Salt Lake City at the time of our interview
and teaching at the University of Utah. His roots, however, are in the
East, and recently he returned to live there with his family. Strand
has received many honors for his poetry, including being named the
poet laureate of the United States. But like most creative people, he
does not take himself too seriously. When asked what was the most
important challenge he was facing at this stage of life, he answered:
"At some moments it's training the puppy not to shit in the house.
At other times, it's trying to get some work done."

Strand does not have a pretentious theory of *ars poetica*. But that
does not mean that he takes his vocation lightly; in fact, his views of
poetry are as serious as any. His writing grows out of the condition
of mortality: Birth, love, and death are the stalks onto which his
verse is grafted. To say anything new about these eternal themes he
must do a lot of watching, a lot of reading, a lot of thinking. Strand
sees his main skill as just paying attention to the textures and rhythms
of life, being receptive to the multifaceted, constantly changing yet
ever recurring stream of experiences. The secret of saying something
new is to be patient. If one reacts too quickly, it is likely that the
reaction will be superficial, a cliché. "Keep your eyes and ears open,"
he says, "and your mouth shut. For as long as possible." Yet life is
short, so patience is painful to the poet.

> Poetry is about slowing down, I think. It's about reading the
> same thing again and again, really savoring it, living inside the
> poem. There's no rush to find out what happens in a poem. It's
> really about feeling one syllable rubbing against another, one word
> giving way to another, and sensing the justice of that relationship
> between one word, the next, the next, the next.

Strand claims that often he starts writing without anything specific in mind. What gets him started is the simple desire to write.
Writing for him—as for the rest of the individuals discussed in this

chapter—is a necessity, like swimming for a fish or flying for a bird. The theme of the poem emerges in the writing, as one word suggests another, one image calls another into being. This is the problem-finding process that is typical of creative work in the arts as well as the sciences.

I'll jot a few words down, and that's a beginning. It can happen when I'm reading something else, I mean, it's different all the time, there's no one way. One of the amazing things about what I do is you don't know when you're going to be hit with an idea, you don't know where it comes from. I think it has to do with language. Writers are people who have greater receptivity to language, and I think that they will see something in a phrase, or even in a word, that allows them to change it or improve what was there before.

I have no idea where things come from. It's a great mystery to me, but then so many things are. I don't know why I'm me, I don't know why I do the things I do. I don't even know whether my writing is a way of figuring it out. I think that it's inevitable, you learn more about yourself the more you write, but that's not the purpose of writing. I don't write to find out more about myself. I write because it amuses me.

But amusement seems like a radical understatement for the way Strand experiences writing. For one thing, it is a never-ending process almost obsessive in its demands on the poet. "I am always thinking in the back of my mind, there's something always going on back there. I am always working, even if it's sort of unconsciously, even though I'm carrying on conversations with people and doing other things, somewhere in the back of my mind I'm writing, mulling over. And another part of my mind is reviewing what I've done." In fact, one of the major problems Strand tries to avoid is a sort of mental meltdown that occurs when he gets too deeply involved with the writing of a poem. At such times, to avoid blowing a fuse, he has developed a variety of rituals to distract himself: playing a few hands of solitaire, taking the dog for a walk, running "meaningless errands," going to the kitchen to have a snack. Driving is an especially useful respite, because it forces him to concentrate on the road and thus relieves his mind from the burden of thought. Afterward, refreshed by the interval, he can return to work with a clearer mind.

Then there is the opposite danger: running into a dry spell. Strand has experienced this, too. After moving to Utah he was unable to write anything serious for many months. A writing block is not merely inconvenient; for someone who defines himself through writing, it is like being in a coma. So writing may be amusing when everything is going well; but between feast and famine the fragile flow is constantly threatened. Moreover, achieving even moderate affluence from writing verse is almost impossible under the best of circumstances. But Strand does not complain about the hardships involved. He feels privileged to be doing what he loves and is impatient with artists who moan and groan about how difficult their lives are. The whine of the victim is absent from his repertoire—as from that of practically all the individuals we interviewed.

In chapter five I quoted Strand's description of the total immersion in the flow of writing. But this state cannot be sustained for long stretches of time: "I could never stay in that frame of mind for an entire day. It comes and goes. If I'm working well, it's there. I will be in a daze, I mean, I will be very disconnected from everything around me. When I'm in a daze I'm creating a space for myself, some psychic space [from] which I can work."

Strand's modus operandi seems to consist of a constant alternation between a highly concentrated critical assessment and a relaxed, receptive, nonjudgmental openness to experience. His attention coils and uncoils, its focus sharpens and softens, like the systolic and diastolic beat of the heart. It is out of this dynamic change of perspective that a good new work arises. Without openness the poet might miss the significant experience. But once the experience registers in his consciousness, he needs the focused, critical approach to transform it into a vivid verbal image that communicates its essence to the reader.

Obsessed as he is with his art, Strand realizes that he could not really work in such a concentrated way longer than he does now, for more than a few hours a day. Besides, the enterprise of writing makes sense only within the context of a broader, more mundane reality. Some artists get so involved in their creations that they lose their appetite for raw experience, but Strand welcomes ordinary life—puttering in the yard, having meals with the family, going on hikes, lecturing, even shopping. These activities:

> take me out of myself. Poetry relocates you in yourself. [When doing ordinary things] you're focused elsewhere. You're not focused on something you're making, something that is entirely formed by you. And you participate in these adventures with other people, which is fun. It's fun to do things with my wife and son and my colleagues; it's fun to visit people.

Mark Strand seems comfortable with his place in the world—with his family, his job, and his dog. He knows that he is good at his craft, which is to express in arresting and accurate language what he has learned from witnessing life. When he was a child, there were no special signs that pointed to his future calling. His parents had struggled to achieve a comfortable middle-class status, and they encouraged their son to be articulate, informed, and well read. But for Strand to become a poet was definitely not a part of their agenda. They were afraid, with good reason, that such a career would never lead to financial self sufficiency. In fact, by age forty five, when some people's thoughts turn to the possibility of retirement, Strand still didn't have a steady job.

> I didn't want one. I realized that everybody else was living better than I was. I had deliberately chosen not to go the university route, not to establish myself in a university, because I thought that I'd fall asleep. I needed a certain existential challenge to keep my mind alert and alive and responsive, at least enough so that I could write poems. But living in New York and scratching around for money, taking this job and that job and commuting, giving too many readings, that sort of thing, was destructive. It destroyed my concentration, so we had to get out.

This is when the Strands moved to Utah, where after another scary spell of poetic drought, his verse began to flow again. Patiently watching and listening to events unfolding around him, alternating between passionate involvement and sardonic detachment, he has found the pattern that best fits the predilections of his consciousness: to be an unassuming yet precise chronicler of life.

THE HAVEN OF WORDS

Of all the writers we interviewed, Hilde Domin most clearly sees literature as an alternative reality, a refuge from the brutish aspects of life. In her seventies, she has achieved a leading position in German letters. Her poems are widely read and are included in official high school textbooks. She has had her share of prestigious prizes, and she is asked to be on numerous literary juries. But hardship and tragedy have marked her life, and it is doubtful she would have survived this long if she had been unable to impose the ordered meter of verse on the chaos of her experiences.

Domin started out studying law at the University of Heidelberg and took courses with the philosopher Karl Jaspers and the sociologist Karl Mannheim. At the university she fell in love with one of the professors, a well-known classical scholar, and, as Hitler was rapidly gaining power in Germany and Domin was Jewish, the two left on a voyage of exile that was to last almost three decades. At first they went to Rome, where Domin's husband had many colleagues. But Jews were vulnerable in fascist Italy as well, so they left for Spain, then for the Dominican Republic, and for short periods they visited the United States. Thanks to her husband's connections and fame, the material hardships they encountered were not as great as they were for many other refugees. But the spiritual pain was hard enough to bear: Having to depend on the charity of hosts, being constantly on the outside of society, having to learn new languages and new skills—and always worrying about the fate of family and friends left behind—caused a chronic state of psychological dislocation.

After World War II, Domin and her husband returned to Germany, and eventually he regained his university chair, this time in Hispanic art, a field of study he had pioneered during his Caribbean exile. Domin, who up to then had been helping her husband as a sort of secretary, translator, and editor, started writing poems herself in 1951. This is how she describes the beginnings of her career: "One evening, I started writing a poem. I didn't have the idea that I wrote, but I started. It happened to me. Like, you know, falling in love. Or like being run over by a car. It happened. I had the language and I needed writing, so I wrote." The precipitating factor was that she felt "annihilated" by her mother's death. All through their life together, she had been protected by her husband; but in this crisis she felt alone and helpless. "And that is why all of a sudden . . . *I flew into language.*"

This flight into a world of symbols saves the writer from the unbearable reality where experience is raw and unmediated. When painful experience is put into words, the poet is relieved of some of her burden:

[The emotion] gets fulfilled, I guess. You know what was in you, and you can look at it now. And it is kind of a catalyst. Wouldn't you say? Yeah, I think so. You are freed for a time from the emotion. And the next reader will take the place of the author, isn't it so? If he identifies with the writing he will become, in his turn, the author. And he then also gets freed. Like the author gets freed. The emotion may not be exactly the same, but it is somehow, how would you say, in harmonic resonance.

Domin's skill with words was not something that manifested itself early or suddenly. She became interested in language after she learned first Greek and Latin, and later Italian, French, English, and Spanish. As she learned to speak these various languages, she became fascinated by the fact that the same word may have a certain set of connotations in one and a very different network of meanings in another. Or that one language could express some emotions or events more accurately than another. She read a great amount and came to cherish especially Shakespeare's sonnets and the works of Goethe. She apprenticed by helping her husband translate some of the classic Spanish poets. But above everything else she felt drawn to German, her native tongue; it was because she could not live where that language was not spoken that she returned to a land where her kin had been killed. "It's normal to find refuge in language," she says. "In music if you are a musician, or if you are a painter, in color."

Struggles with the Field

It took six years before any of her verse was published. These were difficult years, not in the least because her husband, who had been her mentor and protector, bridled at the idea that Domin could have her own voice and independent literary career. At first he patronized her efforts, then he grew resentful, and it took many years for him finally to accept the fact that her fame might surpass his own. But from the very first poem he grudgingly recognized that her verses were true poetry, and this reinforced Domin's resolution. A more ominous obstacle was the politics of the field, which almost succeeded in discouraging her. What kept her going, she thinks, was the fact that she remained oblivious to much of the infighting that took place around her:

I was very naive. I don't know how I could have been, but I was. I did not believe in literary intrigues or any such things, you know, a literary mafia. I mean, for me the work was work, and it has remained so. You know, it was difficult to be a woman, too, at the time. Being pretty. Being pretty is a disadvantage, of course. If you don't want to be kind the way people want you to be kind. But poems make their own way, and also my poems made their way without the support of people who then were in the "mafia."

Like many writers and painters, Domin is torn between endorsing two opposite images of the artist. One is the idealized version, in which genius triumphs no matter what obstacles stand in its way. The second is based on experience, and it recognizes the fact that jealous and antagonistic critics have ways to silence the artist's voice:

Mallarmé says that a poem is like a rocket—it goes up by itself. And that may be true. But then of course it can be cut off. By jealousy. I guess that is the right word. Yeah, of course. But it cannot be ever spoken about and it's long ago past, you know? It is an advantage not to be so young anymore. Nobody wants to sleep with you.

Domin is sensitive to the particular vulnerability of women in the arts. None of the women scientists in our sample hinted at sexual favors being part of the price they had to pay for advancement, but a suspicion of this being the case was not entirely absent from the artists' accounts. It is in part for this reason that naïveté is such a great help for long-term success in the arts. Instead of wasting time hatching plots and counterplots, it allows the focusing of every ounce of energy on painting or writing. Of course, this works only as long as the innocent is also lucky—because it is just as possible to be wiped out by the field, never knowing what happened or why.

Even now, despite her fame, Domin feels like an outsider in the field of literature. When she has to evaluate manuscripts for literary prizes, she concentrates on the merits of the writing instead of the personality and politics of the writer. This is how things should be, of course, but Domin claims they are rarely so. "You know, I am a terrible person when I am on a jury because I have the idea that I am not looking at the person but at the poem. And some people don't. And therefore very soon I am out of the jury." But while she keeps on the periphery of power struggles, she is deeply involved in helping young writers improve their craft. Every week she gets scores of manuscripts from aspiring poets, asking for advice. The poems she thinks are beyond redemption she sends back with a note thanking the writer for his or her confidence. If she sees some promise in the verse, she will spend hours suggesting improvements to the writer—mainly to simplify, to cut out whatever is redundant, flabby, unnecessary. Her own poems read like Japanese haiku, clean to the bone.

Telling It As It Is

Domin is aware that her poetry acts as a catalyst for deep emotions, and usually painful ones—like the depression caused by the death of her mother. Finding words for what is painful begins the healing—through form and style the poet recovers control over tragic events. But for this to work, it is necessary to be absolutely truthful, never pulling punches, always looking at reality without flinching. The ability to do this, Domin thinks, is her strongest claim to being a poet. "I think I am honest, that is why my poems get straight to people regardless of their age or their social situation. Honesty is always touching because so few people are honest, no? I don't make words around it."

Like many other respondents, Domin credits her parents—in this case, her mother—with forming her character.

> It's my nature. I think it depends on my parents. I had such a wonderful childhood because I didn't need to lie. And so I was educated to have confidence in people. If you learn it when you are very young it convinces you, even if later you have bad experiences. On the whole, confidence kind of creates confidence, doesn't it?
>
> I did not have to lie, no. And therefore, I guess possibly, I did

not learn to. I have learned to keep silent, but I did not learn to lie. It was my mother. I could speak lots of intimate things with her, and if other children wanted to not tell where we went to, I told my mother, "We are there and there, but don't tell the other mothers if they call," and my mother always kept face. I think for children the most important thing is whether they can be open at home and have nothing to fear, but can be just straightforward.

Honesty is important to the poet for at least two reasons. The first is that if she lets ideology or undue optimism color the way she reports her experiences, the truth content of the poem will be corrupted. The second is that the poet must be honest with herself, always evaluating what she writes and not letting wishful thinking stop her from improving the evolving work. "In every art, you have to be your own critic," Domin says. "If you get up to a real good standard, you have to be both the one who writes and the one who corrects at the same time. It is paradoxical, but you have to be paradoxical, otherwise you cannot live in this world." The creative individual must reject the wisdom of the field, yet she must also incorporate its standards into a strict self-criticism. And for this one must learn to achieve the dialectical tension between involvement and detachment that is so characteristic of every creative process:

You must always keep distance from yourself. Don't you think? It is the change between being quite close and being quite distant. You must always be in it and always see it from the outside. While you are doing it, you are in it. But you must always keep also a distance. And evidently the more you have the skill, the craft, the more you are able to at the same time be in it and also keep the distance and know what you are doing. Like, for example, when you eliminate a word. In the beginning you eliminate it *after* you have written it. And when you are more skilled you eliminate while you are writing. A schizoid process, is writing. You are the emotional person that kind of furnishes the words, and at the same time you are the rational person that kind of knows which words you want.

But being unflinchingly honest can be dangerous when reality becomes too chaotic, and art can no longer bring order to it. Several of Domin's close friends, Jewish writers who survived the Holocaust, have recently taken their lives, or have lost their minds, in desperation at the renewal of racism and fascism in Europe. She shares their suffering but is not ready to throw in the towel. She still hopes that poetry will help young people to find their way to a better world.

> When you write poetry honestly, and when you read it honestly, then you become an individual and build up a defense against becoming programmed. And if you read poetry to young people, which I very often do—I go to schools, I have even gone to prisons—I feel you can raise in people's minds the wish never to be an opportunist, never to be a mindless follower. To look always at what's happening and not to look away from it. That's the most you can do. You cannot change the world, but you can change the single person, I guess. And a single person who decides not to join the crowd. . . .
>
> You should not look whether you are in or out. You should look into your own heart. Confucius says that you should listen to the silent voice of your own heart. This is what poetry can do.

RELEASED BY STYLE

Anthony Hecht is a lyric poet whose verse has been published in numerous collections and in *The New Yorker* and other leading magazines. He has been awarded fellowships by all the major foundations and has won a great number of prizes for his work, including the Pulitzer in 1968. Hecht's poems are crystalline, elegant to the point of refinement, constructed with a rigorous attention to form. A Vivaldi concerto could provide a passable musical analogy to his writing. He often uses the sonnet, or even earlier *canzoni* of the kind used in the Middle Ages, more than six hundred years ago. The rules of these forms are so rigid that even Dante complained that to write according to them was like hanging chains upon himself and that he never would write in that style again. Yet, paradoxically, it is by following such demanding discipline that poetry can liberate the writer—and the reader—from the jumbled onslaught of raw experience.

It is not coincidental that Hecht's main interests in childhood were first music and then geometry. Both domains are among the most highly ordered symbolic systems, and whoever invests attention in them must follow ordered patterns of thought and emotion. Otherwise Hecht's early years were rather chaotic; his father's business failed three times, and the family not only lost everything but ended up deeply in debt each time. Nor was the emotional atmosphere much more serene; he remembers suffering extraordinary anxiety and loneliness.

Music was the first of these [interests], precisely because it was abstract and therefore could be divorced from all the mess around me. I loved it. I used to listen to it all the time on the radio. I had a little record collection, and played things over and over again until I knew them really by heart. Eventually I would come to know poetry by heart in the same way. I really knew whole symphonies by heart. I knew where every instrument came in and went out and all their figures. I listened with great care, and without being able to read music I did feel I knew these pieces very well indeed. And as I say, the great thing about music is that it is nonreferential so it is completely uncontaminated by anything. Even as a child I scoffed at people whose association with music was always with some sort of sentimental event. You know, "They're playing our tune." That meant nothing to me. A Beethoven sonata was not connected with any emotional event precisely because I didn't want it to be. I wanted it to be pure music.

I had a geometry teacher in high school and I did extremely well, I got honors in geometry. And I loved it. Again, because like music, it's abstract. I think probably music and math were the two things that I liked most as a child.

It is fascinating how the pursuit of artistic domains such as music or poetry, and also of scientific domains like geometry and science, is motivated not so much by the desire to achieve some external goal— a poem or a proof—but by the feeling of freedom from the threats and stresses of everyday life one experiences when completely immersed in the domain. Paradoxically, it is the abstract rules we invent to limit and focus our attention that give us the experience of untrammeled freedom.

Hecht experienced a less temporary and more physical liberation when he went off to college and enjoyed student life for all it was

worth. But the idyllic campus life did not last long: He was con-
scripted and in Europe saw half his infantry company killed or
wounded. The brutality of the war left a deep mark that had to be
exorcised in his work. Again art came to the rescue: After the war
Hecht went back to school, met good mentors and colleagues,
decided that poetry rather than music was his strongest suit, and was
launched on what became a very successful career. One example
illustrates his method of work, as well as the sources of his inspiration:

> There's an awful lot of fussing and fiddling; I feel that the writ-
> ing of a poem is a very conscious act. It's not what it is for some
> people like Ginsberg, and I say this without disrespect of him, but
> his way of writing poetry is altogether different from mine. He is
> annotating the activity of his mind, and I'm trying to make a for-
> mal structure. Once I have the *donnée,* the stuff that I get out of the
> unconscious, it's my job to bring it together.
>
> I can give you an example. It was a poem written about the
> birth of our son, who was born in 1972, in a snowstorm. And in
> 1972 the Vietnam War was still going on. I don't know how this
> evolved—it may have been in that state between wakefulness and
> sleep—I realized that one of the things I was thinking about had to
> do with the sheer randomness of events. How there was a random-
> ness, for example, in the whole process of sexual intercourse and
> conception, there was randomness in the snow as it appeared the
> night of the birth, where it fell and how much it accumulated.
> And there was randomness in the death of soldiers in the field.
> And all of this somehow I knew belonged together if I could find
> the way to get it into a poem.
>
> I do however find that while I'm now talking in concepts, very
> often poems begin, for me, with words. So that very often when I
> leap out of bed in the dark, the thing that I want to jot down is a
> set of words in a certain order, which will be the nucleus of what-
> ever is going to come. I think much more in terms of words than I
> do in terms of other things—concepts, for example.

Like all other writers, Hecht learned to be one by reading extensively. He memorized poems until they "became part of my blood-
stream." Then he spent years writing in the voice of various poets he

admired: John Donne, George Herbert, Thomas Hardy, T. S. Eliot, John Crowe Ransom, Wallace Stevens, W. H. Auden. Assimilating the style of predecessors is necessary before one can develop one's own. Only by immersing oneself in the domain can one find out whether there is room left for contributing creatively to it, and whether one is capable of doing so.

Poetry is whatever poetry has been, with any new inventions that a new poet cares to add to that. But he can't add to it without knowing what it has been. I mean, the only way you decide to become a poet is because you've read a poem. So in some immediate sense poetry depends on the whole poetic tradition of the past. And once you accept that idea then you have to decide, well, out of all that enormous welter of previous poetry, what is most interesting to me? Because there's an awful lot of stuff that nobody really likes or cares about, and there's an awful lot of inferior poetry being written all the time. It takes a very long time to acquire the kind of sensibility that can make intelligent, sound discriminations, what's good and what's not good, what has already been done and what therefore needs now to be done which is different from all the stuff in the past. All that takes time.

As powerful as poetry is, it does not resolve all one's problems. Mastering a symbolic style—be it poetry or physics—does not guarantee one will also bring order to those events that lie outside the rules of the domain. Poets and physicists may bask in the beautiful order of their craft as long as they are working at it, but they are as vulnerable as the rest of us when they step back into everyday life and have to confront family problems, time pressures, illness, and poverty. This is why it becomes so tempting to invest more and more energy in one's work and forget everyday life—in other words, become a workaholic. Developing his poetic skills did not resolve all of Hecht's problems either. His first marriage broke up after seven years, and for a decade afterward he felt that he was floundering and wasting his time, and he was unhappy about it. He credits his second marriage, in 1971, with returning him to an even keel and making "everything seem worthwhile." A fulfilling relationship and a creative profession—what more can one ask? Especially when work consists of adding to the culture's heritage of order and beauty.

A JOYFUL RESPONSIBILITY

Madeleine L'Engle is best known for her children's stories (which are just as interesting to adults), but she has written books of all sorts, to the tune of about one a year for the last few decades. She married at twenty-seven, and her husband, an actor, was her "best editor" for the next forty years. Their three children have been an inspiration in her work, which became really successful only when she herself turned forty, with the publication of *A Wrinkle in Time,* which won the prestigious Newbery Award and became the first volume of a classic trilogy.

L'Engle started writing stories when she was only five years old, and although she also wanted to be an actress and a pianist, she always knew that writing was her true vocation. After college she worked in the theater and started publishing her stories. She still plays the piano, and in a way that is similar to Mark Strand's strategy of driving a car or running errands when the focus on his work becomes excessively absorbing, she uses music to help clear her mind and get back in touch with experiences beyond the compass of rationality:

> Playing the piano is for me a way of getting unstuck. If I'm stuck in life or in what I'm writing, if I can I sit down and play the piano. What it does is break the barrier that comes between the conscious and the subconscious mind. The conscious mind wants to take over and refuses to let the subconscious mind work, the intuition. So if I can play the piano, that will break the block, and my intuition will be free to give things up to my mind, my intellect. So it's not just a hobby. It's a joy.

Her early school years were a dismal experience: "In the middle grades I had terrible teachers, who decided that since I wasn't good at sports, I wasn't very bright. So I did no work for them. I learned nothing in school till I got into high school, and what I learned, I learned at home. I learned absolutely nothing in school. Then I had some good teachers in high school and some really excellent teachers in college." Indirectly, however, the bad school experience and a physical handicap—a bad knee—turned out not to be a total loss. Shunned by peers and teachers, Madeleine spent much of her childhood reading and thinking alone. Now she feels that she couldn't have written her books if she had been happy and successful with her peers. Like most individuals in our sample, she showed her creativity first of all by being able to turn a disadvantage into an advantage. Later in high school, and then at Smith College, she found supportive teachers. It was in a college writing workshop that her literary career became confirmed.

The family environment, on the other hand, seems to have been supportive from the very beginning. Her father had been a journalist, a foreign correspondent. He and her mother married late, and she was the only child of two very busy parents who neither pushed Madeleine nor held her back. They were neither critical of her accomplishments nor overly approving—and L'Engle believes that too much encouragement can be almost as bad as too much discouragement. It was an atmosphere where artistic expression was considered a normal part of life. Then when she was seventeen, her father died, and at that crucial time her mother's unselfishness made a great difference:

> The best thing my mother did, and it was, I think, remarkable, was when my father died. My mother was a Southerner and it was expected by her Southern family that, of course, I'd come home and take care of my poor, widowed mother. And she did everything that she could to free me, to go on to college, and after college, to go back to New York and do my own thing. She in no way held me back. She did not expect me to give up my own work for her. So I was able to go to New York after college without any feeling of guilt and start doing my own thing. Earning my living any way I could and writing my first book, half of which I had written in college.

The situation L'Engle describes is familiar to many creative women but practically unknown to men. Women feel responsible to their families of origin and extended relatives in ways that men do not. Of course, the men in our sample feel a great deal of responsibility to their wives and children, and the depth of guilt they experience if they feel unable to meet their obligations can be overwhelming. But their sense of responsibility is generally limited to the role of husband and father, whereas the women's usually embraces a larger web of kinship.

The Survival of the Human Spirit

The central themes of L'Engle's writing circle around the need for hope. Her fiction, even that aimed at children, typically deals with doomsday scenarios that reach a happy ending because the main characters never lose hope even in the grimmest situation, and they learn from adversity to act with mercy and forgiveness. As in the stories of C. S. Lewis or J. R. R. Tolkien, which share with L'Engle's the assumption that powerful evil forces are always threatening to reduce the world to chaos, innocence wins because it refuses to take the easy way out, because it won't use violence even when it is expedient to do so. She feels that it is especially important to remind readers of these grim realities nowadays, when the media are unable to present a meaningful picture of how things work:

> Television commercials give such a strange view of what life is supposed to be. And a lot of people buy it. Life is not easy and comfortable, with nothing ever going wrong as long as you buy the right product. It's not true that if you have the right insurance everything is going to be fine. That's not what it's really like. Terrible things happen. And those are the things that we learn from. People are incredibly complex. I read a book last winter called *Owning Your Own Shadow*, by Robert Johnson. And one of his theories is that the brighter the light, the darker the shadow. Which is often true.

In her own work, L'Engle feels that the enjoyment of writing comes first, followed by a sense of responsibility for what she writes. Because she knows that her books influence a lot of readers, she is concerned about not passing on a destructive message. Even when the characters in the book suffer and seem at the end of their rope, she believes that "you have to get them out, into some kind of hope. I don't like hopeless books. Books that make you think, 'Ah, life's not worth living.' I want to leave them thinking yeah, this endeavor is difficult, but it is worth it, and it is ultimately joyful."

This sense that despite encroaching darkness there is always a silver lining is not just a rhetorical device for L'Engle; it is a belief that pervades her attitude toward real life as well.

> Oh, I'm a little less idealistic about the world than I might have been thirty years ago. This whole century has been difficult, but

the last thirty years have been pretty awful in many, many ways. I mean, if thirty years ago I had listened to the six o'clock news, I wouldn't have believed it. War is all over this planet. On the other hand, there's a black president in South Africa! Wonderful things happened even while there's terrible things. We wouldn't have believed thirty years ago that the Soviet Union would be dissolved. It's like weather, it's unpredictable. The amazing thing is that despite all the things that happen, the human spirit still manages to survive, to stay strong.

Everything in the Universe Is Interrelated

If the survival of the human spirit is one central theme in L'Engle's work, another is the interrelation of action and reaction, of events at the cosmic and the microscopic levels. A sort of a karmic web pervades her narrative, where violence inside the cells of a body can have repercussions among the stars. Her books are a mixture of science fiction and medieval morality tale. She drew on particle physics and quantum mechanics for *A Wrinkle in Time,* on cellular biology for *Wind in the Door,* and *A Swiftly Tilting Planet* combines the singing magic of the druids with relativity theory. Like most creative individuals, her contribution has been to bring together domains that appear to have nothing in common.

A lot of ideas come subconsciously. You don't even realize where they're coming from. I try to read as widely as possible, and I read fairly widely in the areas of particle physics and quantum mechanics, because to me these are very exciting. They're dealing with the nature of being and what it's all about.

One of the things that we have learned, having opened the heart of the atom, is that nothing happens in isolation, that everything in the universe is interrelated. Physicists have a favorite phrase, "the butterfly effect." That means that if a butterfly should fly in here and get hurt, the effect of that accident would be felt in galaxies thousands of light-years away. The universe is that closely interrelated. And another thing they've discovered is that nothing can be studied objectively, because to look at something is to change it and to be changed by it. Those are pretty potent ideas. I'm reading now a book on the necessity of seeing to light. It's like the tree falling in the forest; it doesn't make a sound if it's not heard? Well, the same thing with sight—light is not there unless it is seen.

L'Engle believes that telling stories is an important way to keep people from falling away from one another and to keep the fabric of civilized life from unraveling. Helping the relationship among people remain harmonious is one of her central tasks. She believes her calling is to reflect on what she has learned from experience and share it with other people, especially children.

In America we no longer value the wisdom of older people. Whereas in so-called primitive tribes, the older people are revered because they have the "story" of the tribe. I think as a country, we're in danger of losing our stories. Planned obsolescence cuts across everything; it doesn't only hit refrigerators and automobiles, it hits people, too. I have wonderful friends of many generations, and I think that's important. I think chronological isolation is awful and chronological segregation is one of the worst of the segregations.

Risking Failure

Like many other creative individuals, L'Engle attributes her success in large part to the ability to take risks. She has been adventurous in her personal life, trying to follow an inner sense of what was right even when it went against the norms and expectations of her social milieu. She flouted popular wisdom by writing in a style that editors and critics thought was too difficult for young people to read, too childish for adults—even though the scientific concepts and philosophical ideas actually were not that easy even for grown-ups to grasp. So it took ten years for her unusual stories to be published. The manuscript of *A Wrinkle in Time* collected rejection slips for two and a half years before a publisher took a chance on it. "You cannot name a major publisher who didn't reject it. They all did." But she was never tempted to compromise her vision in order to play it safe.

One episode she remembers in this context concerns a time early in her career when she was invited to give a talk to a women's group on the West Coast. She prepared a humorous talk but one that adroitly avoided controversial issues. When she showed the draft of the lecture to her husband, he said: "'Well, dear, it's very funny. But they're not paying you to go all that way just to make them laugh. They think you may have something to say. Stick your neck out and say it.' And so I did. Sticking my neck out has been something I have learned to do. And I think it's a good thing."

Her personal credo is well summarized by these few lines, which reflect the stubbornness that has stood her in such good stead so far:

> Human beings are the only creatures who are allowed to fail. If an ant fails, it's dead. But we're allowed to learn from our mistakes and from our failures. And that's how I learn, by falling flat on my face and picking myself up and starting all over again. If I'm not free to fail, I will never start another book, I'll never start a new thing.

ADDING TO THE WORLD

Richard Stern, novelist and professor of literature, recalls three formative stages in his childhood. First when he was exposed to oral narrative, then when he learned to read, and finally when he tried writing himself. Each of these steps enlarged tremendously the limits of his experiential world. His first brush with fiction involved listening to the stories his father told when Richard was practically an infant. This experience is still quite vivid in his mind:

> My first memory, and I think it's a memory, is of lying in the dark. And I swear I have the sense, but I think that's probably imposed, of seeing the slats of my crib. I know that I'm on the right side of the room. On the left side of the same room, in the other corner, is my sister, four years older than I. Somewhere in the middle is my father. And each night he comes in and tells us stories. He was a wonderful storyteller. And his voice and the stories are present to me. I've used the names that he invented in stories later on.

Stern started reading early, and the fairy tales that were his first fare had such an impact on him that his mother, afraid that he would get ill from overexcitement, forbade him to check out any more books from the neighborhood library at Amsterdam Avenue and Eighty-second Street in New York City. Stern found ways to take books out anyway and continued to read voraciously. Reading widely, of course, is how writers learn to master the domain of literature. Stern echoes what everyone else in his field says: "I don't think there are any writers who have not read, who have not been enchanted by books, by stories, by poems."

Finally, during his freshman year at Stuyvesant High School, he experienced his first success as a writer. As is often the case, the success was modest but memorable—it confirmed that he had the ability and provided the first heady taste of admiration:

> A wonderful teacher, Mr. Lowenthal—I can see him now, in his blue suit and high collar, large nose and large Adam's apple, black hair—asked anybody who wanted to write a story. And I had been reading stories so I wrote a story. And the class laughed, and Mr. Lowenthal approved, and I knew this was very important.

Before this episode, Stern wanted to be a Supreme Court justice. As a young Jewish boy who had been inspired by the lives of Justices Brandeis and Cardozo, about whom he had read in a book he thinks was called *The Nine Old Men*, he believed this to be the highest ambition to which he could aspire. But after tasting the exhilaration of authorship in Mr. Lowenthal's class, he sensed that his future direction lay in writing. And he never looked back: He enrolled at the University of North Carolina when he was sixteen years old and almost immediately fell in with a group of poets and writers who remained lifelong friends. They had a literary society, a literary magazine—in short, all the makings of a small field. From college he went on to Harvard, and then to the Iowa Writers' Workshop, where it was possible to get a Ph.D. by writing fiction instead of an academic thesis.

At Iowa he began to publish extensively, and in 1954 one of his stories was included in the prestigious *Best O'Henry Stories*. It is there that he started on the book that made his reputation as a novelist: *Golk*. Equally important perhaps was the fact that at Iowa, and while working on the literary magazine *Western Review*, he met and formed friendships with some of the most influential writers of his generation. Saul Bellow and Philip Roth became particularly close, and eventually both Bellow and Stern went to teach at the University of Chicago. During his travels he became acquainted with some of the most prominent European writers as well; Thomas Mann made a particular impression. Contacts like these are necessary to the creative person for several reasons: They provide benchmarks for evaluating one's own work, they offer competition that spurs one to surpass oneself, they provide helpful criticism, and, last but not least, they open up opportunities and information that can be essential to one's advancement.

The Conversion of the Negative

As we have seen several times already, one typically turns to writing literature in order to restore order to experience. Madeleine L'Engle is concerned with the survival of the human spirit threatened by cosmic chaos; Anthony Hecht was moved by the inanity of war; Hilde Domin by the tragedy of Nazism and the death of her mother. It is not surprising that Stern too uses his writing to exorcise some evil. In his case, the evil seems to be something more private, less dramatic, more related to the normal wear and tear of life. Perhaps one could say that his intent is to explore the damage that psychic entropy causes to our lives—the stunted emotions, the acts of selfishness, the betrayals, the inevitable disappointments that are the conditions of existence. These are the grains of sand that cause the writer to coat them with words to diminish the pain:

> The great thing about this kind of work is that *every* feeling that you have, every negative feeling, is in a way precious. It is your building material, it's your stone, it's something you use to build your work. I would say the conversion of the negative is very important. So I taught to myself what I try to teach my students who are becoming writers: Don't duck pain. It's precious, it's your gold mine, it's the gold in your mine.
>
> Of course there are things in myself which I haven't talked about—and probably won't—which I know are bad, mean, twisted, weak, this, that, or the other thing. I can draw strength from that, without talking about them. I can transform them. They're sources of strength. And as I said earlier, the writer takes those and they're his material.

To overcome the pain of existence, one must be honest with one-self, acknowledging one's faults and weaknesses. Like a surgeon, one must be willing to cut deeply into the festering sores of the psyche. Otherwise too much energy is absorbed in denial, or in ruminating over disappointments. Stern responds to the question about what was the main obstacle he encountered in his life:

> I think it's that rubbishy part of myself, that part which is described by such words as vanity, pride, the sense of not being treated as I should be, comparison with others, and so on. I've tried rather hard to discipline that. And I've been lucky that there has been enough that's positive to enable me to counter a kind of biliousness and resentment—*ressentiment*—which I've seen paralyze colleagues of mine, peers who are more gifted than I. I've felt it in myself. And I've had to learn to counter that.
>
> I would say that the chief obstacle is—oneself.

It is easier to diagnose what's wrong with one's life than to cure it. Like most people who are honest with themselves, Stern is aware that with all the best intentions in the world some bitterness remains, some unrequited ambition rankles, some past choices cause regret. Weakness in others is relatively easy to condone. Stern endorses Pascal's maxim "To understand is to forgive." In fact, one of the most exciting opportunities in being a writer, he feels, is to take a villain or criminal character and make him human by showing what caused him to be so. It is more difficult to forgive oneself, but writing helps to do that, too. After all, the writer is also a part of the human race, and when he explains the failings of a character, to a certain extent he excuses himself as well. And then there is joy in being able to craft a story that will add meaning to the reader's life. The greatest reward of the writer is when the readers

> have enjoyed, had pleasure in the rise and fall, in the symmetry, in the characters, in the situations, so that they feel their understanding deepened. They've felt pleasure, and it's related to something that I've made. I have told my grandchildren stories, and my little nieces and nephew. To see two or four or five faces hanging on things you say is one of the most beautiful things in the world. That kind of attention. You know, I know some actors and

actresses well, and I see through them the connection between their work and mine. It's making human beings comprehensible, so shaping a life in a book or on the stage that an audience suddenly gets a grasp of what's there, what *is*.

As these five cases suggest, the domain of the word is indeed quite powerful. It allows us to recognize our feelings and label them in terms of enduring, shared qualities. In this way both the author and the reader can achieve a certain distance from the immediate raw experience and begin to understand, to contextualize, to explain what otherwise would remain a visceral reaction. Poets and novelists stand up against the chaos of existence. Hilde Domin builds a refuge of words where actions and feelings make sense; Mark Strand chronicles the fugitive experiences that would otherwise fade into oblivion; Anthony Hecht constructs beautiful forms to stem the capricious randomness of fate. Madeleine L'Engle tries to find the connection between events happening within our cells and those happening between the stars; Richard Stern focuses on the fragility of human commitments. Their struggle leaves a record of the human attempt to bring meaning to life. For the most part, it is this struggle that serves as the inspiration for their work.

All of these writers were able to make their contribution only by first immersing themselves in the domain of literature. They read avidly, they took sides among writers, they memorized the work they liked—in short, they internalized as much as they could from what they considered the best work of previous writers. In this sense, they themselves became the forward-moving edge of cultural evolution.

Sooner or later, each of them also became part of the field of literature. They befriended older writers, they gravitated to avant-garde schools and journals, they became intensely involved with other young writers. Eventually they became the gatekeepers by teaching literature and serving on juries, editorial boards, and so on. As opposed to the generally ecstatic relation they had with dead writers, relations with live ones were much more problematic. Domin regrets the infighting of the literary "mafia"; Stern is aware of the potential for bitter jealousy among peers. Somehow or other, however, writers must come to terms with the social organization of their domain if their voices are to be heard.

Another similarity among the writers was the oft-stressed emphasis on the dialectic between the irrational and the rational aspects of the craft, between passion and discipline. Whether we want to call it the Freudian unconscious where childhood repressions linger or the Jungian collective unconscious where the archetypes of the race dwell, or whether we think of it as a space below the threshold of awareness where previous impressions randomly combine until a striking new connection happens by chance, it is quite clear that all the writers place great stock in the sudden voice that arises in the middle of the night to enjoin: "You have to write this."

Everyone agrees that necessary as it is to listen to the unconscious, it is not sufficient. The real work begins when the emotion or idea that sprang from the uncharted regions of the psyche is held up to the light of reason, there to be named, classified, puzzled over, and related to other emotions and ideas. It is here that craft comes into play: The writer draws on a huge repertoire of words, expressions, and images used by previous writers, selects the ones most fitting to the present task, and knows how to make up new ones when needed.

To do so it helps to have a broad base of knowledge that extends beyond the boundaries of literature. Domin draws on her knowledge of many languages; Hecht on music and geometry; L'Engle on quantum physics and microbiology. Being able to braid together ideas and emotions from disparate domains is one way writers express their creativity. Love and death may not have changed for thousands of years; but the way we understand them changes each generation, in part as a result of what we know about other facets of life.

There were many similarities also in the methods these writers follow as they ply their craft. All of them keep notebooks handy for when the voice of the Muse calls, which tends to be early in the morning while the writer is still in bed, half asleep. Most of them have been keeping diaries for many years. They usually start a working day with a word, a phrase, or an image, rather than a concept or planned composition. The work evolves on its own rather than the author's intentions, but is always monitored by the critical eye of the writer. What is so difficult about this process is that one must keep the mind focused on two contradictory goals: not to miss the message whispered by the unconscious and at the same time force it into a suitable form. The first requires openness, the second critical judgment. If these two processes are not kept in a constantly shifting balance, the flow of writing dries up. After a few hours the tremendous concentration required for this balancing act becomes so exhausting that the writer has to change gears and focus on something else, something mundane. But while it lasts, creative writing is the next best thing to having a world of one's own in which what's wrong with the "real" world can be set right.

ELEVEN

THE DOMAIN OF LIFE

We do not know for sure what was the first form of systematic knowledge our ancestors developed. Certainly the attempt to classify plants and animals, to understand health and disease, must have been one of the earliest. The domain that we now call biology, dealing with the forms and processes of life, is one of the fundamental ways humans have tried to make sense of the world in which they lived.

The difference between present knowledge and the knowledge of our ancestors is greater in biology than in any domain except physics. More than four thousand years ago, in all the major centers of civilization—Mesopotamia, Egypt, India, and China—knowledge of medicinal herbs and animal species that had been slowly assembled by preliterate hunters and pastoralists began to be carefully recorded. Fifteen hundred years later Aristotle provided a more scientific classification of animals, and one of his students did the same for plants. But until the last few centuries nobody had any understanding of physiological processes—digestion, breathing, the circulation of blood, the function of the nervous system. No idea of cells, of bacteria and viruses, of genetics and evolution. The difference between what our ancestors could see of the processes of life and what we can see is enormous.

The life sciences now have become so diversified and specialized that we would need several dozen examples to show what the domain consists of. Even a little over a hundred years ago, someone like the German explorer and naturalist Alexander von Humboldt could condense in four volumes the entire spectrum of knowledge that a biologist or earth scientist was then likely to know. Nowadays no single individual can be expected to cover in depth even a small fraction of the content of the discipline. This chapter focuses on three persons who have changed the domain of the life sciences, even though these three case studies represent only a few of the many possible approaches.

A PASSION FOR ORDER

E. O. Wilson is one of the most influential biologists of our time. With more than three hundred technical papers and many books, two of which have won the Pulitzer Prize, he has made important contributions to the classification of ants; to the concept of biodiversity, or the necessity to preserve the variety of life forms; to the study of chemical communication in insects; and to the study of island ecosystems. But he is probably best known as the father of sociobiology, or the ongoing attempt to explain human behavior and social institutions in terms of their selective value over evolutionary time. In the process, he has become involved in deep ideological disagreements that at one point generated a host of enemies both within the field and outside of it. Regardless of fame or adversity, however, Wilson keeps true to his vocation, which is an unusual combination of rigorous fieldwork with inspired insights that bring together facts and principles everyone else thought were unrelated.

His current goal is to achieve the grand synthesis of the social and biological sciences that he initiated with the classic work *Sociobiology*:

> I see a picture forming, one in which I would pay a great deal
> more attention to the fundamentals of the social sciences. And use
> the evolutionary biologist—the biologist's approach, since I'm
> learning some molecular and cell biology too—to winnow and
> reanneal the elements of the social sciences that I think are

required to create a consilience between biology and the social sciences. To the present time, it is still not understood that we need to create a consilience. Many would say it's impossible. And the ones who say it is impossible, they're just a goad to show it is possible. That's what makes this whole domain exciting.

A Naturalist with Steely Ambition

Wilson has been a ceaseless worker all his life. A painful childhood instilled a certain amount of insecurity in him, which he decided to overcome with a relentless drive modeled on an idealized Southern heritage long on pride, sacrifice, and discipline. These were what current psychological jargon calls deficit motives, based on efforts to compensate for undesirable early experiences. But there was also positive motivation: fascination with and love for the living world, and especially for some of its most humble denizens, ants and termites. Wilson wanted to be an entomologist by age ten; some issues of the *National Geographic* and a visit with a friend to the Washington Zoo confirmed that what he wanted most to do in life was to become an explorer and a naturalist.

Like many creative individuals, Wilson was bored in school until he reached college. In the early years, the Boy Scouts provided an environment where he could pursue his own interests and learn at his own speed. One would have thought that Wilson, having impaired vision, would become interested in whales or elephants, but with characteristic obstinacy he chose to focus on the smallest of insects instead. At the age of thirteen, he wrote up the first reports on the mound-nests of the fire ants that were beginning to infiltrate the Southern states, creating a sizable environmental problem. When he was in high school in Mobile, Alabama, a local news editor decided to feature the fire ants and commissioned the young Wilson to write a series of articles. The sudden responsibility, acceptance, and feeling of accomplishment this project provided launched his career.

At about the same age, before entering college, Wilson read Ernst Mayr's *Systematics and the Origins of Species,* which revealed to him that the huge mass of facts in the natural world could be ordered in a meaningful way by adopting the theory of natural selection. Mayr was the first great intellectual influence on Wilson's career; later he became mentor and then valued colleague to the younger naturalist. But many other influences followed, as Wilson retained his youthful curiosity and openness. "I just had one road to Damascus after another, I guess," he says of his intellectual journey. Signposts on that road included the example of James Watson, whose grand reduction of genetics to the double helix of the chromosome and whose boldness and independence he found very appealing. Konrad Lorenz taught him the possibility of explaining animal behavior through ethological observations; the geographer Ellsworth Huntington was responsible for introducing the concepts of evolutionary ecology, which tried to explain why two cultures developing in very similar ecological niches, such as Newfoundland and Iceland, end up being so different. Finally he learned about the principles of kin selection from William D. Hamilton, whose mathematical models of changes in the reproductive rates of populations opened up another door for understanding life processes. From all these very different perspectives, Wilson was about to forge his own great synthesis.

His personal development appears to have been as complex as the intellectual one. We saw in chapter 3 that creative individuals typically alternate between opposite poles on traits that are usually segregated. Wilson mentions several of these polarities—facility vs. persistence, love of subject matter vs. desire to control, selflessness vs. ambition, solitude vs. social acceptance, enjoyment vs. pain—in these reflections about what it takes to be a successful scientist:

> There are a few fields, like pure mathematics and theoretical physics, in which sheer brightness is crucial. It's also interesting that these are the fields in which the best work of the scientist is often, if not usually, done by the age of thirty-five. Harvard's got more than an ample sprinkling of physicists and chemists in the National Academy of Science, and mathematicians whose best work was twenty years ago. They're nice people, but you know they're not going to hit the ball out of the park anymore.
>
> And in the other sciences, persistence and ambition are all-important. I think what is required is a combination of love of the subject—you got into this because you had a self-image and a delight over certain activities and mental operations that you would do regardless of where you went or what your fortunes were. Natural history is like that. You know, you could have forced me to be

a postmaster in Boise, Idaho, and I would have done it and I would have been a very happy man. I would have been out in the early mornings, and the evenings, the weekends, in the mountains. I'd be doing all the same things. Because I loved it and love it.

But the other thing is insecurity, ambition, a desire to control. A scientist—and this is a risky thing for me to confess—wishes to control, and the way to control is to create knowledge and have ownership of it, either by original discovery or by synthesis. I am consumed by a drive to be in more command of broad subjects than anybody else in the world, and it probably is a proprietary instinct that is beyond or separate from my love of the subject. I want to do natural history. I want to be in the field. I could happily spend 360 out of 365 days away from other people, you know, traveling in the rain forest and [in] my library.

But at the same time, I want to feel that I'm in control, that I cannot be driven out of it, that I cannot be stopped, that I will be well regarded for being in it, and that entails control, and control means ambition. It means constantly extending one's reach, renewing, extending, innovating. I think that the combination of those drives is what makes a major scientist.

Being a major scientist, or scholar for that matter, entails, I might add, enormous amounts of work and pain. And you have to accept a certain amount of rejection. You have to tolerate strong rivals. You have to be ignored for periods of time. But the idea of the lone hunter, or the lone voyager or explorer, who's guided by his principles and is going to get there against all odds, that self-image, as romantic and foolish as many people might consider it, is a very powerful force in making a major scientist.

Dodging Bullets

In his own life, Wilson often had to tolerate rivals and rejection. This was due, in part, to a convergence of historical circumstances in the domain and the field of biology in the 1960s that changed the rules of the discipline beyond recognition.

In terms of the domain, it was in that period that knowledge in molecular biology suddenly went through a phase of exponential growth. The naturalist tradition of field work in which Wilson had been trained suddenly looked old and pointless. The great leaders of a former generation were being eclipsed by young experimentalists who could control the chemical processes within cells, decode genetic instructions, and promise to unlock the secrets of biological creation itself. To paraphrase Karl Marx, the point of biology shifted from studying life to actually changing it.

The effect of this revolution in knowledge was that most bright biologists were attracted to its molecular variety, threatening to leave the old guard high and dry, deprived of the necessary recruits. The change in the domain had an immediate impact on the field: Research grants started to go to the laboratories, journals began to publish more experimental articles and fewer fieldwork studies, and the new generation of biologists turned away from the old problems and immersed itself in the seemingly endless but orderly world of cellular processes. In other words, an extreme example of what Thomas Kuhn has called a paradigm shift was sweeping through biology.

In such a situation the most common response for members of the old guard is to resign themselves to the inevitability of "progress," take on an administrative position, or rest on their laurels in some other fashion. But Wilson was still too young, or too determined, to throw in the towel. So he developed a strategy for thwarting historical inevitability that ultimately proved quite successful. He did not try to confront the molecular revolution head-on, or deny its contributions. Instead, by bringing together other current approaches such as mathematical modeling and population studies he was able to resurrect Darwinian natural history in a modern guise. Wilson explains how he pursued his campaign to defend his hold on the domain, and therefore on the field of biology:

It entailed getting together with the brightest people I could find. It meant for me, even though I have limited mathematical gifts, learning mathematics much more than I ever thought I would need, so that I could be literate in model-building, and reeducating myself in my late twenties and early thirties. And it meant, among other things, inventing the term "evolutionary biology." I invented it in 1957 or '58. And then giving a course in it, and in population biology, and as I indicated, presenting a brave front. A lot of it was a Potemkin village, I have to tell you, because so little could be laid out in the new mode of population biology, model-building, and experimentation, and so on, that I had to parade those examples and make the most of them. And this is what I did during much of the sixties in my teaching.

Now there were, here at Harvard especially, a number of very bright young undergraduates, new graduate students, who had considerable mathematical ability, better than mine in most cases, and they listened to me in that course and they saw a career for themselves. They didn't have to go into the milling hordes of molecular biologists and make their way there. They saw a route into biology, a successful career into biology, by way of mathematical modeling and theory and integration and evolutionary biology. And they include very gifted people, a fairly long list of people now in their forties, or even fifties—quite distinguished people.

As Wilson suggests, in order to make a viable creative contribution one must change both the symbolic system and the social system at the same time. It is not enough to come up with new ideas, new facts, new laws. One also must convince young people that they will be able to make a living and a name for themselves by adopting the new perspective.

But the field of biology in the 1960s was being changed by forces other than the molecular revolution, forces that originated outside the domain of biology in the larger sociocultural arena. What biologists did began to be a concern of society at large. Evolutionary theory, with its axiom of the "survival of the fittest," was seen as providing ideological support to entrenched powers. Molecular genetics was raising the specter of scientists deciding what kind of children we should have, and how many. Battle lines were drawn along political lines, and Wilson's effort to achieve a sociobiological synthesis found itself in the cross fire. In these often quite violent confrontations, his early pride and spirit of adventure stood him in good stead:

> I ran into the radical left and had combat experience with the politically correct movement, powered by the last remnants of the counterculture left in the academic world. In the seventies I became so revolted by the dishonesty, including that of some of the respected academics, that it immunized me forever from wanting to curry favor with the people who applauded them.
>
> So, if anything, certain conservative social values that I had any-way from childhood made me much more individualistic and—

what's the word I'm looking for?—independent. You know, as a person I don't think much of the right and I don't think much of the left. My favorite movie is *High Noon*. I don't mind a shoot-out, and I don't mind throwing the badge down and walking away. I sort of have a Hemingwayesque attitude toward life in that regard.

Hunting for Patterns

Wilson typically works on several projects at once, using different methods. This is again a common pattern among creative individuals; it keeps them from getting bored or stymied, and it produces unexpected cross-fertilization of ideas. There are at least four different approaches that Wilson commonly uses. The first is fieldwork in exotic places, which acts as a sort of "nuclear fuel" by providing concrete experiences and data to be elaborated later. The second is attending lectures or meetings, where he absorbs from other experts the latest developments in the domains that interest him. The third is night-work, the serendipitous connection between ideas that unexpectedly arise upon waking up in the middle of the night. And finally there is the systematic work that takes place from morning to early afternoon, which also includes reading, writing, mathematical modeling, and drawing specimens. The crucial insights sometimes occur during the night-work, but more usually they are the result of the systematic work process itself and its combination with the other three approaches:

I think the best eureka ideas I've had are right in the middle of working. Well, for example, a week ago I was sitting having lunch. I do a lot of studying and writing while I'm having lunch. I have a favorite restaurant in Lexington. It's an Italian restaurant. They know me, they let me sit in the corner. I work for up to two hours every lunch period when I can be at home in Lexington. And bring papers. I read books. I make notes.

I was reading an anthropological work, and I was worrying about why there were such great differences among preliterate societies, and things like patriarchy and the transfer of wealth and so on. And then I saw that it was ecological in ways that the author had missed. He was describing it, typical ethnography. He was describing it as though, "Oh, well, human behavior is so flexible.

We have this, and we have that." And I was saying, "No, no. It's ecological. You know, it's this way among the Australian aboriginals because their resources are patchy and unpredictable. It's that way in an African agricultural society because they are not unpredictable and patchy," and so on. And then I started: "But why do these things hold on for such long periods of time? All those fine details of cultural differences hold on?"

And then I thought of the whole notion of ritualization and the need to ritualize and codify and then sacralize some kind of code norm, and that that must be the reason for stasis in cultural differences. In other words, many things will work. But once the society has settled upon something and ritualized and sacralized it, then it becomes very static. And then last night, as I sat and listened to Amartya Sen, the economist, talk about the Nash equilibrium, the steady state of strategies, it occurred to me that they tend to freeze. You know, once they're set. At least theory predicts that they freeze.

It occurred to me that in addition to ritualization, and perhaps as an aid to it, the attainment of Nash equilibrium would be a means of reaching equilibrial solution, as opposed to others, and then holding onto them indefinitely. And that the connection between that concept of strategy equilibrium developed by economists should be related to the notion of cultural stasis and ritualization in anthropology. So I've just given you an example of the way I've been thinking the last several days. Now that last one, Nash equilibrium to cultural stasis, is just the kind of thing that might occur to me as I was falling asleep. As it did, it came to me while I was listening to the talk about the Nash equilibrium, but it easily could have come to me a couple hours later as I was getting ready to go to sleep. OK. And often that sort of thing does come to me, and then I get up, and I write.

But most of Wilson's work does not involve coming up with synthetic insight. It consists, rather, of slow, methodical work. Among his current projects is a monograph on the largest ant genus in existence, which requires identifying and describing six hundred related species of ants scattered around the world—one of the largest genera of any kind of animal. In preparation, Wilson has completed more than five thousand drawings with his own hand. "Now that might sound rather odd," he admits, "but I find it particularly rewarding. I'm doing that on the side. That's sort of like a hobby."

There are few clearer examples of how complex a creative life can be than the one presented by E. O. Wilson. Personal adversities, historical conflicts, and deep changes in the organization of knowledge all clamored for attention and required a positive response. There were many ways that he could go wrong and few that led to acceptable solutions. The way Wilson adapted to the pressing external demands required stubbornness and flexibility, ambition and selfless curiosity. He had to be as pure as the dove while being as cunning as the serpent. In this way, instead of being swept aside by the momentous changes swirling around him, he used the emerging ideas from different domains and created with their help a new way of understanding the intricate web of life.

THE LIFE OF CANCER CELLS

George Klein also hunts small life forms, but his are even smaller, and vastly more deadly, than the fire ants E. O. Wilson pursues. Klein is a pioneer of a recent branch of cell biology known as "tumor" or "cancer biology." It is a domain that has emerged from studies on the chromosomal constitution, genetic changes, immunology, and the role of viruses in the generation of tumor cells. Like other branches of cellular biology, it has exploded this century into a race for knowledge that was made possible by the development of molecular biology, by constant cross-fertilization between the rapidly advancing research laboratories, and by the infusion of funds aimed at conquering cancer. In the most general terms we might perhaps say that cancer biology tries to understand how cancerous cells develop, how they grow, and how they die. Traditionally tumors were viewed strictly as pathological entities to be eliminated by any means available. The new approach also wants to learn how to get rid of them, but it is based on the assumption that this goal can best be accomplished if we think of cancer as populations of cells subject to genetic variation and selection, with their own genetic and environmental history. Then one can ask the crucial question: Why do these cells disobey the growth-controlling instructions that the rest of the organism obeys?

Klein's domain, like that of many other people we interviewed, could scarcely be said to exist until quite recently. The elements of knowledge were there, but they were not put together in a coherent conceptual system. The origins of tumor biology could be traced to the pioneering studies of the American researcher Peyton Rous in the first decade of this century, but like most scientific domains it grew opportunistically by borrowing whatever information was relevant from other expanding disciplines. Science works by putting out conceptual pseudopods that occasionally separate out of the parental field to form an independent discipline; more often than not, however, the shoots are reabsorbed. In this highly charged intellectual atmosphere, research centers compete with each other, as well as complement and stimulate each other's work with their discoveries.

George Klein leads one of the most exciting of these laboratories, at the Karolinska Institute in Stockholm, Sweden. Pre- and postdoctoral fellows from all over the world work in his lab. Klein obtained the funding and helped design the building, and for many years he has been responsible for the fiscal and intellectual life of the lab. One of the dilemmas creative scientists face is that if they wish their ideas to continue into the future, they have to become entrepreneurs; but if they become entrepreneurs, they have to take precious time away from their original research.

In addition to running the institute with all that entails in terms of applying for grants and administration, Klein is involved in many enterprises of a very different sort. He has published several volumes of essays that combine personal reminiscences with philosophical reflections, with titles such as *The Atheist and the Holy City*. His fascination with poetry led him to investigate the life of the great Hungarian poet József Attila and to write about his verse. After reading Benno Muller-Hill's book on the Nazi death doctors, he has become a vocal spokesman for ethical responsibility in science. And finally, at the many international scientific conferences he attends, he has gained the reputation of being the person who can best summarize and integrate the presentations of other specialists.

A Sunny Pessimism

Klein's life began in Hungary under less than auspicious circumstances. His father died before George had a chance to know him, and the loss has remained a constant presence in the son's psyche. On the one hand it gave him an *"incroyable légèreté,"* a great lightness in confronting life without worrying about a paternal censor, a condition Jean-Paul Sartre attributed to those who grew up without a father. "I did not have to carry an Anchises on my back as I swam to a new country," he says, quoting Sartre's metaphor. On the other hand, fatherlessness leaves a burden of a different sort on the son's shoulders: a feeling that as the oldest male, he is now responsible for the welfare of everyone around him.

Klein remained close to his mother, whom he perceived to be very dependent on him emotionally. His main concern became to satisfy her needs, to keep her from being depressed. Even now his greatest fear is that people who depend on him won't be happy, that he will let others down. His greatest source of pride is the ability to control his own emotions so as to preserve harmony in personal and professional relationships.

Klein is Jewish, and the cultural environment of assimilated Hungarian Jewry played a determining role in the formation of his character. An orthodox grandmother was especially important, but what counted even more was the generalized valuing of the sanctity of existence and the expectation that one should achieve excellence in one's life, which he absorbed from the cultural milieu. As he turned fourteen he started doubting the existence of God, and after a two-week spiritual crisis decided that religious beliefs were "absolute nonsense." Even now he "believes absolutely in the nonexistence of God," while retaining his awe at the wonderful mystery of life, which he sees as his task to demystify.

As a teenager, Klein was frustrated in school. Although he was ambitious, he feels he didn't learn anything from the "stupid, oppressive teachers"—except for one, who had a permanent influence on all his students. Kardos Tibor ostensibly taught Italian and Latin, but it is his enthusiasm and love for art and poetry that made him memorable. Klein can still recite Dante's verses, even though he cannot speak Italian. Uninspiring schools did not prevent him from learning important things, however. Like E. O. Wilson, Klein learned self-confidence and love for nature from the Boy Scouts, where he became the youngest patrol leader in the troop. He still remembers fondly the long hikes, the night raids, the pleasant exhaustion after vigorous exertion outdoors. Above all, learning to resist fatigue, hunger, and thirst helped to build up the toughness necessary to confront the future, "when all hell broke loose" toward the end of the war. But on the train ride home from the outings, he was saddened by the empty and boring conversation of his peers.

For intellectual challenges he turned to a different group. He and a few other Jewish students banded together to discuss music, literature, philosophy, the arts, and mathematics during walks on the banks of the Danube—not as a continuation of what went on in school, but in opposition to it. It was the kind of peer group that used to be relatively frequent in Central Europe and is almost unknown in the United States: a group in which the most "serious" boys earned the highest respect, and one demonstrated superiority by being sensitive and having a broad range of knowledge. In that circle one never talked about personal matters, only about abstract ideas and aesthetic experiences. It is thanks to these discussions that his interest in culture is still so lively: "I like Dante more than most Italians, and the Kalevala more than most Finns." And like all the other creative individuals, he spent much of his youth alone. He played the piano and tried to keep his mind in order through music, reading, and thought.

Many decades later Klein developed a new version of the intellectual club. Feeling that specialized scientific interactions were limiting, he started corresponding with kindred spirits, and that correspondence eventually grew into an informal network that spans the globe. All kinds of intellectuals, from physicists to poets, share with him ideas about religion, politics, the arts, and life in general. Occasionally he acts as a clearinghouse for this information by sending copies of letters received from one friend to others he thinks would enjoy reading them. Many of these letters are dictated into a tape recorder in airport waiting rooms and on the subway for later transcription, and a typical letter is four to six pages, single-spaced. The files of this correspondence take up dozens of cabinets near his office.

In a way, it is surprising that Klein ended up choosing a career in medicine. As a child he had been horrified by saliva, vomit, or bodily functions in general. He remembers being both fascinated and frightened by doctors when six or seven years old. But after high school, medicine seemed the only realistic profession to enter. It was not a positive choice, but more of a process of exclusion that made him start on a respectable career in which a Jew would be less likely to be ostracized. It was not until he was twenty-two years old and took a rotation in pathology that he became fascinated by the detective work involved in laboratory research.

In the meantime, World War II was nearing its end, and the fate of the Jews in the formerly protected nations of Central Europe was getting increasingly precarious, as the local governments were caving in under Nazi pressure. Klein worked for the Jewish Council in Budapest, as a secretary to one of its members, and he heard ominous news whispered about atrocities committed in the territories where the German armies had been entrenched. But nobody wanted to accept the truth of these heinous tales, and especially not the comfortably bourgeois, assimilated Jews of Budapest.

The Hungarian government succeeded in protecting the Jews for as long as possible, but on March 19, 1944, German troops occupied the country and installed a fascist regime, which began to assist in the deportation of Jews from the countryside to the extermination camps. Shortly after, Klein read a manuscript circulated underground, which contained the account of Vrba and Wetzler, the first two Jewish prisoners to escape from Auschwitz. The account was horrifying; in objective, nonemotional terms, it gave concrete details of the workings of the death factory. Yet Klein also felt a sense of intellectual satisfaction in learning a truth that was more credible than the disinformation and wishful thinking most of his peers preferred to cling to. The Jewish Council decided to keep the information secret to prevent panic and reprisal from the fascists; but the report reinforced Klein's determination to escape as soon as he had a chance. The thrill of getting at truth, no matter how unpleasant, was to remain a hallmark of his intellectual life.

In October 1944 the Arrow Cross, as the Hungarian fascists were called, escalated their terror. Suspects were rounded up and herded on death marches, or killed outright. Klein was shipped to a labor camp the next month, but he escaped, and, having obtained false papers, he lived in hiding until the Soviet Army arrived on January 10, 1945. Liberated from the Nazi terror, he decided to start medical school as soon as possible. Because the University of Budapest was still in ruins from the war, he and some friends walked to the city of Szeged at the other end of the country, where the university had remained relatively undamaged and courses were starting.

As soon as the University of Budapest reopened its doors, Klein was back in the capital to continue medical studies and started research in histology and pathology. In 1947 two momentous events took place: He met Eva, a fellow student, and they fell in love. Almost immediately afterward he was invited to visit Sweden with a group of students. Given the desolate condition of the country still reeling from the war, such an opportunity would have been a dream come true except that now George regretted leaving the girl he was sure he wanted to marry, even for a short trip abroad.

The visit in Sweden turned out to be a turning point in Klein's life. His research experience in Budapest, slight as it was, happened to fit the needs of Torbjörn Caspersson, the head of the Cell Research Department at the Karolinska Institute in Stockholm, who offered him a job at the lab. He describes his feeling at that time:

> I still remember the mixture of ecstatic happiness and enormous anxiety. My situation appeared totally hopeless. I knew virtually nothing. I was halfway through my medical studies, still far removed from an M.D. I was desperately in love with a girl whom I had only known during a summer vacation of eight days and who was on the other side of an increasingly forbidding political barrier. I did not know a word of Swedish. Still, I was firmly decided to resist the more comfortable possibility of continuing my studies in Hungary.

Before taking up his position, Klein returned to Budapest for a few days, and at the end of the visit he and Eva were secretly married. In the meantime the Iron Curtain was descending between Hungary and the West, ushering in new decades of terror. Fortunately, after a few months Eva was able to follow George to Stockholm, where they both finished their medical studies, and where, after forty-seven years, they are still collaborating on research and pursuing their own independent work—as well as a full married life.

Being a witness to one of the most tragic periods of European history left Klein a "sunshine-colored pessimist." An atheist with a positive outlook, he feels happy even though he is sure that life has no meaning at all. His goal is not to save humanity from disease, or to build a scientific empire, or to be successful. He has identified flow as the moving force of his life. The important thing is not to be bored and not to disappoint those close to him. "Whenever I am concentrating, I am happy," he says. "I am horrified by the very concept of 'taking it easy,' of taking a vacation. I get panic-stricken when at a formal dinner I have to sit next to boring people." But when working on a scientific problem, or involved in anything challenging, Klein feels "the happiness of a deer running through a meadow."

The Synergy of Arrogance and Modesty

The early years in Sweden were not easy for Klein. He had to learn a new language, a new lifestyle, under severe competitive pressures. At first, it took only a cold greeting from a technician at the lab to ruin his entire day. He worked with senior scientists who were bored and alienated, and for a while it seemed that scientific research might be a trap that led to an alienated life. But after a few years he found supportive and inspiring mentors.

A visit to the Institute for Cancer Research (ICR) near Philadelphia was especially memorable in this respect. The U.S. scientific environment was much more friendly and egalitarian than anything comparable in Europe. Despite his youth and inexperience he was treated almost as an equal by older researchers. The description of his boss at ICR is a good model for what a laboratory chief should be, a model that Klein has adopted as his own:

> My own boss was Jack Schultz, a lively man in his sixties. Jack exuded boundless curiosity, joy of life, and great human warmth. He received me as if I were his long lost, finally recovered son. During my stay he often gave me a lift from my rented room to the laboratory. Most of what I know about genetics can be traced to those car rides. But the trip was not over when we arrived. Jack's office was at the far end of a long corridor. Walking down the hallway he would stick his head into every lab and stop and talk with people on the way. He asked them about everything, the health of their kids, mother's broken leg, the weekend excursion, but first and foremost the latest experiment. The people brightened visibly when they saw him. . . . Jack looked, listened, discussed, interpreted, proposed new experiments . . . sometimes half a day passed before we arrived at his office where his secretary waited in despair!

Having tasted the acceptance of the field, Klein lost his "immigrant complex" and started to take the intellectual risks that made his career. In this he was helped by that unusual combination of opposite traits that we see repeatedly characterize creative people. As Klein's friends say, he is a "combination of infinite modesty and a stubbornness bordering on arrogance." Whether because he never had to defer to a father, or because he experienced the ineffectiveness of formal education, or because he saw the blindness of his elders during World War II, or for some still deeper reason, Klein has never been intimidated by established authorities.

One example of how Klein works concerns his early insight into the development of tumors from antibody-forming cells (B-lymphocytes) in different mammals. He had been studying a tumor that affects particularly children in Africa, called Burkitt's lymphoma, which was believed to be caused by a virus. Klein and other researchers found evidence that 97 percent of such tumors contained what came to be called the Epstein-Barr, or EBV, virus. However, this virus alone could not cause the tumor, because most individuals carry it without ever developing the disease. What was the missing piece of the puzzle?

At this point, Klein began to put together information from a variety of sources—cell biology, virology, and immunology. It is this process of connecting seemingly disparate ideas that he finds most enjoyable about his work. He found that for patients who had Burkitt's lymphomas the tips of two chromosomes broke off and changed places. After long and painstaking work aimed at identifying the function of the genes involved in this reciprocal translocation, Klein postulated that the displaced chromosomal fragment contained a growth-controlling gene that upon coming in contact with a highly active immunoglobin gene permanently activated it, driving the cell to the continuous division that results in cancer.

At first his hypothesis was regarded as a "most hair-raising extrapolation" from what was known about chromosomes to the much more minute world of the genes. But only a year after the hypothesis was published in *Nature*, five different laboratories around the world verified the insight that chromosomal translocation plays a decisive role in the development of many forms of cancer by bringing two unrelated genes into close proximity to each other.

Klein sees infinite vistas opening up in his domain; the main challenge is to combine detailed information from the sequencing and splicing of genes, from "the protean foresight of the immune system," and from the understanding of cell pathology, and then put together this information in an understandable scheme of how organisms work. The more one knows about the complexity of the world within the cell, the more wonderful it all seems. "As you go in, it's a jungle," he says, a jungle full of perils and stark beauty.

THE IMMENSE JOURNEY

Few people have had the good fortune to increase substantially human well-being by discovering a new way of healing. One thinks of Edward Jenner and Louis Pasteur, who first made vaccination against disease a feasible cure; of John Snow, who in 1854 discovered that the source of the London cholera epidemic was the Broad Street pump that had been contaminated by sewage, and thus established the link between bacteria and drinking water; of Ignaz Semmelweis, who understood how to avoid mothers' death from infections at childbirth; of Alexander Fleming, whose discovery of penicillin saved untold lives. There are few satisfactions as deep as the knowledge that one has brought such improvements to human welfare.

It is among this fortunate elite that Jonas Salk belongs. As a young medical student he joined a research team studying the tragic disease of poliomyelitis at the University of Pittsburgh. Until that time, polio had been a disease that ruined the lives of tens of thousands of children annually. Every summer, when the rates of the illness peaked, mothers would dread sending their children to camp, or to the movies, or anywhere where they could catch the contagion.

After identifying different strains of the virus in the laboratory, Salk was able to demonstrate first with monkeys, then with humans, that injecting dead viruses induced the formation of antibodies and hence could prevent the disease. The widespread use of what came to be called the Salk vaccine almost completely eradicated an illness that had cast a pall on the lives of every person in the United States.

This breakthrough made Salk a scientific celebrity. Foundations and individual donors vied to offer financial support for his next projects. But Salk, while still interested in continuing laboratory research, had raised his sights even higher: His goal was now to understand the immense journey of evolution from inorganic forms to biological life and finally to the metabiological realm of ideas. To achieve this synthesis it was necessary to bring together people representing every branch of human knowledge. So he planned to use his enormous prestige and financial backing to establish a new interdisciplinary center, a "crucible of creativity" where scientists, artists, and thinkers of different persuasions would come together to stimulate one another's minds. It was to be a physically beautiful space that re-created for our times the intellectual brilliance of Goethe's Weimar, the Medici court, the Platonic academy. In 1960 he teamed up with the visionary architect Louis Kahn, and together they built the splendid structures of the Salk Institute, which stands like a contemporary descendant of ancient Greek temples in a grove above the Pacific Ocean at La Jolla, in southern California. It was in these buildings that Salk's dream of a powerhouse of ideas was to be realized.

Yet history provides ample evidence that even the benefactors of humanity are not immune to the entropy that bedevils ordinary lives. Pasteur had to fight strong criticism against his efforts to use the rabies vaccine; Semmelweis suffered a terminal mental breakdown when all his medical colleagues laughed at his true but too far advanced ideas. It is perhaps not surprising that Salk's second career encountered unexpected obstacles. In order to establish the scientific credentials of the institute, its founder started out by hiring traditional biologists to run its laboratories. Because he wanted to have an institution run along democratic lines, Salk relinquished most of the power to his younger colleagues. Unfortunately, when time came to begin transforming the laboratory into the center of his dreams, Salk found out that traditional scientists had no sympathy for his novel vision. His colleagues preferred to devote all the resources of the institute to pursuing safer, more orthodox biological research. The idea of bringing in astronomers and physicists, not to mention musicians and philosophers, for serious discussions seemed to them mere self-indulgence. The ensuing conflict played itself out along the lines of classical mythology: The creator was dethroned by his offspring. Salk retained an office and ceremonial status but could not implement the ideas that made the institute possible in the first place.

With the resiliency typical of creative individuals, Salk did not let the defeat stop his march toward the synthesis he sought. In several books he developed his thoughts about the evolutionary continuities that stretch from the distant past into the future, where we have to follow them if we wish to survive as a species. As a member of the board of powerful foundations he shaped research and philanthropy.

And with the sudden appearance of AIDS, he rolled up his sleeves once more and returned to the laboratory in the hope of finding a way to prevent this plague by immunological means. But whether in the boardroom or the lab, Salk in his seventies followed a direction discovered very early in life: to reduce human suffering and to become, to paraphrase the title of one of his books, "a good ancestor."

Making Visible the Invisible

A central theme in Salk's life was the effort to see, and to make others see, that which is hidden. At the most obvious level, this has involved bringing to light the viral processes that caused polio. Less directly, his later attempts to assemble men and women from very different domains at his center were also directed at making the invisible visible through conversations that would bring out new ideas that could not arise in the minds of the single individuals but might emerge as a result of the interaction. This is how he described this latter form of creativity:

I find that that kind of creativity is very interesting and very exciting—when this is done interactively between two sets of minds. I can see this done in the form of a collective mind, by a group of individuals whose minds are open and creative and are able to bring forth even more interesting and more complex results. All of which leads me to the idea that we can guide this process—this is in fact part of the process of evolution, and ideas that emerge in this way are equivalent to genes that emerge in the course of time. I see that ideas are to metabiological evolution what genes are to biological evolution.

Q: What needs to be present in the relationship to allow that kind of creativity to emerge?

A: Well, in the first place, minds have to harmonize. There's something of a think-alike quality, an openness, a receptivity, a positive rather than a negative attitude. There's a mutual affirmation; it comes about as kind of a consensus, a reconciliation of differences that exist when you don a new vision or perception.

Any dialogue, such as we're having now, is of that nature. There is a tendency to draw each other out, to bring out the best or the most creative aspect of the mind, or the functioning of the mind.

In this kind of interaction each person helps the others see what they see. That's what is needed in the world today to reconcile differences, resolve conflicts, help us each understand what our belief systems represent, how to reconcile belief and knowledge.

The Human Side of Science

Salk grew up as an overprotected son of a strong and domineering mother. She was an immigrant with little knowledge of English, but as is often the case with the mothers of creative persons she spent a tremendous amount of time with her children and expected a great deal from them. "Whatever we did was never good enough," mused Salk. Childhood was a time of "sweet adversity," with restricted freedom and with great expectations. The folk wisdom contained in ancient proverbs such as "God helps those who help themselves," "The early bird catches the worm," or "Where there's a will there's a way" was also a part of his childhood, and Salk still tends to think aphoristically as a result. Like many of his creative peers, in some respects he does not think of himself as a mature adult: "I'm seventy-six now and I still feel like a child, an adolescent, as if I still have lots to do."

Another strong influence in the early years was the Jewish biblical tradition and the dim awareness of a long line of ancestors who had survived all sorts of adversity. One of his earliest memories was of seeing the soldiers returning from World War I, in the Armistice Day parade, in 1918, when he was only four years old, and wondering what it all meant. Out of these experiences Salk developed a strong sensitivity to human suffering and an unusually heavy sense of responsibility. As a child of ten, he wanted to become a lawyer so he could be elected to Congress and make just laws. He was deterred from these plans in part by his mother's doubts about his ability to win arguments; but even when later he decided on a medical career it was not with the intention of becoming a physician who cared for one patient at a time but with the goal of bringing science into medicine, and so "to make it much more valuable to human beings."

There is a strong sense of responsibility, which I'm aware that I had all my life. And it's been said by others that I seem to have a capacity to take responsibility, to act responsibly, even against odds, even if it's unpopular, if it seems to me important. And that I know is true.

I see much of what we're speaking about as having been innate but also having been actively induced by circumstances, so that throughout my own life I was aware of war and disease and suffering, problems of humanity, and I think I dedicated my life to trying to make the world a better place in which to live, to improve the lot of humanity now and in the future.

This sense of responsibility and sensitivity to suffering helped Salk avoid the mechanistic specialization that many scientists tend to succumb to.

I do see myself as an artist-scientist, scientist-humanist, humanist-scientist. I guess my purpose is different from that of those who are interested in science for science's sake. I'm interested in science as it has relevance to the human condition, so to speak. I try to understand the human side of nature and do something for it. So I have a purpose—a purpose as a humanist somehow, in some innate way. That's why I created this place, to set up this ideal set of circumstances within which scientists would work, I hope being more creative than they would be otherwise. And in fact this does seem to be the case, so it has not failed in a sense, it simply has not yet succeeded in that which would take a little bit longer to emerge.

Patterns of Meaning

Salk's penchant for seeing emergent possibilities often brought him in conflict with those whose clear view of the present blinded them to the future. "Damn it all, Salk," one of his mentors used to say, "why do you always have to do things differently from the way other people do it?" As a medical student, he kept questioning the orthodox opinions of his teachers. In a manner typical of creative individuals, he kept seeing the emperor without clothes while everyone else admired the sovereign's fancy regalia. The basic idea that later resulted in the polio vaccine seems to have already occurred to Salk in the second year of medical school:

We were told in one lecture that you could immunize against tetanus by chemically treated toxins, or toxoids, and in the next lecture we were told that for immunization against virus diseases you had to experience infection itself, you could not use a chemically treated or noninfectious virus. Well, it struck me that both statements couldn't be true, and I asked why that was the case. I guess the reason that was given was, "because." But then two or three years later, I had the opportunity to work on the influenza virus, and I then chose to see whether or not this was true for flu. So I didn't use chemical treatment, I used ultraviolet light to inactivate the virus and found that you could immunize the virus that way. So that was the beginning of a demonstration that one could kill a virus, so to speak, or render it noninfectious, dissociating infectivity and antigenicity or antigenicity and capacity to immunize. And that led to work that eventuated in the influenza vaccine, which is being used today.

And then when I had an opportunity to work on polio, I just evoked the same idea and attempted to see what could be done there, and it proved to be successful. Since then, of course, all of the genetic engineering and the other things that are done to parts of the virus are continuations of this principle. And so I tend to look for patterns. I recognize patterns that become integrated and synthesized and I see meaning, and it's the interpretation of meaning, of what I see in these patterns.

Despite his successes, Salk continued to encounter obstacles in everything he attempted to do; his research on cancer, autoimmune disease, and multiple sclerosis brought him into conflict with various bureaucracies and with peers who saw things differently. "And it was just a matter of persisting and tending to prevail and finding ways around the obstacles."

Salk's best ideas often come to him at night when he suddenly wakes up and after about five minutes of visualizing problems he had thought about the day before he begins "to see an unfolding, as if a poem or a painting or a story or a concept begins to take form." Sometimes when such associations of ideas begin to occur in his mind, Salk claims to feel a palpable physiological response which indicates to him that the right side of the brain has become active. At this point he either falls into a deep sleep, or he sits up in bed, turns on the light, and writes down the thoughts that have occurred to him, for three quarters of an hour to an hour. In this fashion, he has "accumulated a considerable amount of material over the last several years that I'm now beginning to work with, to try to understand or see the themes that have come forth this way."

This tendency to take one's dreams and hunches seriously and to see patterns where others see meaningless confusion is clearly one of the most important traits that separates creative individuals from otherwise equally competent peers. Of course, this fluidity of thought results in something creative only if one has already internalized the rules of a domain. Otherwise, chances are that the dreams will dissolve by morning. And even the most original ideas have little chance to make a difference without the persistence to convince others of their rightness, and without a good dose of luck. Jonas Salk has been blessed once by everything turning out right.

These biologists—Wilson, Klein, and Salk—have led very different lives and contributed to their domain by different means. Yet they share strong similarities, some of which are common to creative individuals across a broad spectrum of disciplines.

All three remember childhoods that were in some way troubling, or even "dysfunctional." One never knew his father, the others never mentioned theirs throughout the interview. All three, however, remember very strong, demanding, or emotionally dependent mothers. Each one felt early on the support of the beliefs and values of a cherished tradition, whether of the American South or of Judaism. None of them was a particularly brilliant student; in fact, school left positive memories with none of them. For Wilson and Klein, the best learning during adolescence occurred in peer groups and the Boy Scouts.

In line with everything else we know about the creative personality, all three men show the complexity we are led to expect. They are selfless and egocentric at the same time, eager to cooperate yet insistent on being in control. They call themselves workaholics, are extremely perseverant, and stubborn when thwarted. They have all taken risks and have defied the dogmas of their fields. At the same time, none is content staying within the limits of his specialization; each is open to a great variety of experiences in art, music, and literature.

In fact, while all three started their careers as specialists in narrow fields—the study of ants, the growth of cancer cells, the control of the polio virus—now that they are past sixty they all see themselves as primarily synthesizers. Their main goal is to connect their specialized knowledge with other domains, or indeed with the evolutionary process itself. How they try to accomplish this synthesis, however, differs quite substantially. Although they all paid attention to developments outside their fields and tried to link their work with other disciplines, Salk seems to do so mostly in terms of intuitive, analogical leaps between widely different processes in the arts and the sciences; Wilson tries to achieve precise "collinearity" between specific biological and cultural processes; and Klein connects biological knowledge that usually proceeds independently, such as virology, genetics, and oncology.

There are also obvious differences in the men's careers. Wilson claims to have known by age six that he was going to become a naturalist; Klein ended up in medicine by default, and was already twenty-two when his interest in cellular pathology caught fire; Salk remembers a generalized wish to help people, but becoming a physician was the second-best choice. Friends and mentors played a very central role in the career of all three, but the kind and the timing of these relationships varied quite a bit.

So far, however, these conclusions could apply just as well to creative individuals in other domains. Is there then no unique component to creativity in biology? Could these three persons have been just as creative if they had become writers, lawyers, physicists, or musicians? Or did they have some trait that attracted them to this specific domain?

It is very difficult to answer these questions with any confidence, but there seems to be something common to these three men that one finds less often in other professions. Over and over, they mention the strong *responsibility* they feel toward other people and the living world in general. Of course, it is possible that a concern for others is the result of having been a life scientist for so many years instead of the reason for entering the profession. Yet Salk claims to have been sorry for the GIs returning from war when he was only four years old. Klein recently visited the village house in the foothills of the Carpathians where he lived with his mother as a child, and as he stepped on the porch he was overwhelmed by the anxiety he used to feel as he tiptoed across the same porch when he was six years old, petrified at the thought of waking his mother who was napping inside—just one of a continuous stream of events in which he had felt that the well-being of others depended on him. Perhaps this kind of guilt, of being burdened with everyone's welfare, is one of the early experiences that predisposes a young person to a career in the life sciences.

But there are indubitably other reasons. All of them enjoy the thrill of venturing into new areas of knowledge; they compare what they do to the work of a detective or an explorer. Wilson describes his professional work as "dodging bullets"; when he talks about his research Klein uses the metaphor of driving a big truck on a slippery road. There is no doubt that the domain of biology offers endless opportunities for flow to those who venture to push back its boundaries. Perhaps it is this combination of empathy with the living world and a predilection for risk and adventure that leads to a creative involvement with the life sciences.

TWELVE

THE DOMAIN OF THE FUTURE

Creativity generally refers to the act of changing some aspect of a domain—to a painting that reveals new ways of seeing, to an idea that explains how stars move and why. But of course there was a time when domains did not exist. The first astronomers, the first chemists, the first composers were not changing a domain but actually bringing one into being. So, in a sense, the most momentous creative events are those in which entire new symbolic systems are created.

To do so, of course, is not easy. The rate of attrition for creativity within domains is very high, and that for new domains must be at least as large. Many people have grandiose ideas about inventing new paradigms, new perspectives, new disciplines. Exceedingly few of them succeed in convincing enough others to form a new field. The four people in this chapter exemplify these hazardous attempts at bringing about a new set of symbolic rules.

Each was successful within an existing scientific domain before trying to establish a new one. None started out on a new course in order to achieve personal advancement, power, or money. A deep concern for the well-being of the world informs their lives. In each case, they addressed a central social problem in an effort to achieve a voluntary reorganization of the human community. Because they could not see how to address these issues adequately from within existing domains, all four struggled to develop new symbolic representations and new social institutions dedicated to the solution of global problems. These are important similarities; but as we shall see, the differences are just as impressive.

THE SCIENCE OF SURVIVAL

The name Barry Commoner has become synonymous with the ecological struggle. He was among the first scientists to realize, in the 1960s, that some of the fruits of technology—from nuclear fallout to pesticides, from oil consumption to solid waste—posed dangers for human health. Trained as a biochemist and biophysicist, Commoner found himself increasingly frustrated by the abstraction and fragmentation of academic science. He tried to influence public awareness through a number of books, and in 1980 through an unsuccessful campaign for the presidency of the United States. For many years now he has directed the Center for the Biology of Natural Systems, now associated with the City University of New York, where he continues to explore the problems posed by runaway technology and their possible solutions.

At War with the Planet

Commoner did not start his career with any specific sense of mission. He had been a fairly good student in high school, and his father, an immigrant tailor, pressured him to become a radio repairman. But then an intellectual uncle pushed him to enroll at Columbia University—not an easy step for a Jewish boy in those days. At the end of his college career, when it had become clear that Commoner had a knack for science and should continue his graduate education, a biology teacher called him in and told him he was going to Harvard. "What do you mean?" Commoner remembers asking. "'I've arranged for you to become a graduate student at Harvard.' I hadn't applied, nothing. 'As a Jew from Columbia, you'll have a very hard time getting a job; I'm sending you to Harvard.'" And so Commoner moved to Harvard, where he received an interdisciplinary education in chemistry, biology, and physics.

After he started his academic career, Commoner was confronted with a number of ominous developments. One was the threat of a nuclear holocaust, which after World War II cast a pall on an entire generation. Two other defining events he describes in the first chapters of his book *Science and Survival*. The first was an electric blackout that shut down power on a 1965 November night across a huge area of the Northeast and Canada. What struck Commoner about this failure was that it was caused by the elaborate computerized controls built into the electric grid, which overcompensated for a surge in demand by closing down the system entirely (a process not dissimilar to what happened more than twenty years later when the computerized programs for buying and selling stocks circumvented human controls and went into a selling frenzy that brokers were unable to stop, thereby causing a market crash).

The second event Commoner describes in his book was the discovery that fallout from nuclear testing in Nevada produced iodine–131 isotopes that were carried by winds to pastures in Utah, where they contaminated the grass cows foraged on. The idoine passed into the cows' milk, and when children drank it, it deposited itself into the cells of their thyroid glands. There the radiation from the iodine occasionally produced goiters and tumors.

Both the blackout and the iodine–131-produced diseases were typical examples of the kind of unintended chain reactions that occasionally occur when technology escapes from human control. Most people chalk up such events as the necessary price to pay for progress and do not worry too much. But Commoner, either because his interdisciplinary training made him think in terms of systemic patterns rather than linear processes, or because of a long personal history as an outsider who has been forced to take a critical perspective, felt that these events were not just side effects but part of the main history of our time.

The main story, according to Commoner, is that we have unintentionally declared war against the planet on which our lives depend. Science started out as a powerful tool for increasing human well-being. But when knowledge within separate domains is pursued without understanding how its applications affect the whole, it unleashes forces that can be enormously destructive. The sorcerer's apprentice, who sets in motion a magic spell that he cannot stop when it begins to get out of hand, is a metaphor that recurs in Commoner's writing.

Of course, he was not alone in this realization. In fact, several groups founded in the sixties helped hone Commoner's ecological consciousness, such as the Committee on Science in the Promotion of Human Welfare of the American Association for the Advancement of Science, and the Committee for Nuclear Information. But with time Commoner developed a personal approach to the problem of helping the environment, one that made it possible for him to envision solutions that were feasible given who he was and what he could do.

Science and Politics

What Commoner realized was that the solution could not come from science alone. To keep runaway technology under control, science and public policy had to work together. When it comes to applying technology, science predictably sells out to the highest bidder. The military ends up controlling the awesome power of radiation; pharmaceutical companies profit from the fruits of chemistry; agribusiness uses biology for its own aims. None of these entrenched interests has any responsibility to preserve the fabric of life on the planet, although each one owns the means for destroying it. So we must step in and regain control in the name of the common interests of continued life on Earth.

Unlike many others who also have perceived the threats of technology, Commoner has kept his faith in science. He realizes that even though science may have gotten us into this mess, we are unlikely to get out of it without its help. So he continues to use the scientific method both to diagnose the problems and to find solutions for them. In doing so, he works with the dedicated humility of a true scholar. For many years now, the efforts of his institute have been focused primarily on solving problems of solid waste disposal. Garbage is not a fashionable topic, but its exponential growth presents real threats that few want to think about. And what is more, it is a problem that can be solved and thus might serve as an example of how to tackle more complex issues. Like all creative individuals we studied, Commoner tends not to waste energy on problems that cannot be solved; he has a knack for recognizing what is feasible and what is not.

Commoner felt that it was not enough just to demonstrate that when you burn trash in incinerators you create dioxin, which is a dangerous pollutant, or that by using too much fertilizer we poison our water supply with nitrates. This was important knowledge, but it would not make any difference as long as special interests benefited from incineration or fertilization. So he concluded that the first priority was to inform the public about these environmental crises and their origins. To do this he used different means: He wrote books and pamphlets, talked to leaders and opinion-makers, gave press conferences, got money from foundations for environmental causes, and developed networks of like-minded people.

In the process he had to break out of the standard scientific domains and from the academic fields that preserve their boundaries. This meant leaving the safe shelter of the universities, a step that few people trained in them have the courage to take:

> I was brought up before World War II, when a number of my professors believed in a duty that the academic has to society generally. But as the generation represented by the World War II scientists began to get older, the academic world became very isolated from the real world. Academic work was discipline-dictated and discipline-oriented, which is really pretty dull, I think. And so the work that we've done has become more and more alienated from the current general direction of academic work, because most people in the university work for the admiration of their peers. The work we do is for the sake of people outside the university.

> Only by crossing disciplinary boundaries is it possible to think holistically, which is necessary if we are to "close the circle" and preserve the organic balance of planetary life forms.

> The prevailing philosophy in academic life is reductionism, which is exactly the reverse of my approach to things. I use the word *holism* in connection with biology and environmental issues. But the academic world has changed a great deal since I was a graduate student. It has become progressively self-involved and reductionistic. And I find that's dull and I'm not interested in doing it.

Instead of letting specialized academic fields dictate how he should approach problems and attempt solutions, Commoner lets "real-world" events dictate where he should turn his attention, and what means he should use to try to control recalcitrant technology. Specific threats, such as the proliferation of toxic waste or the pollution of drinking water by nitrogen isotopes, are what mobilize his energies:

The center has always been directed toward the solution of real-world problems in the environment and energy. Not academic problems. Not problems defined by a discipline. Problems defined by the real world. Particularly people in the community who are confronted by a problem. Our approach to this problem then is to solve it, not to write a paper that will fit into a particular discipline or even a combination of disciplines. That's why I say we are adisciplinary, not interdisciplinary.

This quote has a facile, anti-intellectual ring to it. But Commoner is using science in its most basic, truest sense. What he objects to is not systematic, careful observation, only the irresponsible uses of it. What he objects to is the ritualized worship of domain knowledge for its own sake, instead of the integrated knowledge we actually need to avoid becoming history.

Struggling with Reality

Commoner calls himself "a child of the Depression" who always had to struggle to achieve his goals. This, plus the constant awareness of his marginal status as a Brooklyn Jew in what used to be WASP ivory towers, is probably why he maintained his unorthodox views all his life. Those who are not properly socialized by a field are prime material for the skeptical, divergent thinking approach that often leads to creativity.

Like so many of our respondents, Commoner insists on the importance of maintaining two usually contradictory attitudes toward his work: to keep an emotional link to what he does and at the same time a rigorously objective perspective. There is no doubt that he cares deeply about his topic—the entire pattern of his life is evidence of it. And it is equally clear that he takes the rigor seriously: Among his associates he is famous for writing draft upon draft for each speech or press release, until it is free of ambiguities and weaknesses.

It is not easy to be a maverick and to keep to the narrow path of self-chosen excellence in a nonexisting domain. Commoner ran into various difficulties with university administrators who didn't understand what he was trying to accomplish, with fellow scholars who felt he was trespassing on their turf, with the authorities who wanted to silence his opposition to nuclear weapons and the Vietnam War. His stubborn faith in the necessity of his task kept him from giving up. But he also had to find strategies to keep his mind focused and prevent distractions. As with most other creative individuals, a sort of ascetic discipline orders his attention:

> Well, also, I reject an awful lot. I don't answer letters. I don't do things people ask me to do for the sake of helping them. We help a lot of people in areas where we want to help. But, you know, people call up and say, "I've got this invention." Anything that's commercial, I never touch. I have a whole series of rules like that, to just get rid of things. You have to concentrate on one thing at a time, I think. But, you know, I *can* do two or three things in one day.

SPLICING THE CULTURAL DNA

Hazel Henderson's life theme dovetails almost perfectly with Commoner's. She also is struggling to develop a new interdisciplinary—or adisciplinary—domain to deal with the problems of technology. She also has dedicated her life to keeping our species from destroying the habitat in which it lives. But because she was trained in economics instead of biology, her concern is more with how patterns of consumption affect our uses of resources than with the biochemical consequences of our lifestyles.

Henderson was born and grew up in the United Kingdom, in a loving, traditional family in which gender roles were strictly respected. It is impossible to say why, but Henderson seems to have fallen in love with the world quite early in life:

> When I was five—you know, like where you just open your eyes and you look around and say, "Wow, what an incredible trip this is! What the hell is going on? What am I supposed to be doing here?" I've had that question in me all my life. And I love it! It makes every day very fresh. If you can keep that question fresh and

remember what that was like when you were a child and you looked around and you looked at, say, trees, and you forgot that you knew the word *tree*—you've never seen anything like that before. And you haven't named anything. And you haven't routinized your perceptions at all. And then every morning you wake up and it's like the dawn of creation.

This a good example of Henderson's spirited and open approach to life. It is reminiscent of the American philosopher C. S. Peirce's distinction between what he called "perception" and "recognition"; it is also very similar to the Yaqui sorcerer Don Juan's practice of "stopping the world." But derivative or not, this freshness of perception is entirely consistent with her being.

After high school Henderson made two resolutions: to travel around the world to see how everyone else lived and not to do anything she did not enjoy. For a starter, she wrote to a number of resorts in Bermuda, proposing that she run their hotels in exchange for lodging, good meals, and afternoon lessons with the tennis and golf pros. Her offer was eagerly accepted, and she chose the most glittery resort. This experience greatly improved her tennis game. But what's more important, it showed her the possibility of stepping out of the money economy and of organizing small-scale, mutually beneficial exchange systems. She continued to draw on this experience for the rest of her self-made career. And lack of formal education turned out to be a blessing in disguise. It kept her mind open and allowed her to see freshly the economic system on a global scale.

The Blindness of Nations

The problem Hazel Henderson eventually identified as the issue she was going to invest her life trying to resolve is one that many people feel strongly about: the ruthless exploitation of natural resources and the growing inequalities between the rich and the poor countries. Although we are all aware that there is something dangerously wrong with our way of using energy, the very size and intractability of the problem prevents us even from trying to do anything about it. The most natural reaction is to ignore it, otherwise it would hover in the back of our minds, poisoning each moment with its presence.

What makes Henderson's reaction creative is that she found a way to formulate what is wrong so that she—and others—can do something about it. Like all such conceptual moves, her formulation consists in focusing first on one limited aspect of the problem rather than on the whole intractable mess. Henderson decided to focus on how the seven most industrialized countries—the G-7—and measure their progress and wealth. She concluded that these societies, which represent only about 13 percent of the world's population but use up most of its natural resources, have blinded themselves to reality by measuring their Gross National Product (GNP) without taking into account the social and environmental costs of their so-called progress. As long as this shortsighted accounting continues, she feels, the real economy of the planet will go from bad to worse.

Behind this one problem, Henderson feels, stands another one: the epistemological bias of the last few centuries of Western thought, which has progressed by abstracting bits of reality from their context and then treating each bit as if it existed in isolation from the rest. As long as we keep thinking of progress in this way, we will never see the real implications of our choices.

It's basically linear thinking. Its underlying paradigm is that we're all marching along a time line from the past to the present to the future, and that somewhere along there's lots of assumptions about what progress is, which is normally measured in terms of material abundance, technological virtuosity, and economic growth.

The policy that industrial countries pursue is "OK, top on the agenda is to do this, and second on the agenda is to do that." There's the whole assumption that problems are attacked in that way and solved in that way. I don't think problems are like that. The kind of policy issues that industrial countries are dealing with, maybe you actually have to do ten things at the same time because you're dealing with systems that are all interacting. And if you push the system right there and say, "That's the thing we ought to push on today," all you do is to create six hundred other effects somewhere else in the system that you hadn't noticed. And then you call them, quote, side effects. Whereas there's actually no main effect without, quote, side effects.

The Real Wealth

Having formulated the problem of what is wrong with our dealings with the environment this way, Henderson is able to do something about it. As is usually the case, the formulation of the problem implies its own solution. Formulating the problem is conceptually the most difficult part of the entire process, even though it may seem effortless. In this case, Henderson had two goals: to make people understand the long-term costs of progress and to promote a systemic, instead of a linear, mode of thinking about environmental policies. In terms of the first issue, her position is:

> People are the wealth of nations, you see. The real wealth of nations are ecosystem resources and intelligent, problem-solving, creative people. That's the wealth of nations. Not money, it doesn't have anything to do with money. Money is worthless; everybody knows money is worthless. I do seminars on money. And I start off by burning a dollar bill, saying, "This is good to light a fire but you know it's not wealth. It's a tracking system, to help us track transactions."

And instead of linear thinking:

> My view of the world is systemic and interactive. Unless you have a systemic model of the problem that models all of the interfaces and all of the dynamism—and it probably has to be planetary, within an ecosystem framework—you don't know where to push. When you have a good sense, a good map, of how all of those systems are interacting, maybe the policy will need to be pursued in five places at once in order to have feedback effects, or else your one policy will either dissipate and not change the system, or it will have some bad effect somewhere else, or you may amplify the problem in some other system.

In the most general way, Henderson believes, the problem is to redesign the "cultural DNA," or the set of instructions that keep people motivated—the values and rules of action that direct human energy. The basic question is:

> How do you take natural language and compress it so tightly that it begins to act almost like a mathematical formula? What I'm

interested in is the DNA code of societies and of organizations. That is, the program of rules derived from their values. Every culture is really a high-quality program of software, derived from a value system and a set of goals. And every corporate culture and every institution is like that. And so what I like to do is to write the DNA codes for new organizations.

Midwife of Change

Having identified a general approach to the solution of the problem, one now has to devise a method that will do the job. How does one rewrite the DNA of any organization, let alone the entire planet? It is at this point that the really hard work begins. It would be tempting just to bask in the glory of having found a conceptual model for beginning to solve some of the world's worst problems and let others implement it, if they can. But Henderson's creativity is not primarily at the conceptual level; what makes her work stand out from that of many armchair environmentalists is that she actually tries to carry out her ideas.

How does she do that? Her methods are varied and diverse. She writes articles and op-ed pieces. She writes books about alternative economies. She lectures all around the world. She spends time in potentially sympathetic countries like China or Venezuela, networking with government officials and environmental groups. She tries to influence the G-15 countries to adopt new methods of keeping track of their GNP, methods that take into account the hidden social and environmental costs of technological progress. But the main weapon in her arsenal is the ability to create organizations that will implement parts of her vision. These groups may focus on recycling, or alternative economies, or developing an "alternative GNP" such as her Country Futures Indicators, or questioning the environmental appropriateness of consumerism. This involves finding:

The first people and the first resources to bring in around that DNA code, which will be what you might call the business plan for the organization. And to find these people who really understand what that code is, and then find an initial foundation grant or something. My temptation over the years was, I would hang around too long, because I'd want to make sure that that wonderful little DNA code got etched into the stone tablets of the methodology of that organization so that then I could get back onto the board of directors and generally not worry about it because it was all locked in and everybody agreed on what this organization was. So that it wouldn't be something that had been designed to be a mouse and turn into a hippopotamus.

But with time she discovered that to "hang around too long" was a mistake, because the volunteers who joined her out of idealism would get stifled and dependent on her. Plus her ego would become too tightly bound up with the success of the enterprise. So now she passes on the leadership of the young organization as soon as possible and doesn't worry too much whether her initial design will be followed to the letter.

I learned through the school of hard knocks, actually. I was more ego-driven when I was younger, and I found that I started a lot of social change organizations, through the sixties and seventies, and I learned that if you want social change organizations where there's no money involved, there's no motivation of money, it's just a job to be done in terms of an idealized vision of how the society could be in the future, you'd better not be so ego-driven as to want to take credit yourself for having thought of the idea or founded the organization. Because you're trying to recruit idealistic, wonderful people and you're in a position of having to tell them, "Look, the salary's going to be lousy, or there may be even no salary at first." And so all you have really to offer them is identity and identification with an exciting new organization where they can put their whole energy into it. What I found is that the more I stepped back out of the way, and the quicker I did that, the better the organization took off and the more satisfaction the people whom I brought in to run it had. And I found over a period of years that I learned to jump clear faster and faster. I mean, first I'd be worried, "Oh, is my little baby going to be taken care of properly?"

Making High Mischief

How was Henderson able to implement these methods? It is not easy to pull off the kind of guerrilla warfare she has been waging for three decades against planetary economic mismanagement. Certainly having a high goal helps—there are few projects one can devote one's life to with more self-evident justification. But there are a number of more mundane procedures she had to adopt in order to continue with her work without distraction. One thing she had to resign herself to was doing without a normal family life, and eventually her dedication to the solution of the problem she chose led to an amicable divorce. Another thing she had to give up was the financial security of a good job. But then, as she ruefully admits: "I have always known I was unemployable. Because, you know, I would be fired off any job in the first day for insubordination. Because I'd either tell them how to do it better, or whatever. And so I have always realized that I would have to invent my own job."

And finally, by moving to a small community in north Florida she was able to protect her privacy, to express in her location the maverick values she espoused, and by keeping a low profile, to disarm her political opponents. (These, by the way, are exactly the reasons why Elisabeth Noelle-Neumann also moved her high-tech polling organization to an isolated fifteenth-century farming estate in the rural south of Germany.) This is what Henderson says about her choice of a place to live:

It gives me great delight to be able to interact with a big system like the United States and live in a place that's a backwater, where people would say, "What on earth do you live in the wilds of north Florida for?" To me there's a great delight in that. Because one can be considered by the dominant culture as sort of beneath contempt, you know. I mean, "She's just sort of messing around on the fringes of things." The less people know about your effect on various subsystems, the better.

Being hidden away in north Florida does not mean that Henderson is isolated. Whenever she feels it is worth her time, she travels all around the world. And people who are really concerned with helping solve the problems she cares about come and find her—her house is always full of visitors trying to implement the same "high-level mischief" that characterizes her own enterprises. Her best ideas come either when she is involved in a solitary activity like biking, walking, gardening, or washing dishes, or when talking with interesting visitors. Without the constant dialogue with like-minded people Henderson could not even begin accomplishing her aims.

Henderson's unique career has not been smooth sailing all along the way. Like most creative persons, she had her share of difficulties. At a certain point, twenty years ago, she went through a burnout phase. She had been too involved, too busy, too anxious. The constant traveling and stress were giving her neck pains. She was coming close to a breaking point. So she realized she'd better "make her own mode of sustainable operation." This is when she decided to move to Florida and change her lifestyle. But above all else, she reevaluated her priorities and decided that it wasn't important to get credit for what she had been doing, it wasn't important for her to get anywhere. What mattered was to do the best she could and enjoy it while it lasted, without getting all ego-involved with success. This decision has given her so much peace of mind that now she is busier than before without feeling any stress or pain.

What sustains her instead of the desire for fame is a fundamental feeling for the order and beauty of nature, a calling for creating orderly and beautiful environments around her. In colorful hyperbole she says:

> On one level I feel like an extraterrestrial. I'm here visiting for a while. And I'm also in human form. I'm very emotionally attached to the species. And so I have incarnated myself at this time. But I also have an infinite aspect to myself. It all kind of hangs together quite easily for me. It sounds flippant, but the thing is that this is a spiritual practice for me.

Not many people confess to feeling like extraterrestrials, but one must be able to look at oneself from a certain distance in order to get an objective view of the human condition. And in order to invent new ways of living that are not compromised by past traditions, one must strive to attain such objectivity. Yet at the same time, one must also maintain one's "emotional attachment to the species." This dialectic between rational calculation and passionate involvement was mentioned earlier as one of the traits of creative individuals in general. It is perhaps even more essential for those whose creativity lies outside of traditional domains. This is how Henderson expresses it:

There's a very harmonious continuum of what Zen Buddhists call attachment-detachment. And you should always be in the state where you're both. There's a yin/yang continuum, which we can't understand in Western logic because we have this either/or. But it's "both/and" logic, and it says that there's a constant dance and continuum between attachment and detachment, between the long view, the infinite view, and the incarnated view where we have to learn about limitedness, and finitude, and action.

STEPS TO PEACE

Elise Boulding, married for fifty years to the economist Kenneth Boulding, had a difficult time emerging from the shadow of her famous husband. But after bringing up five children and spending eighteen years as a homemaker, she finished a doctorate in sociology and embarked on a unique career of her own devising. Like the other individuals in this chapter, Elise Boulding discovered her problems in the vicissitudes of real life and tried to solve them first within the boundaries of an existing domain. Upon finding out that this was impossible, she left the security of the academic field and struck out on her own, hoping to develop new approaches to the threats that she saw endangering our future.

No Safe Place Left

The main theme in Boulding's life is peace, peace at all levels—in the home, the community, the nation, the world. It is a concern that matured slowly and now absorbs all her energy. It started when she was a small child in New Jersey:

The fear of war in my childhood was the fear of being gassed, from the stories and movies of World War I. And so the kind of nuclear fears children have today, I had an equivalent fear of populations being gassed. I had a fantasy as a child that if there should be another war I would go to Norway, which is where I was born, and go into the mountains and live in a cabin and be safe. All of my mother's stories were about Norway being the good place. The U.S. was in many ways not such a good place; it was selfish, greedy, corrupt. Even in the twenties. [She laughs.] When I was a senior in college, Norway was invaded. Suddenly there was no safe place to go anymore. And so the internal upheaval, my own coming to terms with that, that I'd lost the safe place. Although I knew that was a childhood fantasy, nevertheless it was very much a part of my own core being.

What Boulding saw was that the world was too interconnected to allow anyone to withdraw to a safe haven. Violence can spread everywhere instantaneously. Just as Commoner and Henderson, Boulding confronted the systemic nature of our mutual dependence. She realized that the only way the world was going to be a safe place was if everyone worked to make it so.

Grounding

Working for world peace is no small task. In fact, it is such a utopian idea that it borders on the naive. Most people, when they realize the dangers global aggression poses to their lives, take psychological shortcuts such as denial or scapegoating. It is so much easier to blame the ills of the world on manageable targets such as the Soviet Union, South Africa, religious fundamentalists, or the liberal establishment, instead of considering the possibility that one's own actions are part of the problem. It is always easier to try to get other people to behave instead of behaving ourselves. Yet when we see the world as a system, it is obvious that it is impossible to change one part of it while leaving the rest unchanged.

Boulding approached the problem of peace from the ground up, so to speak. Part of her talent, like that of other creative individuals, consists in finding a way to deal with a complex problem in a manageable way. The steps are simple and obvious: First, we must raise children to be peacemakers; second, we must understand how families can achieve internal harmony; third, we must link harmonious families into neighborhoods and communities; and finally, people so linked should be made aware of their global identity, of their mutual interdependence:

I discovered international nongovernmental organizations, and that gradual understanding of what it meant that there was a network of eighteen thousand transnational associations, where people had different identities than their national identities. What that meant, and how we could plug into those networks, or use the ones we were in. I spent a lot of my time helping people to understand that whatever they belonged to was in fact a world identity. You see, whatever you're doing locally, whether it's Rotary, Kiwanis, all the service clubs, or whether it's churches, whether it's chamber of commerce, sports, there is no realm of activity [that cannot be done] at a global scale. But always departing from a very strong conviction that unless you understood how your own local community worked, you were useless in working anywhere else. You had to know how things worked locally.

Of course, formulating the problem of peace in terms of these ascending levels of complexity does not make the task easy; but it makes it manageable enough that one can start doing something about it instead of throwing up one's hands in despair. Boulding started out by making sure that she brought up her own children to be "peacemakers." She then took her ideas to the Society of Friends, the Quaker meetinghouse to which the family belonged. From there her influence moved to increasingly large audiences—as chairperson of the Sociology Department at Dartmouth, as a writer and lecturer at both the popular and scientific levels. Like Hazel Henderson, Boulding considers the goal of her writing to change the way people think about world problems: "I'm thinking all the time about the different metaphors we use and how they determine our understanding of how reality works. And how we want to change it." Her activities broadened to include leadership in various organizations, and finally she began to move on the international stage:

Well, right at this moment it's the follow-up of my serving as secretary general of the International Peace Research Association, which duty ended the first of May. But in January, as a result of the Gulf War, we established something called the Commission on Peacebuilding in the Middle East. I undertook to be acting chair just to get it rolling, and I've agreed to do that until October; I would like someone else to take over then. So a lot of my time now is involved in corresponding with people. I'm trying to develop as many Middle Eastern contacts as possible, so that it isn't just people in Europe and the U.S. thinking about what should be happening in the Middle East. I'm committed to trying to gather a lot of background papers and produce an overview document.

No matter how far Boulding's influence extends, her activities stand on the firm foundations of home, family, and community. And even deeper than that, her commitment to peace is rooted in faith: She calls her work "action grounded in God's love." As a Quaker, her conception of God is not tied down to a particular historical interpretation; it is a diffuse and evolving entity. But it is a lively and powerful force that allows her to feel connected to the cosmos in "organic wholeness." She turns to lyrical expression to describe how the relationship to the Godhead affects her: "The bright shaft of longing love that goes into the cloud of unknowing, reaching out to unimaginably distant horizons of creation. And having it inside yourself."

Despite this strong faith and the strong supports of family and community, Boulding's life has not been smooth and without problems. Occasionally she feels depleted, exhausted by the burdens she has chosen to carry. One such crisis occurred on her sixtieth birthday, when "all of a sudden I felt totally surfeited with my own life. I felt I had indigestion from my life, that it was too full, too much. I couldn't stand it. It took me a couple of months to work it through and get back to feeling it was OK, there was room for more experience in my life—you know, I wasn't totally surfeited, wasn't totally clogged up—and to kind of open up again and go on." When the dark night of the soul descends, Elise Boulding retreats to her mountain hermitage. There, surrounded by distant peaks and her cherished objects, her days ruled by ritual prayer and meditation, she can restore inner balance and rediscover her spiritual grounding.

RELEASING POTENTIALITIES

John W. Gardner has had many jobs: He started out teaching psychology in college, was the president of a major philanthropic foundation, was appointed by President Johnson as the first secretary of Health, Education and Welfare, and wrote several influential books. But none of these achievements, each one of which would justify most people's existence, gave Gardner a feeling that he had done enough. Because he was not seeking either money or power, the goal he was striving for remained elusive even though to an objective observer it would have seemed that he had reached it several times over.

The Excellence of Plumbers

What did Gardner try to accomplish in his life? Before answering this question, it helps to know what he identified as the main problem that needed solution, the major goal that was worth investing his energies in. Basically, Gardner became convinced that we don't live up to the potential for excellence that is the birthright of every person.

This has two consequences. Our lives become drab and impoverished. We never experience the feeling of exhilaration that one has when acting at the fullness of one's capacities, the kind of feeling that an Olympic athlete may have when running her personal best, or a poet may have when turning a perfect phrase—what I call flow. The second consequence is that people who are both badly paid and have dull jobs eventually become alienated from the fortunate few. With time, this tension necessarily results in social conflict. The problem, as Gardner saw it, was to implement an ethos of social equality even while recognizing the reality of profound individual differences. In a sense, this would require a concept of excellence that includes plumbers (or, from the point of view of plumbers, a concept of excellence that includes university professors):

> Around my late thirties I came to recognize a very powerful dilemma for the American people. They have an ethos of equality and some words that describe that ethos, and yet people vary tremendously in ability and capacity to reach certain standards. And so the subtitle of the book [Excellence] was "Can We Be Equal and Excellent Too?" It seemed to me that we had to have a conception of excellence that left room for the person who was excellent as a plumber. Excellence at various levels. If you start off and say only these people at the very top are excellent, then you invite a carelessness for all the rest of the society. You're saying it doesn't matter, because they can't be excellent anyway, they're just slobs. That's a terrible way to run the society. Everybody ought to feel that whatever his or her calling, they can be excellent. They can be an excellent mechanic, they can be an excellent kindergarten teacher, as well as being an excellent neurosurgeon or whatever. And it was this that really set me on the road to trying to get some ideas across. But today, thirty years later, those ideas are still very mixed up in people's minds.

Reaching the People

The riots that flared through the major cities of the United States in the late 1960s seemed to confirm Gardner's fears: The segments of society that had been denied a chance to be excellent were beginning to revolt. It was at this point that his creativity really began to surface: He left the comfortable institutional positions where he had been so successful and started moving out of the range where foundations and government bureaus held sway. Basically, he felt that the way to combat alienation was to get people more involved in the decisions that affected their fortunes.

This meant organizing voluntary movements that would inform people of their options and then help them find their voices and their power in the political process. The first such job was heading the National Urban Coalition, which had been a remarkable group of corporate, union, minority, and religious leaders who had come together to address the problems of the cities.

My job was to chair that extraordinary group, and it was a very interesting experience, because I visited the toughest parts of every city. I was certainly deeply into every city where there was a riot. I really got intensive exposure to a side of American life that I knew something about, but I didn't know as deeply. I found it very valuable, and also, it led me to form Common Cause. Because as I studied the things that you might do to correct the situation, I kept running into real ailments of government, shortcomings of the process of government, and concluded that we needed attention to government by citizens. There is lots of attention to government by citizens who are acting as lobbyists for the unions or lobbyists for businesses or lobbyists for all the professional groups, but there wasn't much of a voice for the common good, you know, how do we make this system work, how do we make this city a better city.

Common Cause, which Gardner founded and chaired for many years, was an instant success: In the first six months, it attracted one hundred thousand members. He eventually resigned the leadership of this organization for the same reason that Hazel Henderson passes on leadership as soon as possible to someone else: "With every year that passed, I became more certain that I had the answers. I had a Gardner answer for everything." Knowing all the answers is nice, but it has two disadvantages: It makes the job boring, and it stifles the initiative of one's collaborators.

So Gardner moved on to found another organization called Independent Sector, to provide a forum for all the nonprofit agencies around the country. And he continued lecturing and writing. Approaching his eighties, he also returned to his first career and went back to teaching college with renewed verve. His current interest is the study of community, because he feels that neither the fulfillment of one's potentialities nor the self-organizing power of groups can be achieved if people live in anomic neighborhoods that lack the values and inner rules that make a community an organic, self-correcting system.

Living with a Sense of Responsibility

What made Gardner able to put aside the power and success he had achieved and devote his energies to helping re-create forms of representative government? Obviously, a superior intellect helped; he blazed through school always a few years ahead of his agemates. But being very smart doesn't explain his intrinsic motivation. He could have used the same intelligence to make money on Wall Street, or to advance even higher in government. Instead, he chose to do whatever helped most the common good—not so much out of a sense of obligation, but because he genuinely believed that this was the best thing he could be doing:

> I never did anything that I wasn't strongly motivated to do. I never did anything for a title, for power, for money, unless I was deeply interested in the subject matter. I don't know why I behave that way, but I guess I felt life was short and I wanted to do what I wanted to do. I think the other things can be even more secure if you have that base of motivation, if you stay close to your own values.

Of course, this still does not explain where these values came from and why Gardner accepted their priority over the usual ones with such gusto. By now it should be clear that there is no single explanation for his life choices, but several leads contribute parts of the answer. Gardner himself suggests one obvious reason he has lived with such a strong sense of responsibility: the influence of his parents.

Because his father died when Gardner was only a year and a half old, he could have exerted no direct influence. But as we saw in chapter 7, being orphaned early is a frequent occurrence among creative men, and in such cases the absent father appears to have a life-long effect, asking, as it were, very high standards of achievement from his son. Gardner's mother had a more direct but also powerful influence on his values:

> My mother was a very strong, independent-minded person. She had ideas which, for her time, were very advanced about women's rights and about race relations. She had very strong standards of conduct but they didn't fit some of the conventional hypocrisies of the time. For example, she simply would not allow us to look down on any other race or any other group. It just wasn't permitted in our family. We weren't even conscious of it. Years and years later my brother and I would talk about it and realize that we both had exactly the same attitudes. She had instilled those attitudes early. And she also had a very strong independence. What the group did and thought was not binding for her. In fact, I really had the impression that she felt if you were popular, maybe there was something wrong with you, that, you know, you were accommodating too much and weren't standing up for your views and so forth. There's no question that that had some effect in the way my brother and I grew up.

The implicit demands of an absent father and the mother's uncompromising fairness and independence left their mark on Gardner's character. Another influence was the childhood spent in a booming, optimistic California. While he felt elated at the buoyancy of this environment, he also regretted that, because of frequent moves and his own precocity in school, he didn't develop a sense of community or a network of friends. This early sense of marginality also contributed to his own independence and, perhaps as a compensation, to his later concern with the importance of community:

> It may be, you know, my mother's inclination not to accept what the world might think or what conventional patterns were, or it may be my own tendency to do that, or the fact that I didn't

grow up in a community that set those standards for me, but I never had any trouble doing what I wanted to do.

It is not that Gardner was born with a great sensitivity to social wrongs and grew up with a goody-goody wish to help his fellow men and women. He discovered how enjoyable helping others could be as he discovered he had a knack for doing so:

> I thoroughly enjoy management, but before the age of twenty-nine, when I was thrown into management [in the armed services during World War II], I didn't even know that it was an option, and if someone had said it's an option, I would have said, "It doesn't interest me," because I'd never felt the interest, the sheer interest of helping people put their energies together to get a result.

The same thing happened again twenty-four years later, when as the new secretary of HEW he was thrown in the midst of tumultuous political battles: He discovered that he had the skills of a fighter and that he enjoyed a good fight for a good cause. A few year later, when the urban riots forced him to start the grass-roots organization Common Cause, he found he was able to communicate with a wider public and discovered he enjoyed that. In fact, it was these personal experiences that confirmed Gardner in the belief that we all have much deeper reserves than we know we have and that generally it takes an outside challenge or opportunity to make us aware of what we can actually do. A lot of our potential, he believes, is buried, hidden, imprisoned by fears, low self-esteem, and the hold of convention.

> If I meet with a group of business executives, which I often do, and if we get on this subject, I tell them that my estimate is that when they have completed their careers, they will have realized about half of what is in them. The other half will have remained dormant, because life didn't pull it out of them. Or because they concluded very early that that was not something they were good at. They capped their own abilities. The older they get, the more they avoid the risk that growth involves. You start out early with little failures that lead you to believe, don't try that again. And that list grows and grows. By the time you're middle-aged, there's a

long list of things you will never try again. Some of them you might be very good at but have written them off. You've selected the little area in which you know you can win, you know you're gonna make it. You stay within that safe area. What crises and emergencies do is to lift you out of that little safe area of performing, and you discover you have things in you you hadn't guessed.

Gardner has kept learning and growing. He started out reserved, aloof, and detached. This persona worked well as long as he was an academic researcher, but as the head of a large foundation it was intimidating, so he developed a more friendly demeanor. Similarly, the highly rational approach to problems that is appropriate in academic settings is not as effective when it comes to motivating large groups of people:

> I suppose I was forty before I began to think that I could reach people in other than a rational way, which you have to do if you're going to influence them. If you're going to move them, you have to reach their motivations, you have to get below the surface of their thinking into what moves them, what affects their enthusiasms, their concerns. And I had a number of jobs, several of them self-assigned, in which my capacity to persuade, my capacity to evoke action, was of the essence.

In other words, Gardner realized that to influence the new fields in which he was operating, he needed to develop new strategies and rebuild his own personality in the process. This required a great deal of openness and flexibility on his part. "To get the things done that I had to get done, I had to be more open and more interested. I enjoyed it, and the fruits of it." The ability to discover what one can do well, and enjoy doing it, is the hallmark of all creative people. It is particularly fortunate when this also happens to be something that benefits the community, as it has been in Gardner's case.

THE DOMAIN OF GLOBAL RESPONSIBILITY

What Commoner, Henderson, Boulding, and Gardner have in common is that they have realized the systemic interconnection among the events that happen on the planet, and they are struggling to act on this realization. One way of saying what they are trying to do is that they are attempting to develop a domain of global responsibility and a field to implement it. Commoner emphasizes our uses of energy and resources; Henderson, our lifestyles and consumption patterns; Boulding, violence; and Gardner, the social effects of stunting individual potential. The focus of attention is different in each case, but the causal network they consider is interconnected. Any change in the pattern of energy use, of consumption, of peaceful spirituality, of personal fulfillment affects the others. The central message is that every action has a consequence, that in many important respects the planet is a closed system with fragile boundary conditions, and that unless we take informed action, these conditions may easily be violated.

In a sense, this emerging realization is not so novel. Many simple cultures have developed a systemic view of their cosmos. It is implicit in many of the great world religions. In Judeo-Christian faiths, it is expressed obliquely in the belief of an omniscient God who sees and evaluates even the most minuscule event, such as the fall of a sparrow from a tree branch. It is implicit in the Eastern beliefs in karma, in the endless consequences of each action rippling down the ages toward infinity. According to the ancient Zoroastrian creed, each person was expected to pray forgiveness of the water for having polluted it, of the earth for having disturbed it, of the air for having filled it with smoke. But with the glorious advance of science in the last few centuries, these intuitions of a network of causes and effects binding on individuals were discredited as superstition. The human species was seen as all-powerful, its actions above the laws of nature.

What people like the ones described in this chapter are doing is rediscovering, within the domains of different sciences, the grounds for taking these intuitions seriously. Biochemistry, economics, sociology, and psychology come to the same conclusion: It is dangerous to proceed within the rules of an isolated domain without taking account of broader consequences. It is dangerous to build nuclear devices unless we know that we can dispose of their wastes safely; it is dangerous to waste food and energy when most of the world is cold and starving; it is dangerous to ignore the spiritual needs of people; it is dangerous to underutilize human potentialities.

But how to formulate these isolated bits of knowledge into a coherent symbolic domain? Scientists in the West started to study systems only recently; we still have no way to represent the kind of problems these four creative people are struggling with in any manageable way. To a large extent, we are still at the prescientific, metaphorical stage. The myth of Gaia, which describes the planet as a living, self-correcting organism, is one of them. The anthropic principle, which claims that our thoughts and actions actually make the existence of the universe possible, is another. Commoner, Henderson, Boulding, and Gardner appear to be poised at the threshold between metaphor and natural law; ready to move from poetic insight to systematic understanding.

They share some common traits appropriate to intellectual pioneers. They all felt marginal as they grew up. Commoner because he was Jewish, Henderson because her loyalties were split between a loving mother and a powerful father, Boulding because her Norwegian-American upbringing gave her two different perspectives for interpreting experience, and Gardner because he lost his father, never felt that he belonged to a community, and was always the youngest boy in class. This feeling of marginality caused them never to take orthodox ideas for granted. It helped them break away from domain-bound constraints on their thinking when real-life experience conflicted with them.

All four mentioned repeatedly their constant shifting from action to reflection, from passion to objectivity. In each case, this alternation allowed them to keep learning, to keep adjusting to new situations. Their creativity unfolded organically from idea to action, then through the evaluation of the outcomes of action back to ideas—a cycle that repeated itself again and again.

None of them seems to be motivated by money and fame. Instead, they are driven by a feeling of responsibility for the common good, a feeling that sometimes borders on traditional religious values but more often seems to depend on a spiritual sense for the order and beauty of natural phenomena that transcends any particular creed. It is a contemporary formulation of that most ancient awe that prompted our ancestors to develop images of the supernatural in the first place. But they wear this feeling of responsibility lightly, as a privilege rather than a duty. Although they work hard to help improve our lives, they claim that they never did anything they didn't want to do. Like the other creative persons we studied, flow is the typical state of their consciousness.

THIRTEEN

THE MAKING OF CULTURE

T he world would be a very different place if it were not for creativity. We would still act according to the few clear instructions
our genes contain, and anything learned in the course of our lives
would be forgotten after our death. There would be no speech, no
songs, no tools, no ideas such as love, freedom, or democracy. It would
be an existence so mechanical and impoverished that none of us would
want any part of it.

To achieve the kind of world we consider human, some people had
to dare to break the thrall of tradition. Next, they had to find ways of
recording those new ideas or procedures that improved on what went on
before. Finally, they had to find ways of transmitting the new knowledge
to the generations to come. Those who were involved in this process we
call creative. What we call culture, or those parts of our selves that we
internalized from the social environment, is their creation.

CREATIVITY AND SURVIVAL

There is no question that the human species could not survive, either
now or in the years to come, if creativity were to run dry. Scientists

will have to come up with new solutions to overpopulation, the depletion of nonrenewable resources, and the pollution of the environment—or the future will indeed be brutish and short. Unless humanists find new values, new ideals to direct our energies, a sense of hopelessness might well keep us from going on with the enthusiasm necessary to overcome the obstacles along the way. Whether we like it or not, our species has become dependent on creativity.

To say the same thing in a more upbeat way, in the last few millennia evolution has been transformed from being almost exclusively a matter of mutations in the chemistry of genes to being more and more a matter of changes in memes—in the information that we learn and in turn transmit to others. If the right memes are selected, we survive; otherwise we do not. And those who select the knowledge, the values, the behaviors that will either lead into a brighter future or to extinction are no longer factors outside ourselves, such as predators or climatic changes. The future is in our hands; the culture we create will determine our fate.

This is the evolution that Jonas Salk calls metabiological, or E. O. Wilson and others call biocultural. The idea is the same: Survival no longer depends on biological equipment alone but on the social and cultural tools we choose to use. The inventions of the great civilizations—the arts, religions, political systems, sciences and technologies—signal the main stages along the path of cultural evolution. To be human means to be creative.

At the same time, it does not take much thought to realize that the main threats to our survival as a species, the very problems we hope creativity will solve, were brought about by yesterday's creative solutions. Overpopulation, which in many ways is the core problem of the future, is the result of ingenious improvements in farming and public health. The loss of community and increasing psychological isolation are in part due to the enormous advances in mobility, brought about by the discovery of self-propelling vehicles such as trains and cars. The loss of transcendent values is the result of the success of science at debunking beliefs that cannot be tested empirically. And so on, ad infinitum. This is the reason, for instance, that Robert Ornstein calls human inventions "the axemaker's gift," referring to what happens when a steel axe is first introduced to a preliterate tribe that knows no metals: It leads to easier killing, and it shreds the existing fabric of social relations and cultural values. In a sense, every new invention is an axemaker's gift: The way of life is never the same after the new meme takes hold.

It is not only the clearly dangerous discoveries—distilled alcohol, tobacco, firearms, nuclear reactors—that threaten to wipe out entire populations. Even apparently beneficial inventions have unintended negative consequences. Television is a fantastic tool for increasing the range of what we can experience, but it can make us addicted to redundant information that appeals to the lowest common denominator of human interests. Every new meme—the car, the computer, the contraceptive pill, patriotism or multiculturalism—changes the way we think and act, and has a potentially dark side that often reveals itself only when it is too late, after we have resigned ourselves that the innovation is here to stay.

The development of nuclear energy promised both military and industrial advantages to those countries that were able to seize the opportunity. It was a chance none could refuse. Yet only half a century into the nuclear age, it seems that the toll we must pay for this particular Faustian bargain is so high that it could bankrupt us. Recent estimates are that it will cost the United States more than $300 billion to safely dispose of nuclear waste. Many other countries, like Russia, may not be able to start cleanup operations in time. In a twinkling of planetary history, human ingenuity has succeeded in making a good fraction of Earth unfit for habitation.

There is a basic law of human ingenuity that we try very hard to ignore: the greater the power to change the environment, the greater the chances of producing undesirable as well as desirable results. About four thousand years B.C.E., the discovery of large-scale irrigation in Mesopotamia made that country fruitful and rich beyond anything its neighbors could dream of. But each year the currents of the Euphrates and the Tigris removed a fraction of an inch of the rich topsoil and deposited salty minerals in its place. Slowly the bountiful garden between the two rivers has turned into a desert where almost nothing grows.

To take another example from the other end of the world, the great Maya civilization collapsed about 800 C.E. not because it could not cope with adversity, but because it was destroyed by its own success. There are contending explanations of why that complex culture was reabsorbed by the jungle. Perhaps too many families became wealthy and powerful. These elites felt that they should not work any longer, yet each generation expected to be more comfortable and have a higher standard of living than the one before. With too many chiefs per Indian, inner conflicts finally erupted in murderous civil wars. Another hypothesis is that to build their magnificent temples and palaces, the Maya had limestone stucco, which had to be melted in very hot furnaces. To feed the furnaces they cut down much of the surrounding forest, which resulted in erosion of the soil; the topsoil was washed away and silted up the marshes that the Maya had used to irrigate their terraced fields. Deprived of fertilizer, the fields yielded little food, and the ensuing famines fueled civil disorder that led to chaos and eventual oblivion. The power to create has always been linked with the power to destroy.

A similar pattern of initial success leading to eventual failure holds for memes that shape human energy through ideas. The promises of Nazism, Marxism, and the various religious fundamentalisms give people a simple set of goals and rules. This liberates a wave of psychic energy that for a while makes the society that adopted the creed seem purposeful and powerful. In Germany, Hitler eliminated unemployment when the rest of the industrial world was still in the throes of the Great Depression. In Italy, Mussolini for the first time made the trains run on time. Stalin transformed a backward rural continent into a leading industrial giant. Soon, however, the downside appears: Intolerance, repression, rigidity, and xenophobia usually leading to wars or worse are just some of the usual consequences when social energies are focused by memes that promise superiority to one group at the expense of the others.

But even when the fruits of creativity produce no external damages, their very success can sow seeds that are dangerous to the survival of the culture that has adopted them. The Romans were able to fashion a rich and stable society through the invention of a viable system of laws, administrative arrangements, and military practices. But after a while, Roman patricians saw no reason to exert themselves. The inertia of their success lulled them into a false sense of security. Cheap slave labor made them indifferent to new labor-saving devices. As in the slave-holding American South, fatal complacency appeared as the inevitable dark side of the coin of material comfort.

Similarly, American ingenuity has produced a standard of living and a political stability that are the envy of the world. The result is that Americans—as well as most Europeans—see no reason to work long hours at cheap wages. And who can blame them? But much of the rest of the world is eager to toil hard in undesirable conditions. As a result, productive activity increasingly shifts into the hands of people who have the lowest expectations. When was the last time that you wore clothes made in the USA? Or used domestic TV or video equipment? The reason why the number of immigrants keeps growing is that they are the only ones left willing to do menial jobs.

But even engineers and technically trained workers are steadily getting scarcer in the industrialized nations. Everyone wants to be a professional, or at least a clerical worker sitting behind a desk. The optimists argue that our children are preparing themselves for the jobs of the future, based on information and creative flexibility. But the fact is that the number of new patents being taken out in the United States is also decreasing, and computer literacy is more a question of learning to be a consumer of information than knowing how to generate or use the information acquired. If necessity is the mother of invention, secure affluence seems to be its dysfunctional stepparent.

So through history we see an ironic process that Hegel or Marx would have appreciated: a dialectic whereby the success of a culture develops within itself its own antithesis. The more well-off we become, the less reason we have to look for change, and hence the more exposed we are to outside forces. The result of creativity is often its own negation.

It is true that in the past a society that had advanced far in creating complex memes could survive for hundreds or even thousands of years more or less unchanged, living on its initial cultural capital. The Egyptians were able to do so, and so were the Chinese. But such a luxury is no longer available, in part because of the very advances made in the past few centuries. Communications have improved to the point that information, technology, and access to capital are almost evenly distributed across the globe. Those who use these resources most efficiently and with greater determination are likely to control the future. No society can any longer enjoy the splendid isolation of the Nile's empire, or even that of Victorian England.

So what is the verdict in this tangled tale? The current fashion is still to acclaim creativity without reservation. People deemed to be creative can do no wrong; they will save us from past mistakes and lead to a bright future. Of course, occasionally there are dissenting voices. The psychoanalyst Géza Róheim wrote that the whole enterprise of life, and especially its latest conscious episodes, amounts to a huge mistake. The ideal state of matter is inorganic; life is just a feverish sickness, a passing cancer on the serene stage of a crystalline universe.

More to the point, the general public also seems to be getting second thoughts about the value of the culture that our ancestors have created. It is not only in Russia, Iran, India, or Brazil that people's faith in science, democracy, and many of the other good things humankind has fought so hard to achieve is shaken. Spasms of traditionalism run through gleamingly modern Japan, and forces groping toward a return to simpler times are gathering strength in the United States. Recovering shared values, a sense of community, and a more serene lifestyle would be great accomplishments. Unfortunately, turning back is more likely to involve a renewed belief in magic, astrology, the supernatural, and the superiority of one's ethnic traditions relative to all others.

Neither uncritical acceptance nor wholesale dismissal of human creativity will lead us far. It would be so nice if we could look at culture and determine objectively: This is good, that is bad. But history does not unfold in black and white. Each great advance contains within it a new vulnerability. Some memes are indispensable today but a hindrance tomorrow. It is as absurd to believe that progress is always desirable as to reject it out of hand.

Creativity in the Context of Human Evolution

The argument so far has tried to establish two points: that creativity is necessary for human survival in a future where the human species plays a meaningful role and that the results of creativity tend to have undesirable side effects.

If one accepts these conclusions, it follows that human well-being hinges on two factors: the ability to increase creativity and the ability to develop ways to evaluate the impact of new creative ideas. Let's focus first on the second requirement.

Why can't we leave the evaluation of new ideas to their respective fields, or to the "invisible hand" of the marketplace? Unfortunately, neither of these two institutions is well equipped to cope with the task. Almost by definition, the members of a field are devoted to advancing the hegemony of their domain, without much regard for the rest of the culture. Although a few physicists banded together after World War II to alert society to the dangers of nuclear proliferation, the field as a whole could not resist lobbying for expanded research and applications of high-energy physics. Similarly, a few physicians have sounded the alarm about high-tech medicine interfering with progress in public health, but the majority of the field, led by the American Medical Association, sees its duty as endorsing the proliferation of expensive equipment and procedures.

Left with carte blanche, every field naturally wants to control as many of the resources of society as possible, and more. The American Psychological Association would be happy if every school, business firm, and family had its own resident psychologist. The interest of artists is to convince the rest of society that things would be better if everyone became a collector of art, while the interest of dentists is to assure us that we would be happier if we devoted most of our free time to oral hygiene. Each field welcomes any new idea that promises to expand its hold on societal resources.

In addition, even if there were no selfish, material reasons involved, each field would still push for the implementation of new ideas in its domain, regardless of long-term consequences. A person who has worked for years within the limits of a narrow specialization naturally believes that new developments in his or her domain are the most important and therefore take precedence over developments elsewhere. It is difficult to convince a physicist who has devoted a lifetime to high-energy physics that advances in nuclear technology should not be supported all the way.

Each field is understandably proud of its achievements and is quick to invoke academic freedom, free speech, free thought, or any other serviceable ideology to defend itself against attempts to evaluate its contributions in terms of the common good, as opposed to criteria internal to the field. Within a liberal worldview, to challenge an artist's right to exhibit whatever he or she pleases—a desecrated flag, a vase of urine, a mutilated body—amounts to anathema. Scientists recoil in horror at the thought that anyone else should decide what is or is not good science. A person who has been awarded a Nobel Prize in physics has almost no choice but to believe that he or she is heir to the only possible way of studying the world. To paraphrase Voltaire, he naturally believes that his is the best of all possible sciences, and therefore that any attempt to question the inevitable unfolding of physics by physicists is an anti-intellectual attack on the integrity of science. Each field expects society to recognize its autonomy, yet each feels in the last analysis accountable only to itself, according to the rules of its own domain. For all of these reasons, it is useless to expect fields to monitor their own creative ideas in terms of the long-range public good.

The other alternative is for the market to determine the value of novelties. As in many other social processes, our tendency is to trust the wisdom of the marketplace and tacitly to endorse its priorities. But of course by now everyone suspects that the so-called free market is as real as Santa Claus or the Easter bunny. When the World Bank loans untold millions to Brazil to build nuclear reactors it cannot either use or pay for, the transaction is not a response to free-market forces but to the interests of a few big American firms that build reactors. To use another example: Every nation, from France to Finland, from Japan to the United States, tries to protect its agricultural base by paying farmers what the free market will not deliver.

But even if the free market were a reality, it is doubtful that its decisions would be wise as far as our future well-being is concerned. In the first place, market decisions tend to be oriented to the present. Given a choice, consumers choose a product or process that provides an edge right now, with little concern for consequences. I am going to buy the can of deodorant that saves me a few seconds each morning regardless of the hypothetical effects of its spray on the ozone layer. If I were to buy a handgun, I would probably buy one that shot more bullets faster than its competitors, even though that more efficient gun might be the cause of more accidents.

Mass-produced commodities are especially vulnerable to being chosen on the basis of short-term benefits. Fast food is more profitable when it satisfies the most basic taste needs, which were established in our genetic past when fat and sugar were in short supply. A hamburger with fries and a milkshake would make an exquisite banquet for a caveperson but is not particularly healthful for sedentary citizens. Private-sector television is similarly vulnerable to criticism. The kind of spectacles we are genetically programmed to watch have not changed much since the Romans flocked to the arena to see gladiators disembowel each other on the sand. It is difficult to imagine beneficial contributions to evolution from watching soap operas and MTV on the home screen.

Yet, as we have seen earlier, we cannot ignore evolution. The culture that survives to direct the future of the planet will be one that encourages as much creativity as possible but also finds ways to choose novelty on the basis of the future well-being of the whole, not just of the separate fields. What is needed is a self-conscious effort to establish priorities and to use something like an "evolutionary impact analysis" as one of the bases for the social endorsement of new ideas.

A policy of this type should not result in any kind of philistine thought-policing. Artists should be encouraged to follow their muse, scientists should be respected for following a hunch wherever it leads them. On the other hand, why expect society to support novelties that are valued within a given field but may harm the commonwealth?

The greatest art, East or West, was not produced when the artists set the agenda, but when patrons insisted on certain standards that benefited them. Patrons wanted primarily to be admired by the public, so the art they demanded had to appeal to and impress the entire community. In this sense, medieval and Renaissance art, commissioned by popes and princes, was in reality more democratic than it has become since the art world gained the power to separate itself from the rest of society—as a field with its own peculiar tastes and criteria of selection.

It admittedly would be more difficult to achieve a public evaluation of scientific creativity. In most scientific domains the frontiers of knowledge have moved so far beyond the grasp of laypersons that only those within the respective fields can be expected to make any sort of informed decision. But it is probably the case that within each field there are enough individuals with both expertise and a sense of the public good who could be deputized to serve the interests of society.

Currently research grants are evaluated in terms of either the priorities set by the field or the political agenda of the administration disbursing the funds. Perhaps it could be possible to establish a sort of civil service above party politics and disciplinary fashions, composed of those who aspire to be "good ancestors," as Jonas Salk called them, and who would be willing to represent the claims of evolution when assessing whether scientific advances should receive social support. Inevitably such a group would be composed mainly of older individuals, and therefore it would be open to criticism from younger colleagues who are more concerned with advancing their own scientific careers. On the other hand, the probability for dispassionate wisdom is greater among those who have had more, and more varied, experience and who can see their expertise in a broader context—and these in turn are likely to be older persons. Yet our society expects very little from its elders. This might be one important contribution of seniors that will benefit everyone.

WAYS TO INCREASE CREATIVITY

For billions of years, evolution has proceeded blindly, shaped by random selective forces. We were created by chance. Now, however, humans have become one of the most powerful, and therefore the most dangerous, forces operating on the planet. Therefore, if we wish evolution to continue in a way that corresponds with our interests, we must find ways to direct it. And this involves developing mechanisms for monitoring new memes, so that we can reject those that are likely to be harmful in the long run and encourage alternatives that are more promising.

But before selection can begin operating, novelty must be generated. In other words, there have to be new ideas to choose from. So it is now time to turn to the question, What ways are there to increase the frequency of novel ideas worthy of being adopted by the culture? To answer that question, I consider strategies that apply at each of the three levels that define the components of a creative system: the person, the field, and the domain.

More Creative Individuals

We have seen that central among the traits that define a creative person are two somewhat opposed tendencies: a great deal of curiosity and openness on the one hand, and an almost obsessive perseverance on the other. Both of these have to be present for a person to have fresh ideas and then to make them prevail. Is it possible to increase the number of people who have these characteristics?

We don't know for sure. In part we don't have the answer because it is not clear to what extent these traits might be genetically controlled. Of course, it is unlikely that our chromosomes have a single location for an openness gene, and that, depending on which of several alternatives fills each spot, one person might be born with an inclination to be curious, while another will be born indifferent. But it is quite possible that a combination of instructions issued from a number of genes might interact to predispose a person to be more or less open.

But biological inheritance is only part of the story, as we discussed before. Early background has a significant effect. Interest and curiosity tend to be stimulated by positive experiences with family, by a supportive emotional environment, by a rich cultural heritage, by exposure to many opportunities, and by high expectations. In contrast, perseverance seems to develop as a response to a precarious emotional environment, a dysfunctional family, solitude, a feeling of rejection and marginality. Most people experience either one or the other of these early environments, but not both of them.

However, creative individuals seem more likely to have been exposed to both circumstances. John Hope Franklin grew up in a very supportive and stimulating family, but suffered from discrimination because of his race. Isabella Karle grew up in a socioeconomically marginal family, but her parents were warm, stimulating, and supportive.

Of course, many children with similar backgrounds never became creative, and several creative persons in our sample had early experiences that did not conform to this type. It is impossible to argue that one must have a certain kind of family background in order to become creative. But there definitely seems to be an increased likelihood that bimodal early experience is related to later creativity. And this kind of weak relationship is probably the best we can expect to get when trying to assess a causal link between two such heterogeneous concepts as "early experience" and "creativity." But a weak link is better than none. At least we might hope that by providing elements of both experiences, the proportion of people showing the traits of creativity could be increased.

The same argument applies to the other trait-pairs mentioned in chapter 3. Parents and educators should know that a milieu that encourages both solitude and gregariousness may add, even if infinitesimally, to the chances of a child being able to express his or her creativity. Children who have not learned to tolerate solitude are especially at risk in terms of never developing enough in-depth involvement in a domain and lacking opportunities to reflect and incubate ideas. On the other hand, children who are too shy and reclusive need selfless intermediaries, such as van Gogh or Kafka had, lest their contributions disappear from the culture.

Similarly, a certain flexibility about gender roles is likely to help. If a child is too strongly socialized to act in terms of a strict gender stereotype, his or her creativity is likely to be inhibited. In other words, the traits that distinguish a complex personality are likely to add a higher statistical probability of creative expression. The contribution of each trait may be very small, and none is likely to be indispensable. Yet when all of them are present, the prognosis should be more favorable.

In addition to these motivational and personality factors, there are, of course, important cognitive variables. Here too, genetic inheritance might play an important role. Each one of us has particular strengths and predispositions that make us sensitive to some dimension of reality more than another. But again, early exposure and opportunity to engage in a particular domain is essential to developing the inherited potential. A child who is encouraged to question is likely to develop a problem-finding attitude. A child who is introduced to inductive reasoning may have an advantage in making sense of the world.

Above all else, it helps to become involved in a domain early. E. O. Wilson, who probably knows more about ants than any other individual in the world, started his studies when he was six years old. Linus Pauling became fascinated with the way chemicals combined at about the same age. Ravi Shankar was playing music professionally as a child, and György Faludy knew he was a poet in grade school. Vera Rubin was less than ten years old when she decided she had to become an astronomer. It is important to realize that in none of these cases did the parents push their children to study chemistry, music, poetry, or astronomy—the child's spontaneous interest led to the involvement. The role of the parent was limited to providing opportunities, taking seriously the child's interest after it showed itself, and then supporting the child's involvement, as when Rubin's father helped his daughter to build a telescope. If the parents had been more directive, it is unlikely that the child's involvement would have progressed very far.

But most of the individuals in our study did not start that early; in fact, many embarked on their eventual careers in college or later. However, they were all directed by curiosity to master some sym- bolic form to a degree rare in other children. Elisabeth Noelle-Neu- mann played intensely with make-believe villages and loved to write; Mark Strand painted; and Jacob Rabinow took apart any piece of machinery he could lay hands on.

So while specializing in a particular domain can wait until late adolescence, an intense involvement in *some* domain might be neces- sary if a person is to become creative. Without developing a skill he or she is confident in, without having the experience of acquiring a knowledge base, a young person may never get up enough nerve to change the status quo. Hilde Domin didn't write her first poem until late in life, but she had learned and studied half a dozen languages. Sooner or later, however, it becomes essential to master the special- ized knowledge of a particular domain. Here, knowing the basics is essential. Acquiring the foundations of math and physics for a scien- tist, of drawing for an artist, of the classics for a writer is the starting point for any further innovation.

Yet it is important to keep in mind that most breakthroughs are based on linking information that usually is not thought of as related. Integration, synthesis both across and within domains, is the norm rather than the exception. Madeleine L'Engle is inspired by molecu- lar biology to write her stories; Ravi Shankar finds ways of harmo- nizing the music of India and Europe; and almost all scientists cross and recross the boundaries of physics, chemistry, and biology in the work that turns out to be creative.

Even when not directly integrated in one's work, other domains contribute to the overall mental life of creative individuals to a degree that belies the stereotype of the sterile, narrowly trained spe- cialist. Music enriches the life of many, and so do the arts and litera- ture. Scientist Manfred Eigen plays in a chamber orchestra, politician Eugene McCarthy writes poetry. Ceramist Eva Zeisel in her seven- ties started researching and writing a history of race relations in New York City. Business leader Robert Galvin collects antique maritime maps and studies constitutional history.

This breadth, this interest that overflows the limits of a given domain, is one of the most important qualities that current schooling and socialization are in danger of stamping out. If nothing else, this study should renew our determination that narrow specialization shall not prevail. It is not only bad for the soul but also reduces the likelihood of making creative contributions that will enrich the culture.

What the Field Contributes

Most of us deep down believe that a person who is creative will prevail regardless of the environment. The Romantic idealization of the solitary genius is so solidly lodged in our minds that to state the opposite—that even the greatest genius will not accomplish anything without the support of society and culture—borders on blasphemy.

But the reality appears to be different. Favorable convergences in time and place open up a brief window of opportunity for the person who, having the proper qualifications, happens to be in the right place at the right time. Benjamin Spock was one of the first pediatricians with psychoanalytic training, and therefore he was in a good position to write an authoritative and popular child-care book incorporating the latest Freudian ideas. A few years earlier the task would have been impossible; a few years later it would have been redundant. Ravi Shankar learned music from the musical group run by his family, Robert Galvin inherited his business, and practically all the women scientists in this cohort benefited from the opening up of laboratory jobs due to young male scientists being drafted to fight in World War II.

The point is not that external opportunities determine a person's creativity. The claim is more modest, but still extremely important: No matter how gifted a person is, he or she has no chance to achieve anything creative unless the right conditions are provided by the field. And what might these conditions be?

In terms of what we have learned from this study, it is possible to single out seven major elements in the social milieu that help make creative contributions possible: training, expectations, resources, recognition, hope, opportunity, and reward. Some of these are direct responsibilities of the field, others depend on the broader social system. If our argument is correct, then creativity can be substantially increased by making sure that society provides these opportunities more widely.

Let us take these elements one at a time. Clearly, the availability of training is crucial for developing any kind of talent. If Michael Jordan had been born in a country where basketball was not practiced, he would never have been able to refine his skills and would not have been recognized. A society that can match effectively opportunities for training with the potentials of children has an impact on the frequency of creative ideas its members produce.

Of course, training is expensive, and therefore hard choices must be made. Which domains should be taught, and how widely? Currently American public schools try to save costs by eliminating instruction in the arts, music, athletics, and all other areas that the public considers nonessential. On the whole, however, trying to save by cutting opportunities for learning is one of the most benighted solutions a society can adopt. Perhaps only Jonathan Swift's solution to the Irish famine is more objectionable.

Expecting high performance is a necessary stimulus for outstanding achievement and hence for creativity. High expectations should start within the family, continue in the peer group, in the school, and in the community at large. Having high expectations is not a comfortable thing. Asian youth in the United States have internalized very high academic goals from their culture, and consequently have relatively low self-esteem, because it is so difficult for them to live up to expectations. Young African-Americans generally have lower academic goals, and hence their self-esteem tends to be higher.

Certain families have long traditions of artistic, scientific, or professional accomplishment that set high standards for the young person. Nobel Prizes ran in the families of Subrahmanyan Chandrasekhar and Eva Zeisel; Heinz Maier-Leibnitz followed in the footsteps of a distant ancestor. Of course, excessive or unrealistic expectations do more harm than good. In our study, parents and mentors usually conveyed their faith in the young creator's abilities indirectly, almost taking excellence for granted, rather than nagging, pushing, or insisting.

Presumably it is best when not only the family and the school but also the entire community and society expect high performance of a young person. Ethnic traditions were often cited as having influenced the motivation to achieve. Jewish, Southern, and Mormon beliefs about one's exceptional vocation were just some of the examples. In the mainstream U.S. society, excellence in academic domains is not expected. What we do expect more than perhaps any other society in history is that children should grow up happy and well adjusted.

But while Japanese parents, for instance, believe that their children can and should learn calculus, most American parents are content with minimal scholastic performance. It is difficult to see how young people are to take academic domains seriously if they sense their elders don't really care.

Resources are crucial for creativity to develop, but their role is ambiguous. It is true that having access to the best examples of the past helps, and so does being able to afford the necessary materials. About thirty years ago, I remember reading about one of the emerging African nations that decided to institute a space research program. They selected some healthy young men as astronaut candidates. To get used to the gravitational forces involved in launching a space probe, a would-be astronaut would crouch inside a barrel, which his companions twirled around in circles at the end of a rope. Clearly it is extremely difficult to contribute useful new ideas to space exploration if all one has is a barrel and a rope.

Yet too many resources also can have a deadening effect on creativity. When everything is comfortable and better than anywhere else, the desire for novelty turns to thrills and entertainment instead of trying to solve basic problems. When Florence exploded with creativity in the fifteenth century it was one of the richest cities in Europe, a center of learning and information. At the same time it was a city racked with internal political turmoil, threatened from the outside, literally fighting for its continued existence. What can we learn from these contradictory trends? Certainly, if we wish to encourage creativity, we have to make sure that material and intellectual resources are widely available to all talented and interested members of society. Yet we should realize that a certain amount of hardship, of challenge, might have a positive effect on their motivation.

At some point in their careers, potentially creative young people have to be recognized by an older member of the field. If this does not happen, it is likely that motivation will erode with time, and the younger person will not get the training and the opportunities necessary to make a contribution. The mentor's main role is to validate the identity of the younger person and to encourage him or her to continue working in the domain. The guidance of an older practitioner is important also because there are hundreds of ideas, contacts, and procedures that one will not read in books or hear in classes but that are essential to learn if one hopes to attract the attention and the approval of one's colleagues. Some of this information is substantive, some is more political, but all of it may be necessary if one's ideas are to be noticed as creative.

In some fields, like science, math, or music, it is possible to measure extraordinary talent through standard tests. Thus testing has been an important feature of many successful cultures, from ancient China to the current United States. While impersonal recognition through testing might be an important step in some domains, it can only be the first one in the development of creative persons, for whom a close master-apprentice relationship is of great importance. In our study we found that a few individuals were taken in hand by competent adult practitioners very early in life, many were recognized during high school, and most of the remaining had an important mentor by the time they were of college age. Again, recognition by a mentor is not strictly necessary, but it must definitely contribute to the realization of creative potential.

Training, expectations, resources, and recognition are to no avail, however, if the young person has no hope of using his or her skills in a productive career. In our culture, a huge number of talented and motivated artists, musicians, dancers, athletes, and singers give up pursuing those domains because it is so difficult to make a living in them. In a study of American adolescents, we found that almost 10 percent of thirteen-year-olds wanted to be architects when they grew up. At a rough guess, this is probably a thousand times what the field of architecture can accommodate. It is not realistic to expect a great deal of talent to be attracted to a domain, no matter how important it is, if there is little chance of practicing it. The people who succeed in the smaller fields are like Vera Rubin, to whom not being an astronomer was "unthinkable."

After hope, one also needs to have real opportunities to act in the domain. It has been said that the great musical creativity that blossomed in Germany in the eighteenth and nineteenth centuries was in large part due to the fact that each aristocratic court that ruled the many principalities had to have an orchestra to amuse itself and to show its superiority over the others. There was constant interest in and competition for new musical talent. A Bach, Handel, or Mozart had no difficulty in having his music performed and then evaluated by an eager crowd of connoisseurs. If there are fewer creative classical composers now, it is probably not due to a lack of talent but to a dearth of opportunities to display it.

The problem is especially severe in fields that require long and specialized training and then suddenly run out of opportunities. Many young physicians who have trained in some of the more high-tech and well-paying disciplines, such as anesthesiology or radiology, are finding themselves unemployed as insurers cut costs and force hospitals to release patients earlier than they used to do. There are growing numbers of excellently trained but unemployed mathematicians and physicists as well, and several disciplines, like marine biology, which appeal to a great number of young people, continue to have relatively few openings.

It is true that there are many instances of creative individuals who seem to *make* their own opportunities. After all, Albert Einstein was a lowly clerk in the Swiss patent office when he wrote up his ideas about relativity. Next thing we know, he was being offered several professorships. No doubt other such cases exist. But even in the case of Einstein we might perhaps argue that his chances of being recognized would have been much fewer, or nonexistent, if physics had not achieved such prestige at the start of the century, thus inflating the demand for novelty. In any case, the fact that some individuals prevail even when opportunities are few does not imply that there could not be even more creative achievers if the opportunities were greater.

Finally, rewards—both intrinsic and extrinsic—help the flowering of creativity. There is no question that at the beginning of the Renaissance an infusion of golden florins into ambitious projects attracted many young Florentines to the arts. Brunelleschi was a member of the first cohort of artists in the Quattrocento who would almost certainly not have taken up such a career even a generation earlier. He came from a respectable professional family that considered artists despicable craftsmen. But with the sudden infusion of money and prestige, it was possible for him as well as many other talented young men of good families to envision careers in architecture, painting, or sculpture.

Probably very few creative persons are motivated by money. On the other hand, very few can be indifferent to it entirely. Money gives relief from worries, from drudgery, and makes more time available for one's real work. It also enlarges the scope of opportunities:

One can buy necessary materials, hire help if needed, and travel to meet people from whom one can learn. Artists are supposed to be above financial concerns, but in reality they can use money just as much as anyone else: first, in order to buy supplies, and, second, to evaluate their own success.

It is enough to read the autobiography of the famed Renaissance goldsmith Benvenuto Cellini to realize how important money can be to an artist as a gauge of self-worth. In the four and a quarter centuries since Cellini died, money has become increasingly the main measure of a person's success. The importance of honor, respect, or a good conscience keeps diminishing in comparison to the rewarding power of money. Presumably creative individuals respond to financial incentives to a lesser extent than most people, but they do so nevertheless.

Similarly, public recognition and acclaim are certainly not necessary to truly creative persons, yet they are not rejected either. Creative persons are often arrogant and egocentric, but they are also insecure and can benefit from approval. Being at the cutting edge isolates a person from his or her fellows, and it helps to feel appreciated. In one of the most high-powered research institutes in the country, where many a Nobel Prize was won, there used to be an associate director whose main job was to pay a daily visit to each scientist's lab and marvel at his or her latest accomplishments—even though he often had little idea what they were. This practice was based on the strong belief that a pat on the back does wonders for creative productivity, and apparently not without cause.

Intrinsic rewards also can help or hinder a talented person's commitment to a domain. There are times when a dull discipline becomes suddenly exciting, or when the reverse happens. Every scientist talks with starry-eyed nostalgia about the glory days of physics in the first third of this century; computer sciences or molecular biology today attract the same enthusiasm from bright young people. Not because these domains promise wealth and fame, but because they are so interesting, so intellectually challenging—and therefore rewarding.

Intrinsic motivation can be easily stifled. Boring schools, insensitive mentoring, rigid work environments, too many pressures and bureaucratic requirements can turn an exciting intellectual adventure into a chore and extinguish the sparks of creativity. Alan Kay, whose inventions were central to the development of personal computers, claims semi-seriously that the firm he worked for lost tens of millions of dollars by refusing to install a $14,000 shower in a corner of his office, because most of his new ideas came while showering. Perhaps the most immediate improvement in the flow of creativity is to make the pursuit of a given domain more intrinsically rewarding. Relatively easy and inexpensive interventions are possible, and the anticipated results could be great.

But many will argue that nothing the field can do will make a difference. A creative person is precisely the one who despite all obstacles prevails. This equation may be true, but its converse is not. There is no evidence that training and reward do not increase creative contributions.

In my view of the situation, if the systems model of creativity is accurate, then it follows that creativity can be enhanced just as much by changing the field—by making it more sensitive and supportive of new ideas—as by producing a greater number of creative individuals. Better training, higher expectations, more accurate recognition, a greater availability of opportunities, and stronger rewards are among the conditions that facilitate the production and the assimilation of potentially useful new ideas.

CONTRIBUTIONS OF THE DOMAIN

It is easy to see how creative contributions might increase if there were more people acting creatively, and it is also relatively easy to comprehend how the field might help in this regard. It is less clear what the role of the domain could be. Does the way information is coded and preserved have anything to do with how easy or difficult it is to make a creative change in a discipline?

The Accessibility of Information

For many centuries European science, and knowledge in general, was recorded in Latin—a language that no one spoke any longer and that had to be learned in schools. Very few individuals, probably less than 1 percent, had the means to study Latin enough to read books in that language and therefore to participate in the intellectual discourse of the times. Moreover, few people had access to books, which were handwritten, scarce, and expensive. The great explosion of scientific creativity in Europe was certainly helped by the sudden spread of information brought about by Gutenberg's use of movable type in printing and by the legitimation of everyday languages, which rapidly replaced Latin as the medium of discourse. In sixteenth-century Europe it became much easier to make a creative contribution not necessarily because more creative individuals were born then than in previous centuries or because social supports became more favorable, but because information became more widely accessible and easier to add to.

This historical example is just one of many that have influenced the rate of creativity at different times. Often intellectual or power elites hide their knowledge on purpose, to keep to themselves the advantages that go with the information. To accomplish this they develop arcane languages, mysterious symbols, and secret codes that are meaningless to those not initiated into the guild. The priestly castes of Mesopotamia and Egypt, the Chinese bureaucrats, the clerical hierarchies of Europe were not particularly interested in sharing their knowledge with all comers. Thus they were not motivated to make the representation of their knowledge transparent.

Some of this desire for exclusive control of knowledge survives. And even those who have the most selfless and democratic views about the information they control often unwittingly make what they know inaccessible by using a language, a style, or a method of exposition that a layperson cannot understand. Sometimes such obscurantism is inevitable, but often it is an unnecessary habit left over from the past, or a shortcut that makes one's thoughts more accessible to the initiated while putting them out of anyone else's reach.

A colleague in the English Department of our university regularly consults with some of the large law firms in the city, whose senior partners retain him to teach young lawyers how to communicate in English instead of lawyerese. It is easy in law school to slip into a technical jargon that stupefies even other lawyers—and cannot be understood at all by those who are not trained in law. The same applies to other domains: Graduate students in psychology are taught to write in the awkward prose of the specialized journals. This helps to make communication within the field faster and clearer—if arguably less rich and evocative. In any case, the speed and clarity thus gained make the information almost inaccessible to those who are not initiated into the language of the domain.

Linguistic obfuscation is only one means by which domains become isolated. The more general problem is that each domain is becoming increasingly specialized not only in its vocabulary but also in the conceptual organization of its rules and procedures. Recently a professor of chemistry sent an article dealing with some of the broader implications of the second law of thermodynamics to a philosophical journal. The editor, in turn, sent it out to two referees for evaluation; both referees thought that the article did not deserve publication. Then the editor, who liked the piece, called up the author to give him the bad news: "I really cannot publish your article, because the two physicists I sent it to for review both advised against it." "You sent my article to two *physicists?*" asked the author in disbelief. "Physicists don't understand thermodynamics. You should ask some chemists for advice." And, in fact, when chemists were asked, the negative opinion was reversed.

The laws of thermodynamics are of course central to both physics and chemistry. Yet the processes denoted by these "laws" appear sufficiently different so that if one looks from the perspective of physics one might derive consequences that are trivial or even wrong from the perspective of chemistry, and vice versa. What makes this breakdown in communication among disciplines so dangerous is that, as we have repeatedly seen, most creative achievements depend on making connections among disparate domains. The more obscure and separate knowledge becomes, the fewer the chances that creativity can reveal itself.

It is also true, however, that some recent technological advances help trends moving in the opposite direction. The availability of personal computers might yet level the field of play as much as the printing press did five centuries ago. When everyone can access immediately scholarly references, unpublished scientific articles, news reports, multimedia presentations of works of art, and personal ideas in progress through information networks, a great variety of new voices might join the specialized discourse of the disciplines. And, presumably, creativity will benefit from it.

The Organization of Knowledge

Whether it is easy or difficult to recognize novelty in a domain depends in large part on how the memes and the rules of the domain are organized. It was easier to reach a consensus on whether a given painting was or was not an improvement on the art of the period when communities shared common criteria of beauty. It is easier to recognize creativity in music when one can compare each new composition to an established canon. Conversely, when aesthetic criteria become fragmented and largely idiosyncratic, as they have become since World War I, it is more difficult to be sure whether a new painting or piece of music deserves to be remembered and passed on to the next generation or whether it is just a novelty to be forgotten as soon as possible.

Similarly, it should be easier to tell whether a new way of doing things is better than the old in mathematics, which is an extremely coherent domain; it would be slightly more difficult in physics, and even more so in biology and economics; it would be most difficult in the other social sciences and philosophy, which are not as tightly connected by an internal network of laws. When the domain is not strictly integrated by logical rules, it is difficult for the field to judge whether novelty is valuable, and thus whether it should be included in the domain. (Of course, the fact that a domain is more integrated does not necessarily mean it is *better*. Chess is a very logical domain, and if anyone were to discover a new opening combination or effective endgame, the discovery would be instantly adopted by players all around the globe. This does not mean that chess is preferable to philosophy just because it is potentially easier to be creative in it.)

Domains wax and wane in their ability to generate novelty. A century ago many scientists believed that there was not much new one could say about physics. Most physicists believed that all they could do was help tidy up a neat Newtonian universe. This, of course, was just before a sequence of new discoveries and perspectives ushered in the most dramatically creative period of physics in the first three decades of the twentieth century; a period after which all the old physics had to be rewritten from a different perspective.

A domain generates novelty only when there is a convergence between an instability within it and the mind of a person who is able to cope with the problem. Therefore, even the most creative persons usually contribute only a few, sometimes only one, great new idea— the one they were prepared for, the one for which the timing was right. Because of the impact of his early papers on relativity, Einstein was expected to keep astonishing the world as long as he lived. But the great convergence between Einstein's mind and the domain of physics was effectively over before he was forty years old, and in the second half of life his contributions made little difference to it.

Sometimes the domain is changed by a new way of thinking, by better measurements, or by new instruments that allow better observations. Usually all of these are involved. The Ptolemaic view of the universe was replaced by the current one in part because Tycho Brahe spent untold hours in his observatory charting the path of stars, in part because Copernicus found an elegant model to represent the movement of the planets, and in part because Galileo improved the telescope enough to be able to see the moons of Jupiter. Whenever a better way of representing reality is found, it opens up new paths of exploration and discovery.

The organization of knowledge is especially important when it comes to passing it down to the next generation. To be creative, a person must first understand the domain. If the knowledge in the domain is nearly incomprehensible, few young people will bother learning it, and thus the chances of creative innovations will be less. But sometimes there are equally valid conflicting claims about how knowledge should be transmitted. The Suzuki method of teaching music results in impressive performance by children, but some claim that its rigidity discourages musical expression and innovation. Anyone who has seen the before-and-after works of children taught by the methods sponsored by the Getty Center for Education in the Arts must marvel at the sudden maturity and professionalism of the drawings; again, however, critics wonder if the fidelity in the transmission of the craft will reduce innovation. Conversely, the many new versions of math taught in U.S. schools claim to emphasize mathematical thinking and understanding at the expense of memorizing rigid rules and focusing on a single way of solving a problem. To more traditional parents and teachers, these efforts only serve to "dumb down" math and further erode our children's comparative standing in this important domain.

Who is right? Which method is more likely to pass on the requisite knowledge? Which is more likely to lead to creative achievements? The likely answer to these questions is to be found in the unglamorous middle ground. To cope well with numbers it is essential to automate as many mental operations as possible—and this requires some memorizing and practicing. On the other hand, to use numbers effectively in real life one must also have a good intuitive grasp of how to approximate, how to round, when and how to use different operations. Perhaps the most important thing to remember in this debate is that there is no single right way to teach a domain and that the way knowledge is transmitted should be appropriate to the skills of the learner. It would be ridiculous to teach math to a four-year-old who has learned calculus on his own—and apparently there are such children—the same way one teaches the rest of the class.

If there is more than one right way to pass on knowledge, there are many more wrong ways of doing it. Whenever the information is untrue, illogical, superficial, redundant, disconnected, confusing, or—especially—dull, the chances of its getting across to students is diminished, and so is the likelihood of a creative response.

Flow and Learning

The origins of culture can easily be explained by necessity. Technology, science, even the arts were defensive adaptations our ancestors discovered to improve their chances of survival, or in order to increase their comfort. While sharks developed stronger teeth and antelopes faster legs, we built weapons and cars. Some birds use colorful plumage or elaborate nests to impress the competition and woo the opposite sex; we display our desirability through fashionable clothes, expensive homes, and refined manners. In this sense, it is perfectly true that necessity is the mother of invention.

These primitive reasons for having a culture are still operative. We are motivated to learn, to become experts, to innovate and strike out in new directions in large part because to do so promises very real material advantages. We no longer compete, as our ancestors did, primarily in terms of physical prowess or simple skills. The ability to run fast, to kill a wolf, or to bring down a buck are of marginal significance. What counts more is the ability to do well in the cultural arena, where the relevant skills are defined by complex domains. And success in a creative cultural endeavor—a Nobel Prize or a best-selling novel—brings with it wealth and respect, admiration and power.

With time, other reasons for creating culture have emerged, and in many ways they are now more important, at least for some people some of the time, than the ancient reasons based on competition and material advantage. Operating within a domain can become rewarding in and of itself. To find the right words for a poem, the secret of a cell's behavior, or a way to make better microchips for less money is an exhilarating experience in its own right, even if no one else knows about it, and no rewards follow. Almost all of our respondents spoke eloquently and spontaneously about the importance of these intrinsic rewards. If they did not feel this joy, external rewards would not have been sufficient to motivate them to extend their efforts into uncharted regions.

But whereas experts in a discipline usually love what they do, this emotion is generally not available to students or young practitioners. Especially in the sciences, beginners see only the drudgery of the discipline. Teachers rarely spend time trying to reveal the beauty and the fun of doing math or science; students learn that these subjects are ruled by grim determinism instead of the freedom and adventure that the experts experience. Not surprisingly, it is difficult to motivate young people to master aspects of the culture that seem cold and alienating. As a result, knowledge in these areas might become eroded and creativity increasingly rare.

So one obvious way to enhance creativity is to bring as much as possible of the flow experience into the various domains. It is exhilarating to build culture—to be an artist, a scientist, a thinker, or a doer. All too often, however, the joy of discovery fails to be communicated to young people, who turn instead to passive entertainment. But consuming culture is never as rewarding as producing it. If it were only possible to transmit the excitement of the people we interviewed to the next generation, there is no doubt that creativity would blossom.

FOURTEEN

ENHANCING PERSONAL CREATIVITY

The major purpose of this book was to describe how creativity works, how culture evolves as domains are transformed by the curiosity and dedication of a few individuals. But another goal was to learn, from the lives of such men and women, how everyone's life could be made more creative. How can our days, too, be filled with wonder and excitement? To answer this question I move from objective description to prescription. I present my own reflections on what we have learned so far and try to derive from it some practical advice. Just as a physician may look at the physical habits of the most healthy individuals to find in them a prescription that will help everyone else to be more healthy, so we may extract some useful ideas from the lives of a few creative persons about how to enrich the lives of everyone else.

You probably already have formed some ideas about how to experience life more creatively. At the very least, you have learned about the obstacles that creative individuals have to surmount and the strategies they use to increase the likelihood that they will accomplish original work. In this chapter I will distill these insights and present them as explicit suggestions for how to apply them to everyday life.

These suggestions hold no promise for great creative achievement. As is clear by now, to move from personal to cultural creativity one needs talent, training, and an enormous dose of good luck. Without access to a domain, and without the support of a field, a person has no chance of recognition. Even though personal creativity may not lead to fame and fortune, it can do something that from the individual's point of view is even more important: make day-to-day experiences more vivid, more enjoyable, more rewarding. When we live creatively, boredom is banished and every moment holds the promise of a fresh discovery. Whether or not these discoveries enrich the world beyond our personal lives, living creatively links us with the process of evolution.

Most of the suggestions derived from the study of creative lives can be implemented by anybody regardless of age, gender, or social condition. Some of the steps, however, are more appropriate to parents or other adults who want to provide optimal conditions for developing the creativity of children. We cannot change conditions in our own childhood that would make us more curious and hence enhance creativity; but we can change conditions for the next generation. Instead of pointing out each time which suggestions are for adults and which for children, I trust the reader's judgment to make the appropriate distinctions.

I am assuming that each person has, *potentially*, all the psychic energy he or she needs to lead a creative life. However, there are four major sets of obstacles that prevent many from expressing this potential. Some of us are exhausted by too many demands, and so have trouble getting hold of and activating our psychic energy in the first place. Or we get easily distracted and have trouble learning how to protect and channel whatever energy we have. The next problem is laziness, or lacking discipline for controlling the flow of energy. And finally, the last obstacle is not knowing what to do with the energy one has. How to avoid these obstacles and liberate the creative energy we all have is what I review in this chapter.

THE ACQUISITION OF CREATIVE ENERGY

With our present knowledge, even an expert neuroanatomist could not tell Einstein's brain from yours or mine. In terms of the capacity for processing information, all brains are extremely alike. The limits on how many bits of information we can process at any given time are also similar. Nor is the speed of information processing noticeably different from one brain to the next. In principle, because of the similarity in cerebral hardware, most people could share the same knowledge and perform mental operations at similar levels. Yet what enormous differences there are in how people think and what they think about!

In terms of using mental energy creatively, perhaps the most fundamental difference between people consists in how much uncommitted attention they have left over to deal with novelty. In too many cases, attention is restricted by external necessity. We cannot expect a man who works two jobs, or a working woman with children, to have much mental energy left over to learn a domain, let alone innovate in it. Einstein is supposed to have written his classic papers on the kitchen table of his small apartment in Berne, while rocking the pram of his baby. But the fact is that there are real limits to how many things a person can attend to at the same time, and when survival needs require all of one's attention, none is left over for being creative.

But often the obstacles are internal. In a person concerned with protecting his or her self, practically all the attention is invested in monitoring threats to the ego. This defensiveness may have very understandable causes: Children who have been abused or who have experienced chronic hunger or discrimination are less likely to be curious and interested in novelty for its own sake, because they need all the psychic energy they have simply to survive. Taken to the extreme, a sense of being too vulnerable results in the form of neurosis known as paranoia, where everything that happens is interpreted as a threatening conspiracy against the self. A paranoid tendency is one obstacle to the free deployment of mental energy. The person who suffers from it usually cannot afford to become interested in the world from an objective, impartial viewpoint, and therefore is unable to learn much that is new.

Another limitation on the free use of mental energy is an excessive investment of attention in selfish goals. Of course, we all must first and foremost take care of our own needs. But for some people the concept of "need" is inflated to the point that it becomes an obsession that devours every waking moment. When everything a person sees, thinks, or does must serve self-interest, there is no attention left over to learn about anything else.

It is difficult to approach the world creatively when one is hungry or shivering from cold, because then all of one's mental energy is focused on securing the necessities one lacks. And it is equally difficult when a person is rich and famous but devotes all of his or her energies to getting more money and fame. To free up creative energy we need to let go and divert some attention from the pursuit of the predictable goals that genes and memes have programmed in our minds and use it instead to explore the world around us on its own terms.

Curiosity and Interest

So the first step toward a more creative life is the cultivation of curiosity and interest, that is, the allocation of attention to things for their own sake. On this score, children tend to have the advantage over adults; their curiosity is like a constant beam that highlights and invests with interest anything within range. The object need not be useful, attractive, or precious; as long as it is mysterious it is worthy of attention. With age most of us lose the sense of wonder, the feeling of awe in confronting the majesty and variety of the world. Yet without awe life becomes routine. Creative individuals are childlike in that their curiosity remains fresh even at ninety years of age; they delight in the strange and the unknown. And because there is no end to the unknown, their delight also is endless.

At first, curiosity is diffuse and generic. The child's attention is attracted to any novelty—cloud or bug, grandfather's cough or a rusted nail. With time, interest usually becomes channeled into a specific domain. A ninety-year-old physicist may retain childhood curiosity in the realm of subatomic particles but is unlikely to have enough free attention left over to marvel at much else. Therefore, creativity within a domain often goes hand in hand with conformity in the rest of life. Einstein at the peak of his breakthroughs in physics played traditional music on his violin. But narrowing attention to a single domain does not mean limiting the novelty one is able to process; on the contrary, complex domains like poetry, history, physics, or politics reveal constantly expanding perspectives to those who venture to explore them.

So how can interest and curiosity be cultivated, assuming that you feel the desire to do so? Some specific advice may help.

Try to be surprised by something every day. It could be something you see, hear, or read about. Stop to look at the unusual car parked at the curb, taste the new item on the cafeteria menu, actually listen to your colleague at the office. How is this different from other similar cars, dishes, or conversations? What is its *essence?* Don't assume that you already know what these things are all about, or that even if you knew them, they wouldn't matter anyway. Experience this one thing for what it is, not what you think it is. Be open to what the world is telling you. Life is nothing more than a stream of experiences—the more widely and deeply you swim in it, the richer your life will be.

Try to surprise at least one person every day. Instead of being your predictable self, say something unexpected, express an opinion that you have not dared to reveal, ask a question you wouldn't ordinarily ask. Or break the routine of your activities: Invite a person to go with you to a show, a restaurant, or a museum that you never visited before. Experiment with your appearance. Comfortable routines are great when they save energy for doing what you really care about; but if you are still searching, they restrict and limit the future.

Write down each day what surprised you and how you surprised others. Most creative people keep a diary, or notes, or lab records to make their experiences more concrete and enduring. If you don't do so already, it might help to start with a very specific task: to record each evening the most surprising event that happened that day and your most surprising action. This is a simple enough assignment and one you will find is fun to do. After a few days, you can reread what you have written and reflect on those past experiences. One of the surest ways to enrich life is to make experiences less fleeting, so that the most memorable, interesting, and important events are not lost forever a few hours after they occurred. Writing them down so that you can relive them in recollection is one way to keep them from disappearing. And after a few weeks, you may begin to see a pattern of interest emerging in the notes, one that may indicate some domain that would repay exploring in depth.

When something strikes a spark of interest, follow it. Usually, when something captures our attention—an idea, a song, a flower—the impression is brief. We are too busy to explore the idea, song, or flower further. Or we feel that it is none of our business. After all, we are not thinkers, singers, or botanists, so these things lie outside our grasp. Of course, that's nonsense. The world *is* our business, and we can't know which part of it is best suited to our selves, to our potentialities, unless we make a serious effort to learn about as many aspects of it as possible.

If you take time to reflect on how best to implement these four suggestions, and then actually start putting them into effect, you should feel a stirring of possibilities under the accustomed surface of daily experiences. It is the gathering of creative energy, the rebirth of curiosity that has been atrophied since childhood.

Cultivating Flow in Everyday Life

The rebirth of curiosity doesn't last long, however, unless we learn to enjoy being curious. Entropy, the force behind the famous Second Law of Thermodynamics, applies not only to physical systems but to the functioning of the mind as well. When there is nothing specific to do, our thoughts soon return to the most predictable state, which is randomness or confusion. We pay attention and concentrate when we must—when dressing, driving the car, staying awake at work. But when there is no external force demanding that we concentrate, the mind begins to lose focus. It falls to the lowest energetic state, where the least amount of effort is required. When this happens, a sort of mental chaos takes over. Unpleasant thoughts flash into awareness, forgotten regrets resurface, and we become depressed. Then we turn on the TV set, read listlessly the advertising supplement of the newspaper, have pointless conversations—anything to keep our thoughts on an even keel and avoid becoming frightened by what is happening in the mind. Taking refuge in passive entertainment keeps chaos temporarily at bay, but the attention it absorbs gets wasted. On the other hand, when we learn to enjoy using our latent creative energy so that it generates its own internal force to keep concentration focused, we not only avoid depression but also increase the complexity of our capacities to relate to the world.

How can we do this? How can we relearn to enjoy curiosity so that the pursuit of new experiences and new knowledge becomes self-sustaining?

Wake up in the morning with a specific goal to look forward to. Creative individuals don't have to be dragged out of bed; they are eager to start the day. This is not because they are cheerful, enthusiastic types. Nor do they necessarily have something exciting to do. But they believe that there is something meaningful to accomplish each day, and they can't wait to get started on it.

Most of us don't feel our actions are that meaningful. Yet everyone can discover at least one thing every day that is worth waking up for. It could be meeting a certain person, shopping for a special item, potting a plant, cleaning the office desk, writing a letter, trying on a new dress. It is easier if each night before falling asleep, you review the next day and choose a particular task that, compared to the rest of the day, should be relatively interesting and exciting. Then next morning, open your eyes and visualize the chosen event—play it out briefly in your mind, like an inner videotape, until you can hardly wait to get dressed and get going. It does not matter if at first the goals are trivial and not that interesting. The important thing is to take the easy first steps until you master the habit, and then slowly work up to more complex goals. Eventually most of the day should consist of tasks you look forward to, until you feel that getting up in the morning is a privilege, not a chore.

If you do anything well, it becomes enjoyable. Whether writing a poem or cleaning the house, running a scientific experiment or a race, the quality of experience tends to improve in proportion to the effort invested in it. The runner may be exhausted and aching, yet she also is exhilarated if she is putting all her strength into the race. The more activities that we do with excellence and style, the more of life becomes intrinsically rewarding.

The conditions that make flow possible suggest how to transform everyday activities so that they are more enjoyable. Having clear goals and expectations for whatever we do, paying attention to the consequences of our actions, adjusting skills to the opportunities for action in the environment, concentrating on the task at hand without distractions—these are the simple rules that can make the difference between an unpleasant and an enjoyable experience. If I decide to learn to play the piano or speak a foreign language but feel frustrated or bored doing so, the chances are that I will give up at the first opportunity. But if I apply the flow conditions to the learning task, then it is likely that I will continue to expand my creative potential, because doing so is fun.

It is easier to start with the most mundane activities all of us have to take care of. How can you get more enjoyment from brushing your teeth? Taking a shower? Dressing? Eating breakfast? Getting to work? Take the simplest of these routines and experiment with engineering its flow potential. How do you apply flow conditions to loading the dishwasher? If you take this question seriously and try to answer by testing various alternatives, you will be surprised at how much fun brushing teeth can be. It will never be as enjoyable as skiing or playing in a string quartet, but it might beat watching most television programs.

After you have practiced improving the quality of experience in a few everyday activities, you might feel ready to tackle something more difficult—such as a hobby or a new interest. Eventually you will master the most important skill of all, the metaskill that consists in being able to turn any activity into an occasion of flow. If the autotelic metaskill is developed enough, you should be able to enjoy any new challenge and be on the way to the self-sustaining chain reaction of creativity.

To keep enjoying something, you need to increase its complexity. As Herodotus remarked, we cannot step in the same river twice. Nor can you enjoy the same activity over and over, unless you discover new challenges, new opportunities in it. Otherwise it becomes boring. Brushing teeth cannot stay enjoyable for very long—it's an activity that just does not have enough potential for complexity. True, one can preserve the challenge of even the simplest activity by combining it with something else—for instance, while brushing teeth you might plan the coming day or reflect on what happened yesterday. But generally it is more satisfying to become involved in activities that are inexhaustible—music, poetry, carpentry, computers, gardening, philosophy, or deep personal relationships.

Most domains are so complex that they cannot be exhausted in a lifetime, not even the lifetime of the human race. It is always possible to learn a new song, or to write one. It is always possible to find a better way to do anything. That is why creativity—the attempt to expand the boundaries of a domain—makes a lifetime of enjoyment possible.

Habits of Strength

After creative energy is awakened, it is necessary to protect it. We must erect barriers against distractions, dig channels so that energy can flow more freely, find ways to escape outside temptations and interruptions. If we do not, entropy is sure to break down the concentration that the pursuit of an interest requires. Then thought returns to its baseline state—the vague, unfocused, constantly distracted condition of the normal mind.

It is often surprising to hear extremely successful, productive people claim that they are basically lazy. Yet the claim is believable. It is not that they have more energy and discipline than you or I; but they do develop habits of discipline that allow them to accomplish seemingly impossible tasks. These habits are often so trivial that the people who practice them seem strange and obsessive. At first many people were mildly shocked that the great Alfred Einstein always wore the same old sweater and baggy trousers. Why was he being so weird? Of course, Einstein wasn't trying to upset anybody. He was just cutting down on the daily effort involved in deciding what clothes to wear, so that his mind could focus on matters that to him were more important. It may seem that choosing slacks and shirts takes so little time that it is pretentious to worry about it. But suppose it takes only two minutes each day to decide how to dress. That adds up to 730 minutes, or twelve hours a year. Now think of the other repetitive things we have to do throughout the day—comb hair, drive cars, eat, and so on. And then think not only of the time it takes to do each of these things but of the interruption in the train of thought they cause, both before and after. Having to choose a tie could derail a whole hour's worth of reflection! No wonder Einstein preferred to play it safe and wear the same old clothes.

At this point, some readers may smell a contradiction. On the one hand I am saying that to be creative you should be open to experience, focus on even the most mundane tasks—like brushing teeth—to make them more efficient and artistic. On the other hand I am saying that you should conserve creative energy by routinizing as much of everyday life as possible so that you can focus entirely on what really matters. Isn't this contradictory advice? Not really—but even if it were, you should by now expect a certain amount of paradox in creative behavior.

The reason it is not a contradiction to be open and focused at the same time is that these contrary ways of using psychic energy share a similarity that is more important than their differences. They require *you* to decide whether at this point it is better to be open or to be focused. They are both expressions of your ability to control attention, and it is this, not whether you are open or focused, that matters. Before you have discovered an overriding interest in a particular domain, it makes sense to be open to as much of the world as possible. After you have developed an abiding interest, however, it may make more sense to save as much energy as you can to invest in that one domain. In either case, what is important is not to relinquish control over creative energy so that it dissipates without direction.

A few more words may be needed here concerning the concept of "control" as applied to attention. It should be realized that one way of controlling is to relinquish control. People who meditate expand their being by letting go of focused thought. This way they aim to achieve a spiritual union with the energy behind the world of appearances, the force that drives the universe. But this way of giving up control is itself directed, controlled by the mind. It is very different from just sitting and gossiping, passively consuming entertainment, or letting the mind wander without purpose.

What can you do to build up habits that will make it possible to control attention so that it can be open and receptive, or focused and directed, depending on what your overall goals require?

Take charge of your schedule. Our circadian rhythms are to a large extent controlled by outside factors: the rising sun, the commuter train schedule, a job's deadline, lunchtime, a client's needs. If it works for you, it makes perfect sense to abandon yourself to these markers so that you don't have to decide what to do when. But it is also possible that the schedule you are following is not the best for your purposes. The best time for using your creative energies could be early in the morning or late at night. Can you carve out some time for yourself when your energy is most efficient? Can you fit sleep to your purpose, instead of the other way around?

The times when most people eat may not be the best for you. You might get hungry earlier than lunchtime and lose concentration because you feel jittery; or to perform at the top of your potential it may be best to skip lunch and have a midafternoon snack instead. There are probably best times to shop, to visit, to work, to relax for each one of us; the more we do things at the most suitable times, the more creative energy we can free up.

Most of us have never had the chance to discover which parts of the day or night are most suited to our rhythms. To regain this knowledge we have to pay attention to how well the schedule we follow fits our inner states—when we feel best eating, sleeping, working, and so forth. Once we have identified the ideal patterns, we can begin the task of changing things around so that we can do things when it is most suitable. Of course, most of us have inflexible demands on our day that cannot be changed. Even John Reed has to keep to an office schedule, and Vera Rubin has to adapt her curiosity to times when telescopes are available for observation. The needs of children, spouses, and bosses must often take precedence. Yet time is more flexible than most of us think.

The important thing to remember is that creative energy, like any other form of psychic energy, only works over time. It takes a certain minimum amount of time to write a sonnet or to invent a new machine. People vary in the speed they work—Mozart wrote concerti much faster than Beethoven did—but even Mozart could not escape the tyranny of time. Therefore, every hour saved from drudgery and routine is an hour added to creativity.

Make time for reflection and relaxation. Many people, especially those who are successful and responsible, take the image of the "rat race" seriously and feel uncomfortable, even anxious, if they are not busily at work. Even at home, they feel they must be always cleaning, working in the yard, or fixing things. Keeping constantly busy is commendable and certainly much better than just lounging around feeling sorry for yourself. But constant busyness is not a good prescription for creativity. It is important to schedule times in the day, the week, and the year just to take stock of your life and review what you have accomplished and what remains to be done.

These are times when you should not expect any task to be done, any decision to be reached. You should just indulge in the luxury of reflection for its own sake. Whether you intend it or not, new ideas and conclusions will emerge in your consciousness anyway—and the less you try to direct the process the more creative they are likely to be. It may be best to combine these periods of reflection with some other task that requires a certain amount of attention, but not all of it. Preferably this should involve some physical or kinesthetic component. Typical activities that facilitate subconscious creative processes are walking, showering, swimming, driving, gardening, weaving, and carpentry.

Neither constant stress nor monotony is a very good context for creativity. You should alternate stress with periods of relaxation. But remember that the best relaxation is not doing nothing. It usually involves doing something very different from your usual tasks. Some of the most demanding activities like rock climbing, skiing, or skydiving are relaxing to people who have desk jobs because they provide opportunities for deep involvement with experiences that are completely different from the usual.

Learning to control one's sleep patterns can also be very important. Some very effective businesspeople and politicians pride themselves on sleeping very few hours each night, and they claim that short sleep makes them feel more energetic and decisive. But creative individuals usually sleep longer and claim that if they cut down on sleeptime the originality of their ideas suffers. It is impossible to come up with a single amount that is ideal for everyone. Like everything else, the important thing is to find the length of time that best fits your own requirements. And don't feel guilty if you sleep a few hours more than is considered normal. What you lose in waking time will probably be made up in terms of the quality of experience while you are awake.

Shape your space. We saw in chapter 6 that surroundings can have an influence on the creative process. Again, it is not what the environment is like that matters, but the extent to which you are in harmony with it.

At the macro level, the question may be whether you feel you would be happiest if living at the seashore, surrounded by mountains or plains, or in the bustle of a big city. Do you like the change of seasons? Do you hate snow? Some people are physically affected by long sunless periods. There can be many reasons why you might feel trapped in the place you live, without a choice to move. But it's a great waste to spend your entire life in uncongenial surroundings. One of the first steps in implementing creativity at the personal level is to review your options of life contexts and then start thinking about strategies for making the best choice come true.

At the midlevel, determine what sort of community you want to sink roots in. Every city and rural area have neighborhoods that are stratified in terms of price, centrality, type of activities available, and so forth. As with everything else, we are limited in the choice of exactly where to live—very few can afford to live near the Pebble Beach golf course in California, the Lionshead ski run in Vail, Colorado, or on Park Avenue in Manhattan. But even as redundant housing developments are slowly swallowing the landscape, there is still much choice left of where one can live, more than most of us care to exercise. And it is important to live in a place that does not use up a lot of potential energy either by lulling the senses into complacency or by forcing us to fight against an intolerable environment.

At the micro level the choices are much more readily available to everyone. We all can decide what kind of environment to create in our home. As long as there is a roof overhead even the poorest among us can organize space and collect things that are meaningful and conducive to the use of creative energies.

The house of a Hindu Brahman or a traditional Japanese family is likely to be bare of almost all furniture and decoration. The idea is to provide a neutral environment that does not disturb the flow of consciousness with distractions. At the other extreme, a Victorian home bursts with dark, heavy furniture and knickknacks. In this case, the owner's sense of control is bolstered by lavish possessions. Which is the best way to go? Obviously none of these environments is better in an absolute sense. What counts is which solution allows you to use attention most effectively. It is easy to find out which microenvironment best fits your self: Try different kinds and pay attention to your feelings and reactions.

Another way space can help creativity is by following the maxim "A place for everything, and everything in its place." Developing a routine for storing such things as car keys and eyeglasses repays itself more than a hundredfold in time saved. If you know your home and office so well that you can find anything even if blindfolded, your train of thought need not be continuously interrupted to look for something. This doesn't mean that your desk or living room should always be neat. In fact, the work space of creative individuals is often messy and it tends to put off more ordered souls. The important thing is that they know where everything is, so they can work without too much distraction. Many can find papers and organize their work better when their desk is covered with clutter than when things are properly filed away. But if a clean desk makes you feel and work better, then by all means keep it clean.

The kind of objects you fill your space with also either help or hinder the allocation of creative energies. Cherished objects remind us of our goals, make us feel more confident, and focus our attention. Trophies, diplomas, favorite books, and family pictures on the office desk are all reminders of who you are, what you have accomplished, and therefore what you are likely to achieve. Pictures and maps of places you would like to visit and books about things you might like to learn more about are signposts of what you might do in the future.

And then there are the objects that we carry and that help create a personalized, portable psychic space. In most traditional societies, people always took along a few special objects that were supposed to increase the power of the owner. This "medicine bundle," or talisman, might include the claws of a bear killed in the hunt, some clamshells found on the beach, or some herbs that revived the wearer from a difficult illness. Having these objects hanging from one's neck provided a feeling of strength and identity. We also tend to keep in our purses and wallets items that represent our self and its values. Pictures of our children, friends' addresses, a book or movie title scribbled on a napkin—these all remind us of who we are and what we like. Choosing carefully what to carry with us makes it easier to be comfortable with ourselves and therefore to use psychic energy effectively when the opportunity arises.

Another space that is important to personalize is our car. Cars have become important extensions of the self; for many people, the car is more like a castle than the home is. It's in the car that they feel most free, most secure, most powerful. It is where they can think with the greatest concentration, solve problems most efficiently, and come up with the most creative ideas. That is why it is so difficult to get people to use public transportation instead of their cars. Of course, it is possible that in the near future cars will be as obsolete as eating beef, and for the same reason—shortage of fuel. In the meantime, however, it makes sense to learn to use one's vehicle in the way that is most conducive to the environment as well as to the expression of creative potential.

Find out what you like and what you hate about life. It is astonishing how little most people know about their feelings. There are people who can't even tell if they are ever happy, and if they are, when or where. Their lives pass by as a featureless stream of experience, a string of events barely perceived in a fog of indifference. As opposed to this state of chronic apathy, creative individuals are in very close touch with their emotions. They always know the reason for what they are doing, and they are very sensitive to pain, to boredom, to joy, to interest, and to other emotions. They are very quick to pack up and leave if they are bored and to get involved if they are interested. And because they have practiced this skill for a long time, they need to invest no psychic energy in self-monitoring; they are aware of their inner states without having to become self-conscious.

How can you learn the dynamics of your emotions? The first thing is to keep a careful record of what you did each day and how you felt about it. This is what the Experience Sampling Method accomplishes—pagers are programmed to signal you at random times during the day, and then you fill out a short questionnaire. It is possible, after a week, to have a good idea of how you spend your time and how you feel about various activities. But you don't need an elaborate experiment to find out how you feel. Be creative and invent your own method of self-analysis. The basis of ancient Greek philosophy was the injunction to *know thyself.* The first step toward self-knowledge involves having a clear idea of what you spend your life doing and how you feel while doing it.

Start doing more of what you love, less of what you hate. After a few weeks of self-monitoring, sit down with your diary or your notes and begin to analyze them. Again, it takes some creativity, but it should not be very difficult to draw out the main patterns of daily life. It's not more convoluted than planning comparison shopping or studying stock market graphs. And it is so much more important in the long run.

You may find that, contrary to what you had thought, the few times you were with your spouse during the week you had great conversations and felt relaxed. That at work, despite stress and hassles, you felt better about yourself than when watching television. Or, conversely, that most of the time when you were at work you felt listless and bored. Why were you so irritated with your children? So impatient with the people you work with? So cheerful when walking down the street?

You may never find out the deep reasons that answer these questions. Perhaps there *are* no deep reasons. The point is that once you know what your daily life is like and how you experience it, it is easier to begin getting control over it. Perhaps the pattern of feelings shows that you should change your job—or learn to bring more flow to it. Or that you should be outdoors more often, or find ways to do some more interesting things with your children. The important thing is to make sure that you spend your psychic energy in such a way that it brings back the highest returns in terms of the quality of experience.

The only way to stay creative is to oppose the wear and tear of existence with techniques that organize time, space, and activity to your advantage. It means developing schedules to protect your time and avoid distraction, arranging your surroundings to heighten concentration, cutting out meaningless chores that soak up psychic energy, and devoting the energy thus saved to what you really care about. It is much easier to be personally creative when you maximize optimal experiences in everyday life.

INTERNAL TRAITS

The next step, after learning to liberate the creative energy of wonder and awe, and then learning to protect it by managing time, space, and activity, is to internalize as many of these supporting structures into your personality as possible. We can think of personality as a habitual way of thinking, feeling, and acting, as the more or less unique pattern by which we use psychic energy or attention. Some traits are more likely than others to result in personal creativity. Is it possible to reshape personality to make it more creative?

It is difficult for adults to change personalities. Some of the habits that form personality are based on temperament, or the particular genetic inheritance that makes one person very shy, or aggressive, or distractible. Temperament then interacts with social environment—

parents, family, friends, teachers—and some habits are strengthened, others weakened or repressed. By the time we are out of our teens, many of these habits are strongly set, and it is difficult to invest attention—to think, feel, or act—in any other way than what our traits allow.

It is difficult, but it is not impossible. Strangely, in our culture we spend billions of dollars trying to improve our looks, but we take a fatalistic attitude toward our personal traits—as if it was beyond our abilities to change them. If all the energy expended on dieting, cosmetics, and dressing up were turned to other uses, we could easily solve the material problems of the world. Yet most of that energy is wasted because how we look, or how much we weigh, is more difficult to change because it is more dependent on genetic instructions than are personality traits. And, of course, improving who we are is a great deal more important than improving how we look.

To change personality means to learn new patterns of attention. To look at different things, and to look at them differently; to learn to think new thoughts, have new feelings about what we experience. John Gardner was by temperament extremely introverted. He was shy and retiring, undemonstrative and unemotional. This worked well for him up to a certain point, but when in his forties he became a foundation officer he realized that he was intimidating the applicants who were coming to ask him for support. As they described their projects, they were hoping to get some reaction, some signal from him, and all they got was noncommittal silence.

At that point he decided to become more extroverted. He forced himself to smile, to make small talk, to show some vulnerability in conversation. It wasn't easy to change these deeply ingrained habits, but every little success made him a much more effective leader and communicator—the domains in which his creativity eventually asserted itself most strongly. He never became an out-and-out extrovert, but he now impresses an interlocutor as warm and caring—which was always potentially a part of his personality, but he had been unable to show it.

If we go through life with habits that are very rigid, or inappropriate to the kind of job we do, the creative energy gets dammed up or wasted. Thus it helps to consider how to apply what we learned about the personalities of creative individuals to the traits that may be useful in everyday life.

Develop what you lack. All of us end up specializing in some traits, which usually means that we neglect traits that are complementary to the ones we developed. For example, if someone learns to be very competitive, he or she probably has a hard time cooperating; an intuitive, subjective person usually ends up mistrusting objectivity. Even though Aristotle figured out twenty-five centuries ago that virtue consists in the golden mean between such opposite traits as courage and prudence, we still take the easy way out, which is to be one-dimensional.

As we know, creative individuals tend to be exceptions to this rule. In chapter 3 I presented the ten main dialectic poles that describe their personality. The point here is that everyone can strengthen the missing end of the polarity. When an extrovert learns to experience the world like an introvert, or vice versa, it is as if he or she discovered a whole missing dimension to the world. The same happens if a very feminine person learns to act in what we consider a masculine manner. Or if an objective, analytic person decides to trust intuition for a change. In all of these cases, a new realm of experience opens up in front of us, which means that in effect we double and then double again the content of life.

To start, it makes sense to identify your most obvious characteristic, the one that your friends would use to describe you—such as "reckless" or "stingy" or "intellectual." If you don't trust your own assessment, you can ask a friend to help. When you have identified a central trait, you can begin to try its opposite. If you are basically reckless, take a future project, or relationship, and instead of rushing into it plan your moves carefully and patiently. If you are stingy, splurge. If you are an intellectual, get someone to explain to you why football is such a great sport and try watching a ball game in light of this knowledge. Keep exploring what it takes to be the opposite of who you are.

At first it won't be easy and will seem like a waste of time. Why try to save money when you enjoy being spendthrift? Why trust intuition when you are so comfortable being a rational person? Breaking habits is a little like breaking your own bones. What should keep you trying is the knowledge that by experiencing the world from a very different perspective, you will enrich your life considerably.

Shift often from openness to closure. Perhaps the most important duality that creative persons are able to integrate is being open and receptive on the one hand, and focused and hard-driving on the other. Good scientists, like good artists, must let their minds roam playfully or they will not discover new facts, new patterns, new relationships. At the same time, they must also be able to evaluate critically every novelty they encounter, forget immediately the spurious ones, and then concentrate their minds on developing and realizing the few that are promising.

Because this is such a central trait, it is particularly important to practice it. Take some task you often do at your job—for instance, writing a weekly report on a project you are involved in. Start with relaxing your mind; look out the window if you can, or let your eyes roam unfocused over the desk and the office. Now try to grasp what are the most important issues about the project. Grasp not only intellectually but also at a gut level, emotionally. What's *really* important? What gives you a good feeling about it? What scares you? Or try to get images in your mind, like scenes in a film. Picture the people involved in the project. What are they doing? What are they saying to each other?

Then start jotting down some words on a pad, or on the computer. Any word that comes to mind concerning your feelings about the project or the movie in your mind. Words that describe facts, or events, or persons. When you have a few words down, see if you can string them together into a story—it should not be too difficult. The story you glimpse at this stage represents your strongest feelings about what is happening on the project.

It is at this point that the emphasis might shift from openness to discipline. Begin to choose words carefully, keeping in mind the goals of your department, division, or the corporation as a whole, as well as the interests, tastes, and prejudices of the bosses who will read the report. You want to be effective and convincing. So muster all your skills to write a report that conveys your beliefs as clearly and succinctly as possible. If you manage to be intuitively receptive at the beginning, and rationally critical later on, the report will be considerably more creative than if you relied exclusively on one of these strategies.

Shifting from one of these poles to the other is important also in relationships—between friends, spouses, or parents and children. For a relationship to work, it is essential to listen to the other person, to try to imagine why she says what she says, what she feels, how she sees the world. It is essential to change perspectives when necessary, to compromise, to understand the world and to act differently, because this is what the other person's reality requires. Yet it is just as important to remain in touch with our own beliefs and perspectives. In a relationship we should be able to shift moment by moment from our own viewpoint to that of the other. We can see depth only because looking with two eyes gives us slightly different perspectives. How much deeper can we see when instead of two eyes we rely on four! This dual vision again doubles the riches of the world we experience and makes it possible to react creatively to it.

Aim for complexity. The ability to move from one trait to its opposite is part of the more general condition of psychic complexity. Complexity is a feature of every system, from the simplest amoeba to the most sophisticated human culture. When we say that something is complex we mean that it is a very differentiated system—it has many distinctive parts—and also that it is a very integrated system—the several parts work together smoothly. A system that is differentiated but not integrated is complicated but not complex—it will be chaotic and confusing. A system that is integrated but not differentiated is rigid and redundant but not complex. Evolution appears to favor organisms that are complex; that is, differentiated and integrated at the same time.

Complexity also is a feature of human personality. Some people are integrated but not very differentiated: They hold on to a few ideas, opinions, or feelings. They are predictable. They come across as boring, one-dimensional, rigid. There are others who express many opinions, who are changeable and constantly striving to accomplish something new and different, but who give the impression that they have no center, no continuity, no ruling passion. They have a differentiated consciousness that is not well integrated. Neither of these ways of being is very satisfying.

As we have seen, creative individuals seem to have relatively complex personalities. Neither the centrifugal nor the centripetal force prevails—they are able to keep in balance the contrary tendencies that make some people turn inward until each becomes a hard shell,

and others fly outward at random. A creative person is highly indi-
vidualized. She follows her own star and creates her own career. At
the same time, she is deeply steeped in the traditions of the culture;
she learns and respects the rules of the domain and is responsive to
the opinions of the field—as long as those opinions do not conflict
with personal experience. Complexity is the result of the fruitful
interaction between these two opposing tendencies.

But psychological complexity is not just a luxury reserved for creative individuals. Every person who wants to realize fully the poten-
tiality of what it is to be human, and who wants to take part in the
evolution of consciousness, can aim for a more complex personality.
To do so we need to explore and strengthen those traits that are now
lacking, to learn to shift from openness to discipline, within a context of curiosity and awe for the miracle of life. The notion of complexity adds a deeper layer of understanding of why it is important to
achieve this. By fully expressing the tendencies of which we are
capable, we become part of the energy that creates the future.

THE APPLICATION OF CREATIVE ENERGY

Up to now I have said nothing about the role of thinking in personal
creativity. The reason is that if motivations, habits, and personality
traits are in place, most of the job is done. It is inevitable that one's
creative energies will start to flow more freely. Nevertheless, it is also
useful to consider what kind of mental operations expedite novel
solutions to problems in the domain of daily life.

Problem Finding
Creative people are constantly surprised. They don't assume that they
understand what is happening around them, and they don't assume
that anybody else does either. They question the obvious—not out
of contrariness but because they see the shortcomings of accepted
explanations before the rest of us do. They sense problems before
they are generally perceived and are able to define what they are.

The reason we consider the artists of the Renaissance so creative is
that they were able to express the emancipation of the human spirit
from the shackles of religious tradition before the humanist scholars
or anyone else did. The use of perspective in painting broke down
the flat hierarchical order of Byzantine composition. The introduc-

tion of expression, movement, and everyday subject matter into pictorial art lifted human experience to the level of importance previously occupied by static representations of religious ideas. Without expressly intending to, without a clear understanding of the consequences of their actions, the Renaissance artists changed our perspective on the world.

The creativity of artists in this century also consisted in formulating a new visual perspective on the human condition, albeit a much more pessimistic one this time. The experiments with cubism, abstraction, and expressionism in the visual arts, in music, and in literature were precursors of relativism in the social sciences and deconstructionism in philosophy. They expressed in visible form the problems of our age: The lack of a common set of values, the suspicion of ultimate beliefs, the loss of faith in progress brought about by two world wars and their horrors—these were prefigured in the distorted, anguished, and random representations that populate modern art.

If you learn to be creative in everyday life you may not change how future generations will see the world, but you will change the way *you* experience it. Problem finding is important in the daily domain because it helps us focus on issues that will affect our experiences but otherwise may go unnoticed. To practice this skill you might try the following suggestions.

Find a way to express what moves you. Creative problems generally emerge from areas of life that are personally important. We have seen that many individuals who later changed a domain were orphaned as children. The loss of a parent has a huge impact on a young person's life. But what, exactly, is this impact? Does the sadness include a feeling of relief? Of heightened responsibility? Of freedom? Of increased closeness to the surviving parent? Unless one finds words, ideas, or perhaps visual and musical analogies to represent the impact of the loss on one's experience, it is likely that the parent's death will cause violent pain at first, a generalized depression later, and with time its effects will disappear or work themselves out unconsciously, outside the range of rational control.

Other problematic issues in early life include poverty, illness, abuse, loneliness, marginality, and parental neglect. Later in life the main reasons for unease may involve your job, your spouse, or the state of the community or of the planet. Lesser concerns may derive from a temporary threat: the scowl of a boss, the illness of a child, the change in the value of your stock portfolio. Each of these is likely to interfere with the quality of life. But you will not know what ails you unless you can attach a name to it. The first step in solving a problem is to find it, to formulate the vague unease into a concrete problem amenable to solution.

Look at problems from as many viewpoints as possible. When you know that you have a problem, consider it from many different perspectives. How you define a problem usually carries with it an explanation of what caused it. Our first impulse is to label problems by relying on tried-and-true prejudices. If we have a disagreement with our spouse we immediately assume that we are innocent and the fault is with the other party. This may be true some of the time but certainly not always. The most realistic assumption is that both parties are at fault, and the question is to understand what motivated each partner to take his or her position in the argument.

Also, although the argument may be ostensibly about one thing, for instance, money, don't assume that appearances are true. The disagreement really may be about financial decision making and hence about power; or it might be about lack of respect or about inequality in the amount of psychic energy invested in the relationship. How you identify the nature of the problem is critical for the kind of solution that will eventually work.

Creative individuals do not rush to define the nature of problems; they look at the situation from various angles first and leave the formulation undetermined for a long time. They consider different causes and reasons. They test their hunches about what really is going on, first in their own mind and then in reality. They try tentative solutions and check their success—and they are open to reformulating the problem if the evidence suggests they started out on the wrong path.

A good way to learn problem finding in everyday life is to stop yourself when you sense you have a problem and give it the best shot at a formulation. If someone has been promoted ahead of you, you might define the problem as "This happened because the boss dislikes me." As soon as you do this, *reverse the formulation*: "It happened because I dislike the boss." Does this way of looking at the problem make sense? Could it be at least partly true? And then immediately consider a few more alternatives: "It happened because I haven't kept up with the changing job as much as I should have" or "Lately I have been too distraught by what happens at home, and it affected my performance." Which formulation comes closest to representing the problem? Perhaps each is true to a certain extent, and your colleague's promotion was overdetermined by several unrelated causes.

It is possible that you eventually decide that the fact that you didn't get promoted is no problem at all. Being passed over may give you more time to spend at home, to learn something new, to devote your psychic energy to some other task. You may come to realize that the problem was your competitiveness, your ambition, the fact that you invested all your energies in advancing on the job rather than doing a good job for its own sake, or living more fully. So the failed promotion, instead of being the problem, is really the first step to the solution of a more fundamental problem.

Perhaps none of these formulations is "right" in the sense that it identifies correctly the causes of the event. Nevertheless, it is extremely important to identify the nature of the problem, because what you will do next depends on it. By naming the problem and attributing a cause to it you will shape not only the past but, more important, the future. It is in this sense that the lives of creative individuals are less determined than most of our lives. Because they pause to consider a greater range of possible explanations for what happens to them, they have a wider and less predictable range of options to choose from.

Figure out the implications of the problem. Once you have created a formulation, you can begin to entertain possible solutions. Of course, solutions even to a simple problem like "Joe was promoted ahead of me" vary incredibly depending on how you formulated it and therefore what causes you attributed to it. Solutions might include finding interests outside the job, or learning to understand and to like the boss, or catching up on job skills—or a little of each.

At this stage, too, it pays to consider a variety of solutions, to entertain different possibilities. Creative individuals experiment with a number of alternative solutions until they are certain that they have found the one that will work best. Again, as soon as you think of a good solution, it is useful to think of an opposite one. Even the most experienced person is often unable to tell in advance, just by thinking, which solution will do the trick. So first trying one way of going about the problem, then trying another tack for a while, and then comparing results often yields the most creative result. It is good to be quick and consistent. But if you wish to be creative you should be willing to run the risk of sometimes seeming indecisive.

Implement the solution. Solving problems creatively involves continuous experimentation and revision. The longer you can keep options open, the more likely it is that the solution will be original and appropriate. Artists who do more original work change their technique as they are painting, and their paintings develop on the canvas in less predictable ways than those of less original artists. This is because the original artist is more ready to learn from the emerging work; he or she is alert to the unexpected and is willing to go with a better solution if one presents itself. Similarly, creative writers often start a story without knowing how it will end; the ending emerges as they follow the logic of the evolving story.

How does this apply to creativity in the domain of everyday life? To take an absurdly trivial example, if you are giving a party and want to make sure that the seating arrangement around the dinner table is the most appropriate for a good mixing of the guests, it makes sense to prepare a seating plan. But if by the time dinner is ready you notice that some of the guests whom you had scheduled to sit side by side seem cool toward each other, you may want to change the plans at the last moment. And if the dinner turns out to be dull, you should try to match up people in different combinations for coffee and dessert.

Such flexibility works only if you keep paying close attention to the process of solution and if you are sensitive enough to the feedback so that you can correct the course as new information becomes available. The reason most people prefer routine, tried-and-true solutions to their problems is that this requires less psychic energy. In fact, we could not afford to be creative all the time because we would soon stretch the limits of attention and collapse. Routine results in great savings. But it makes good sense to know how to come up with a creative solution when we need one and can spare the effort.

Divergent Thinking

Not all thinking involves the solution of problems. Sometimes we are asked to respond to what other people say, or to produce ideas in response to events, without having a particular problem that needs to be formulated and solved. There are more or less creative ways to pursue these less focused mental tasks. In talking to a friend I can use trite phrases or I can try to say things in a fresh, topical way that more closely represents what I feel at the moment. I can use stock images or try for more vivid ones, based on common experiences.

Most commercial programs designed to increase individual creativity focus on this particular aspect. They try to enhance three dimensions of divergent thinking that are generally held to be important to creativity: fluency, or the knack for coming up with a great number of responses; flexibility, or the tendency to produce ideas that are different from each other; and originality, which refers to the relative rarity of the ideas produced. Brainstorming programs are ways to stimulate people to increase the fluency, flexibility, and originality of their ideas and responses. You can obtain the same results by taking things in your hands and following these suggestions.

Produce as many ideas as possible. If you have to write a thank-you note, a report, or a letter, identify a key word and then try to generate as many synonyms for it as possible. If you get stuck, turn to a thesaurus. Or instead of words that mean the same thing, shift to meanings that are similar but lead in different directions. At first go for quantity; later you will be critical and edit for quality.

If you are planning a weekend or a vacation, do the same thing: First, come up with as many options as you can think of, even if they are not all very sensible. A crazy suggestion may jolt you into thinking in new directions and lead to more acceptable alternatives you would not have considered otherwise. If you are shopping for clothes in a department store, don't just go straight to the familiar floor but try on the greatest variety time allows. Browse for books outside the accustomed categories. If your boss asks for an opinion, don't give only the predictable pet viewpoint based on your interests. Surprise her with a whole range of ideas, options, and possibilities—how wild you can afford to be depends on how conservative she is.

Have as many different ideas as possible. Quantity is important, but try to avoid redundancy. Variety in conversation, in the selection of music, in a menu, is generally appreciated. It pays off to learn how to alternate topics of conversation, types of restaurants, kinds of shows, ways of dressing. Robert Galvin of Motorola trained himself to do a simple mental exercise: Whenever someone says something, he asks himself, What if the opposite were true? Imagining alternatives to what others hold to be true is probably going to be useless 99 percent of the time. But that one other time the practice of flipping to a divergent perspective might generate an insight that is not only original but also useful.

Try to produce unlikely ideas. Originality is one of the hallmarks of creative thinking. If asked to come up with names for a baby, or ways to use a paper clip, or things to do at a party, a creative person is likely to give answers that are different from the answers of the majority. But these answers won't be bizarre. Once people hear them, they are likely to say, "Of course! Why didn't I think of it myself?"

It is more difficult to learn how to think in original ways than to learn how to be fluent and flexible. It requires cultivating a taste for quality that is not necessary for the other two. One exercise involves taking a random paragraph from the paper each day and seeing if you can find unique, more memorable ways of expressing the same ideas. If the paragraph is too dull or obscure, substitute another. Or you can look at your office or your living room, and ask yourself whether it reflects your personal taste, and if not, what you could do to bring it in closer harmony with your unique self.

If your job involves frequent meetings and conferences, you might cultivate the habit of jotting down brief summaries of what the others around the table have said. Then you can quickly generate alternative positions to those that have been expressed, or integrate the various perspectives in a more comprehensive perspective. Instead of stating views that are based on your previous positions, use the lines of force emerging in the meeting to suggest new ways of thinking about the issues.

To think in a divergent mode requires more attention than thinking in the usual convergent style. As usual, it takes more energy to be creative than to be a routine thinker. Therefore, you must choose when to try for creativity and when not to; otherwise you might burn yourself out in a blaze of intense originality.

Choosing a Special Domain

If creativity consists in changing a particular domain, then personal creativity consists in changing the domain of personal life. We call a physicist creative if he or she changes the way physics is practiced; a person who can change his or her own life we call personally creative. The domain of personal life consists of the rules that constrain psychic energy, the habits and practices that define what we do day in, day out. How we dress, how we work, how we conduct our relationships define this domain, and if we can improve on it, the quality of life as a whole is improved. The suggestions in this chapter have been about how to increase creativity in the domain of everyday life.

But even though personal life can be very complex, it is also limited in scope. Much of what makes life interesting and meaningful belongs to special domains: Music, cooking, poetry, gardening, bridge, history, religion, baseball, and politics are symbolic systems with their own special rules, and they exist outside any individual's life. They and thousands of other such systems make up culture, and we become human by seeing the world through the lenses they provide. A person who learns to operate by the rules of one of these domains has a chance to expand enormously the range of his or her creativity.

Too many people assume that most of the world is off-limits to them. Some consider art as being beyond the realm of possibility, others sports or music. Or dancing, science, philosophy—the list of things that are "not for me" can be endless. And it is true that some domains just don't agree with some people. But generally the problem is that cultural resources are underutilized. Either because of ignorance, low self-esteem, or habits of thought established early, we discount the possibility that we could enjoy and be good at many of the things that make others happy. It took several years of jail for Malcolm X to realize the power of religion and of politics, and to discover that he had gifts for both.

Few of us know in advance what domains we may have an affinity for. Prodigies are children who from an early age show a definite gift in some direction, but most of us are not prodigies and it takes us decades of trial and error to find out what we are best cut out for. Even in our sample, some individuals did not realize what their vocations were until they were middle-aged. And often the realization was forced on them by outside factors, such as a war or the necessity to do something that then turned out to be just right.

It is important to try as many domains as possible. Start with things you already enjoy and then move to related domains. If you like to read biography, you might try history. Swimming may lead to skin diving, to scuba, and then—why not?—to skydiving. Learning to operate within a new domain is always difficult, and love at first sight is rare. A certain amount of persistence is necessary. On the other hand, it makes no sense to persevere in an activity that gives no joy, or the promise of it.

Eventually you should be able to find one or more domains that fit your interests, things that you enjoy doing and that expand your life. Ideally we should be able to do so in as many domains as possible. But in practice the limits on psychic energy make it impossible to take on more than a few discrete activities seriously.

There are two dangers as you become involved in a domain. The first is addiction; some domains are so seductive that you may invest so much attention into it that you have none left for your job and family. Some chess players become so taken by the game that to all intents and purposes they become zombies; the same can be true of betting on horses, collecting art, studying the Bible, or cruising the Internet.

The other danger is the opposite: You can become so diffuse, so eclectic, that what you feel in different domains ends up being the same superficial experience. Like the traveler who goes everywhere and is still the same boring, provincial soul he was before he left, many people seem to gain nothing from sampling the best that the culture has to offer. As is usually the case, the best solution does not lie with the extremes.

As you learn to operate within a domain, your life is certainly going to become more creative. But it should be repeated that this does not guarantee creativity with a capital c. You can be personally as creative as you please, but if the domain and the field fail to cooperate—as they almost always do—your efforts will not be recorded in the history books. Learning to sculpt will do wonders for the quality of your life, but don't expect critics to get ecstatic, or collectors to beat a path to your door. The competition among new memes is fierce; few survive by being noticed, selected, and added to the culture. Luck has a huge hand in deciding whose c is capitalized. But if you don't learn to be creative in your personal life, the chances of contributing to the culture drop even closer to zero. And what really matters, in the last account, is not whether your name has been attached to a recognized discovery, but whether you have lived a full and creative life.

APPENDIX A

BRIEF BIOGRAPHICAL SKETCHES OF THE RESPONDENTS WHO WERE INTERVIEWED FOR THIS STUDY

Adler, Mortimer J. Male. b. 12/28/02. Philosopher, author. American. Recipient, Aquinas Medal, American Catholic Philosophical Association (1976). Honorary Trustee, Aspen Institute for Humanistic Studies (1973–). Author, *How to Read a Book* (with Charles Van Doren, 1940); *Six Great Ideas* (1981); *The Paideia Program* (1984). Chairman of the board of editors, *Encyclopædia Britannica* (1974–). Associate editor, *Great Books of the Western World* (1945–; editor in chief, 2d ed., 1990), *Syntopicon* (1952, 1990); editor in chief, *The Annals of America* (21 vols., 1968). See his *Mortimer J. Adler: Philosopher at Large* (1977). Interviewed by Kevin Rathunde (1/17/91). Age 88.

 Anderson, Jack. Male. b. 10/19/22. Journalist, author, writer. American. Recipient, Pulitzer Prize for national reporting (1972). Author, *The Anderson Papers* (with George Clifford, 1973), *Fiasco* (with James Boyd, 1983); others. See his *Confessions of a Muckraker* (with James Boyd, 1979). Interviewed by Kevin Rathunde (5/6/91). Age 68.

 Asner, Edward. Male. b. 11/15/29. Actor. American. Recipient, five Golden Globe awards; seven Emmy awards. President, Screen Actors Guild (1981–1985). Roles in theater, motion pictures, and television, including *The Mary Tyler Moore Show* (TV series, 1970–1977); *Roots* (TV miniseries, 1977); *Lou Grant* (TV series, 1977–1982). Interviewed by Kevin Rathunde (4/30/91). Age 61.

Bardeen, John. Male. b. 5/23/08; d. 1/30/91. Physicist, teacher. American. Recipient, Nobel Prize in physics (1956; with Walter Brattain and William Shockley) for research in semiconductors and the discovery of the transistor effect; Nobel Prize in physics (1972; with Leon Cooper and J. Robert Schrieffer) for their jointly developed theory of superconductivity. Author of many scientific papers. Interviewed by Mihaly Csikszentmihalyi, with Kevin Rathunde (6/14/90). Age 82.

Baskin, Leonard. Male. b. 8/15/22. Sculptor, graphic artist (printmaker, painter). American. Recipient, medal of merit for graphic arts, National Institute of Arts and Letters (1969). Represented in the permanent collections of the Metropolitan Museum of Art; Museum of Modern Art; Library of Congress; National Gallery of Art; others. Founder, Gehenna Press. Author, *Iconologia* (1988); others. See *Baskin: Sculpture Drawings & Prints,* by George Braziller (1970). Interviewed by Sean Kelley and Grant Rich (4/8/95). Age 72.

Bethe, Hans. Male. b. 7/2/06. Physicist, teacher. American (b. Germany). Recipient, Nobel Prize in physics (1967) for his work on stellar energy; National Medal of Science (1976); Albert Einstein Peace Prize (1992). Author, *Basic Bethe: Seminal Articles on Nuclear Physics 1936–1937* (with Robert F. Bacher and M. Stanley Livingston—the "Bethe Bible" to generations of nuclear physicists, 1986); others. See *Hans Bethe, Prophet of Energy,* by Jeremy Bernstein (1980). Interviewed by Jeanne Nakamura (3/29/93). Age 86.

Blackwood, Easley. Male. b. 4/21/33. Composer, eductor. American. Appeared as soloist with the Indianapolis Symphony Orchestra at age 14; studied with Oliver Messiaen, Berkshire Music Center (1949); Paul Hindemith, Yale (1950–54); and Nadia Boulanger, Paris (1954–1957); Appointed to faculty of University of Chicago (1958). Recipient: Fulbright Fellowship (1954); first prize, Koussevitzky Music Foundation (1958, for Symphony No. 1); Brandeis Creative Arts Award (1968); commissions from the Chicago Symphony Orchestra and the Library of Congress. Composer: four symphonies; Symphonic Fantasy (1965); 3 Short Fantasies for Piano (1965); *Un Voyage à Cythère* for Soprano and 10 Players (1966); 12 Microtonal Etudes for Synthesizer (1982). Interviewed by Grant Rich (5/23/95). Age 62.

Booth, Wayne. Male. b. 2/22/21. Literary critic, teacher. American. Recipient, Christian Gauss Prize, Phi Beta Kappa (1962); David H. Russell Award, National Council of Teachers of English (1966). President, Modern Language Association (1981–1982). Author, *The Rhetoric of Fiction* (1961); *The Company We Keep: An Ethics of Fiction* (1988); others. See his *The Vocation of a Teacher* (1988). Interviewed by Mihaly Csikszentmihalyi and Kevin Rathunde (6/7/90). Age 69.

Boulding, Elise. Female. b. 7/6/20. Sociologist, activist, teacher. American (b. Norway). Recipient, Ted Lentz Peace Prize (1977); National Woman of Conscience Award (1980); Jessie Bernard Award, American Sociological Association (1981). Author, *The Underside of History* (1976); *Building a Global Civic Culture: Education for an Interdependent World* (1988); others. Interviewed by Kevin Rathunde (8/1/91). Age 71.

Boulding, Kenneth. Male. b. 1/18/10; d. 3/19/93. Economist, philosopher, teacher, writer (poet). American (b. England). Recipient, John Bates Clark Medal, American Economic Association (1949); Ted Lentz International Peace Research Award (1976). President, American Economic Association (1968); Peace Research Society (1969–1970); others. Founder (with others), *Journal of Conflict Resolution* (1957). Author, *The Economics of Peace* (1945); *The Image* (1956); *Beyond Economics: Essays on Society, Religion, and Ethics* (1968); others. See *Creative Tension,* by Cynthia Kerman (1974). Interviewed by Kevin Rathunde (8/1/91). Age 81.

Burbidge, Margaret. Female. b. 8/12/19. Observational astronomer, professor. American (b. England). Research on physical properties, energy sources, and radiation mechanisms of quasistellar objects and active galaxies. Director, the Center of Astrophysics and Space Sciences (1978–1984). Recipient, numerous prizes and awards, including the Helen B. Warner Prize (1959); Bruce Gold Medal, Astronomical Society of the Pacific (1982); Russell Lectureship Award (1984); National Medal of Science (1984); Albert Einstein Medal (1988). Author, *Quasi-Stellar Objects* (with G. Burbidge, 1967); also more than 300 research articles. Interviewed by Carol A. Mockros (10/3/95). Age 76.

Butler, Margaret. Female. b. 3/27/24. Mathematician, computer scientist. American. As staff mathematician in the early 1950s, she assisted in the development of one of the first digital computers. First woman elected a fellow of the American Nuclear Society. Executive Board, Association of Women in Science. Interviewed by Mihaly Csikszentmihalyi, Carol A. Mockros, and R. Keith Sawyer. Age 77.

Campbell, Donald. Male. b. 11/20/16. Psychologist, teacher. American. Recipient, Distinguished Scientific Contribution Award, American Psychological Association (1970); award for distinguished contribution to research in education, American Educational Research Association (1980). President, American Psychological Association (1975). Author, *Methodology and Epistemology for Social Science: Selected Papers* (1988); *Experimental and Quasi-Experimental Designs for Research* (with Julian C. Stanley, 1966); others. Interviewed by Mihaly Csikszentmihalyi (4/21/91). Age 77.

Chandrasekhar, Subrahmanyan. Male. b. 10/19/10; d. 8/21/95. Astrophysicist, author, teacher. American (b. India). Recipient, Nobel Prize in physics (1983; with William A. Fowler); Royal Astronomical Society

Gold Medal (Great Britain, 1953); National Medal of Science (1966). Author, *An Introduction to the Study of Stellar Structure* (1939); *Radiative Transfer* (1950); *The Mathematical Theory of Black Holes* (1983); others. Author (general science), *Truth and Beauty: Aesthetics and Motivations in Science* (1987); others. See *Chandra,* by Kameshwar C. Wali (1991). Interviewed by Kevin Rathunde (3/26/91). Age 80.

Coleman, James. Male. b. 5/12/26; d. 3/25/95. Sociologist, teacher. American. Recipient, Paul Lazarsfeld Award for Research, American Evaluation Association (1983); American Sociological Association Publication Award (1992). Author, *Introduction to Mathematical Sociology* (1964); *Equality and Achievement in Education* (1990; includes a summary of the 1966 "Coleman Report" on equality of educational opportunity); *Foundations of Social Theory* (1990); others. See his "Columbia in the 1950s" in *Authors of Their Own Lives: Intellectual Autobiographies of Twenty American Sociologists,* edited by Bennett M. Berger (1990, pp. 75–103). Interviewed by Mihaly Csikszentmihalyi (4/20/90). Age 63.

Commoner, Barry. Male. b. 5/28/17. Biologist, teacher, activist. American. Recipient, Newcomb Cleveland Prize, American Association for the Advancement of Science (1953); Phi Beta Kappa Award (1972). Author, *The Closing Circle* (1971); *The Politics of Energy* (1979); *Making Peace with the Planet* (1990); others. Presidential candidate, Citizens Party (1980). See "Barry Commoner: The Scientist as Agitator" in *Philosophers of the Earth,* by Shirley Chisolm (1972, pp. 122–39). Interviewed by Kevin Rathunde (5/7/91). Age 73.

Davies, Robertson. Male. b. 8/28/13; d. 12/03/95. Writer, journalist. Canadian. Recipient, Louis Jouvet Prize for directing, Dominion Drama Festival (1949); Lorne Pierce Medal, Royal Society of Canada (1961); Governor-General's Award for Fiction (Canada, 1973). Editor and publisher, *Peterborough* (Ontario) *Examiner* (1942–1962). Author, Deptford Trilogy (1970, 1972, 1975); *What's Bred in the Bone* (1985); others. See *Robertson Davies: An Appreciation,* edited by Elspeth Cameron (1991). Interviewed by Mihaly Csikszentmihalyi (5/11/94). Age 80.

Davis, Natalie. Female. b. 11/8/28. Historian, teacher. American. Decorated chevalier, l'Ordre des Palmes Académiques (France, 1976). President, American Historical Association (1987). Author, *Society and Culture in Early Modern France* (1975); *The Return of Martin Guerre* (1983); *Fiction in the Archives* (1987). See her interview in *Visions of History,* edited by Henry Abelove (1983, pp. 99–122). Interviewed by Kevin Rathunde (6/28/91). Age 62.

Domin, Hilde. Female. b. 7/27/12. Poet, essayist, translator. German. Recipient, Rilke-Preis (Germany, 1976); Bundesverdienstkreuz (1983). Author (poetry), *Nur eine Rose als Stütze* (1959); *Ich will dich* (1970); others.

See her *Von der Natur nicht vorgesehen* (1974). Interviewed by Mihaly Csikszentmihalyi (9/9/90). Age 78.

Dyson, Freeman. Male. b. 12/15/23. Physicist, teacher, author. American (b. England). Recipient, Max Planck Medal (Germany, 1969); Enrico Fermi Award (1994); National Book Critics Circle Award (1984). Author of many scientific papers. Author (general science), *Weapons and Hope* (1984); *From Eros to Gaia* (1992); others. See his *Disturbing the Universe* (1979). Interviewed by Kevin Rathunde (9/1/91). Age 67.

Eigen, Manfred. Male. b. 5/9/27. Chemist. German. Recipient, Nobel Prize in chemistry (1967; with Ronald Norrish and George Porter) for work on rapid chemical reactions; Otto Hahn Prize (Germany, 1962). Author, *Laws of the Game* (with Ruthild Winkler, 1981); *Steps Towards Life* (with Ruthild Winkler-Oswatitsch, 1992); others. Interviewed by Mihaly Csikszentmihalyi (9/17/90). Age 62.

Faludy, György. Male. b. 9/22/10. Poet, translator. Canadian (b. Hungary). Recipient, honorary doctorate, University of Toronto (1978). Author (in English translation), *Selected Poems* (1985), others; (in Hungarian) *Villon Ballads* (1937, burned by the Nazis in 1944; 1947 edition pulped by the Communists in 1948); *A Keepsake Book of Red Byzantium* (1961); others. See his *My Happy Days in Hell* (1962). Interviewed by Mihaly Csikszentmihalyi (6/5/91). Age 80.

Franklin, John Hope. Male. b. 1/2/15. Historian, teacher. American. Recipient, Clarence L. Holte Literary Prize (1986); Sidney Hook Award, Phi Beta Kappa (1994). Author, *From Slavery to Freedom: A History of Negro Americans* (1947; 7th ed., 1994); *George Washington Williams: A Biography* (1985); *Race and History: Selected Essays 1938–1988* (1990); others. See his "John Hope Franklin: A Life of Learning" in *Race and History* (pp. 277–91). Interviewed by Kevin Rathunde (11/7/90). Age 75.

Galvin, Robert. Male. b. 10/9/22. Electronics executive. American. Recipient, Golden Omega award, Electronic Industries Association (1981); National Medal of Technology (1991); Bower Award for Business, Franklin Institute (1993). With Motorola, Inc. 1940–present (president, 1956–1990; CEO, 1964–1986). Motorola, Inc. received the Malcolm Baldrige National Quality Award (1989 [the first year awarded]) Author, *The Idea of Ideas* (1991). Interviewed by Mihaly Csikszentmihalyi and Kevin Rathunde (9/10/91). Age 68.

Gardner, John W. Male. b. 10/8/12. Psychologist, writer, teacher. American. Recipient, honorary degrees from various colleges and universities; USAF Exceptional Services Award (1956); Presidential Medal of Freedom (1964); National Academy of Sciences Public Welfare Medal (1966); U.A.W. Social Justice Award (1968); AFL-CIO Murray Green Medal (1970); Christopher Award (1971). Chairman, Urban Coalition

(1968–1970). Founder and chairman, Common Cause (1970–1977). Member, Task Forces on Education under Presidents Kennedy and Johnson. Director, Time, Inc. (1968–1972). Author, *Excellence* (1961, 1984); *Self-Renewal* (1964, 1981); *On Leadership* (1990). Interviewed by Mihaly Csikszentmihalyi (8/18/91). Age 78.

Gordimer, Nadine. Female. b. 11/20/23. Writer. South African. Recipient, Nobel Prize for literature (1991); Booker Prize (England, 1974); Grand Aigle d'Or (France, 1975). Author, *The Conservationist* (1974); *Burger's Daughter* (1979); *Something Out There* (1984); others. See *Nadine Gordimer,* by Judie Newman (1988). Interviewed by Jeanne Nakamura (11/21/94). Age 71.

Gould, Stephen Jay. Male. b. 9/10/41. Paleontologist, geologist, science historian, author, teacher. American. Recipient, close to twenty honorary degrees; Medal of Excellence, Columbia University (1982); Silver medal, Zoological Society London (1984); Edinburgh medal, City of Edinburgh (1990); Britannica award and gold medal (1990). MacArthur Foundation prize fellow (1981–1986). Named Humanist Laureate, Academy of Humanism (1983). Author, *Ontogeny and Phylogeny* (1977); *The Panda's Thumb* (1980, award-winning); *The Mismeasure of Man* (1981, award-winning); *Hen's Teeth and Horse's Toes* (1983, award-winning); *Bully for Brontosaurus* (1991); others. Interviewed by Grant Rich (4/10/95). Age 53.

Gruenenberg, Nina. Female. b. 10/7/36. Journalist, editor. German. Columnist, political reporter, associate editor, *Die Zeit* (Hamburg). Listed as the 41st most influential woman in Germany. Interviewed by Mihaly Csikszentmihalyi (9/18/90). Age 56.

Harris, Irving Brooks. Male. b. 8/4/10. Business executive, philanthropist. American. Recipient, several honorary degrees; Chicago UNICEF World of Children award (1985); honorary membership award, Chicago Pediatric Society (1986). Director, Gillette Safety Razor Co. (1948–1960); chairman of the board, Science Research Associates (1953–1958); president, Michael Reese Hospital and Medical Center (1958–1961); president emeritus, Erikson Institute; president and cofounder, The Ounce of Prevention Fund (1982–). Clifford Beers lecturer, Yale University (1987). Interviewed by Mihaly Csikszentmihalyi (5/21/91). Age 80.

Hart, Kitty Carlisle. Female. b. 9/3/15. Arts administrator, actress, singer. American. Recipient, National Medal of Arts (1991). Appointments, Special Consultant to the Governor on Women's Opportunities (1966); Independent Commission to review the National Endowment for the Arts (1990); vice chairman (1971) and chairman (1976–present), New York State Council on the Arts. More than twenty-five principal stage credits (musical comedy, opera, operetta, and drama), including: *Champagne Sec* (1933); *The Rape of Lucretia* (1948); *Die Fledermaus* (1966–1967); *On Your Toes* (1984). Principal film credits: *She Loves Me Not* (1934); *Here Is My*

Heart (1934); *A Night at the Opera* (1935). Principal television credits: *Who Said That?* (1948–1955); *I've Got a Secret* (1952–1953); *What's Going On?* (1954) *To Tell the Truth* (1956–1967). See her *Kitty: An Autobiography* (1988). Interviewed by Nicole Brodsky (2/8/95). Age 79.

Hecht, Anthony. Male. b. 1/16/23. Poet, critic, teacher. American. U.S. Consultant in Poetry, Library of Congress (1982–1984). Recipient, Pulitzer Prize for poetry (1968); Bollingen Prize in Poetry (1983); Ruth B. Lilly Poetry Prize (1988). Author, *The Hard Hours* (1968); *The Venetian Vespers* (1977); others. See *Anthony Hecht,* by Norman German (1989). Interviewed by Mihaly Csikszentmihalyi (12/10/93). Age 70.

Henderson, Hazel. Female. b. 3/27/33. Economist, author. American (b. England). Named Citizen of the Year by the New York Medical Society (1967) for her role in founding Citizens for Clean Air. Author, *Creating Alternative Futures: The End of Economics* (1978); *The Politics of the Solar Age: Alternatives to Economics* (1981); *Paradigms in Progress: Life Beyond Economics* (1991); others. Interviewed by Kevin Rathunde (6/19/90). Age 57.

Holton, Gerald. Male. b. 5/23/22. Physicist, historian of science, teacher. American (b. Germany). Recipient, Oersted Medal, American Association of Physics Teachers (1980); George Sarton Medal, History of Science Society (1989); Andrew Gemant Award, American Institute of Physics (1989). Author, *Thematic Origins of Scientific Thought: Kepler to Einstein* (1973, 2d ed., 1988); *The Advancement of Science, and Its Burdens* (1986); others. Interviewed by Kevin Rathunde (2/25/91). Age 68.

Holton, Nina. Female. b. 1924. Sculptor. American (b. Austria). Studied and apprenticed with Mirko Basadella at Harvard's Carpenter Center for the Visual Arts, and with Dmitri Hadzi in Rome. About thirty group and one-person exhibitions in Boston, San Francisco, Washington, DC. Works in the Fogg Art Museum, the Van Leer Jerusalem Foundation collection; others. Articles published in *Leonardo*. Interviewed by Kevin Rathunde (2/25/91). Age 66.

Honig, William. Male. b. 4/23/37. Educational administrator, lawyer. American. Clerk, California Supreme Court; superintendent of public instruction for the State of California. Regent, University of California. Author, *Last Chance for Our Children* (1985); others. Interviewed by Keith Sawyer (9/29/92). Age 55.

Johnson, J. Seward, Jr. Male. b. 4/16/30. Sculptor, businessperson. American. Collections and public placements throughout the United States (more than twenty-five states), Bermuda, Canada, West Germany. Sculptures are generally life-size, cast in bronze, and follow the genre of hyperrealism. Founder of foundry (Johnson Atelier). See his *The Sculpture of J. Seward Johnson, Jr.: Celebrating the Familiar* (1987). Interviewed by Kevin Rathunde (8/13/90). Age 60.

Karle, Isabella. Female. b. 12/2/21. Experimental chemist, crystallographer. American. Recipient, Superior Civilian Award USN (1964); Annual Achievement Award, Society of Women Engineers (1967); Chemical Pioneer Award (1984); Lifetime Achievement Award Women in Science and Engineering (1986); The University of Michigan (1987); award for distinguished past president, American Crystallographic Association (1987); Gregori Aminoff Prize, Swedish Royal Academy of Sciences (1988); Bijvoet Medal (1990). Author, more than 250 scientific articles, book chapters, and reviews. Interviewed by Carol A. Mockros (5/8/92). Age 70.

Karle, Jerome. Male. b. 6/18/18. Theoretical chemist, crystallographer. American. Head, laboratory for structure matter, Naval Research Laboratory. Research associate, Manhattan Project (1943–1944); president, International Union of Chrystallography (1981–1984); Nobel Prize in chemistry (1985). For nineteen years, member, National Research Council. Chair, chemistry section of the National Academy of Sciences (1988–). Author, numerous scholarly articles. Interviewed by Carol A. Mockros (5/8/92). Age 73.

Klein, George. Male. b. 7/28/25. Biologist, author. Swedish (b. Hungary). Recipient, Prix Griffuel, Association pour la Recherche sur le Cancre (France, 1974); Harvey Prize, Technion–Israel Institute of Technology (1975); Dobloug Prize, Swedish Academy of Literature (1990). Author, more than 800 scientific papers. Author (philosophy), *Pieta* (1992 in English; original work published 1989). See his *The Atheist in the Holy City* (1990 in English; original work published 1987). Interviewed by Mihaly Csikszentmihalyi (5/9/90). Age 64.

Konner, Joan Weiner. Female. b. 2/24/31. University administrator, broadcasting executive, television producer. American. Professor and dean, Graduate School of Journalism, Columbia University. Executive producer, Bill Moyers' *Journal* (1978–1981). Recipient, twelve Emmy Awards; Edward A. Murrow Award; others. Interviewed by Kevin Rathunde (5/19/92). Age 61.

Kurokawa, Kisho. Male. b. 4/8/34. Architect, author, town planner. Japanese. Recipient, Gold Medal, French Academy of Architecture; Japan Grand Prize of Literature (1993). Architect, Nakagin Capsule Tower (1972); Hiroshima City Museum of Contemporary Art (1986); others. Author, *Metabolism '60* (with others, Tokyo, 1960); *The Philosophy of Symbiosis* (London, 1994); others. See his *Kisho Kurokawa—From Metabolism to Symbiosis* (London, 1992). Interviewed by Jeanne Nakamura (10/12/94). Age 60.

Lanyon, Ellen. Female. b. 12/21/26. Artist, professor. Founder, Chicago Graphics Workshop (1952–1955). Recipient, F. H. Armstrong Prize (1946, 1955, 1977); Fulbright Fellowship (1950); M. Cahn Award (1961); Casandra Foundation Grant (1971); National Endowment for the

Arts grant (1974, 1987); Herwood Lester Cook Foundation (1981). More than ten retrospective shows. More than forty one-woman exhibitions. Numerous group shows. Represented in many permanent collections, some of which include The Art Institute of Chicago, The Museum of Contemporary Art; the Library of Congress, Denver Art Museum. See *Art: A Woman's Sensibility*, by A. Adrian. Interviewed by Carol A. Mockros (3/19/93). Age 66.

Lederberg, Joshua. Male. b. 5/23/25. Biologist, teacher. American. Recipient, Nobel Prize in physiology and medicine (1958) for research in the genetics of bacteria; National Medal of Science (1989). President, Rockefeller University (1978–1990). Author of many scientific papers. See his "Genetic Recombination in Bacteria: A Discovery Account" (*Annual Review of Genetics*, 1987, vol. 21, pp. 23–46). Interviewed by Keith Sawyer (6/15/92). Age 67.

L'Engle, Madeleine. Female. b. 11/29/18. Writer. American. Recipient, Newbery Medal (1963); Sequoya Award (1965); Regina Medal (1985); Alan Award, National Council of Teachers of English (1986); Kerlan Award (1990). Author of more than forty works, including: *A Wrinkle in Time* (1962); *The Arm of the Starfish* (1965); *A Wind in the Door* (1973); *The Irrational Season* (1977); *A Swiftly Tilting Planet* (1978); *A Severed Wasp* (1982); *An Acceptable Time* (1989); *Certain Women* (1992); *Troubling a Star* (1994). See her *A Circle of Quiet* (1972), *The Summer of the Great-grandmother* (1974), and *The Irrational Season* (1977) for autobiographical information. Interviewed by Nicole Brodsky (5/19/94). Age 75.

Levertov, Denise. Female. b. 10/24/23. Writer. American (b. England). Recipient, Harriet Monroe Memorial Prize (1964); Lenore Marshall Poetry Prize (1975); Lannan Literary Award (1993). Author, *Here and Now* (1956); *The Freeing of the Dust* (1975); others. See *Understanding Denise Levertov*, by Harry Marten (1988). Interviewed by Jeanne Nakamura (2/27/95). Age 71.

LeVine, Robert A. Male. b. 3/27/32. Anthropologist, teacher. American. Fellow, American Academy of Arts and Sciences. Recipient, Research Career Scientist Award, National Institute of Mental Health (1972–1976). Author, *Culture, Behavior, and Personality* (1973); *Child Care and Culture: Lessons from Africa* (with others, 1994); others. Interviewed by Kevin Rathunde (2/22/91). Age 58.

LeVine, Sarah. Female. b. 8/14/40. Author, anthropologist. American (b. England). Currently working on a doctorate in Sanskrit and Pali. Author, *Mothers and Wives: Gusii Women of West Africa* (1979); *Dolor y Alegria: Women and Social Change in Urban Mexico* (1993). Novels under the name of Louisa Dawkins: *Natives and Strangers* (1985); *Chasing Shadows* (1988). Interviewed by Kevin Rathunde (2/22/91). Age 51.

Livi, Grazia. Female. b. 1932. Writer, journalist. Italian. Recipient, Premio Viareggio for the Essay (1991). Author of several novels and books of essays, including *La distanza e l'amore* (1978); *L'approdo invisibile* (1980); *Le lettere del mio nome* (1991); others. Interviewed by Mihaly Csikszentmihalyi (5/15/91). Age 58.

Loevinger, Jane. Female. b. 2/6/18. Research psychologist, professor. American. Professor, Washington University, St. Louis, Missouri. Created an influential theory of personality development. Author of numerous books and scholarly articles, including *Ego Development* (1976); "On the Self and Predicting Behavior," in R. Zucker, J. Arnoff, and A. Rabin, editors, *Personality and the Prediction of Behavior* (1984). Interviewed by Carol A. Mockros (11/6/92).

MacCready, Paul. Male. b. 9/29/25. Aeronautical engineer. American. Recipient, Edward Longstreth Medal, Franklin Institute (1979); Reed Aeronautics Award, American Institute of Aeronautics and Astronautics (1979). Engineer of the Century, American Society of Mechanical Engineers (1980). Leader of the team that won the Kremer Prize (1977) for human-powered flight. Author, scientific papers. Interviewed by Jeanne Nakamura (6/13/93). Age 67.

Mahfouz, Naguib. Male. b. 12/11/11. Writer. Egyptian. Recipient, Nobel Prize in literature (1988); State Prize for Literature (Egypt, 1957). Author, Cairo Trilogy (1956, 1957); *Zuqaq al Midaqq* (1947; translated as *Midaq Alley*, 1981); *Miramar* (1967). See *Naguib Mahfouz: From Regional Fame to Global Recognition,* edited by Michael Beard and Adnan Haydar (1993). Interviewed by Sherafoudin Malik (6/94). Age 82.

Mahoney, Margaret. Female. b. 10/24/24. American. Foundation president, Commonwealth Fund. Formerly executive associate, Carnegie Corporation. Trustee, John D. & Catherine T. MacArthur Foundation (1983–); Dole Foundation (1984–). Board of directors for Alliance for Aging Research (1987–). Overseas Development Council (1988–). Recipient, Alpha Omega Alpha Award (1985); Women's Forum Award (1989); Frank H. Lahey Award (1992); Walsh McDermott Award (1992). Author, numerous professional articles. Interviewed by Carol A. Mockros (4/13/94). Age 69.

Maier-Leibnitz, Heinz. Male. b. 3/28/11. Physicist, teacher, author. German. Director of first European research reactor, in Grenoble, France. Recipient, Otto Hahn Prize; Grosses Verdienstkreuz mit Stern; Pour le Merit. Author of many scientific publications. Author, *Zwischen Wissenschaft und Politik* (1979); others. See *Wie Kommt man auf einfaches Neues? der Forscher, Lehrer, Wissenschaftspolitiker und Hobbykoch Heinz Maier-Leibnitz,* edited by Paul Kienle (1991). Interviewed by Mihaly Csikszentmihalyi (8/29/90). Age 79.

Mayr, Ernst. Male. b. 7/5/04. Zoologist, curator, teacher, writer. American (b. Germany). Recipient, honorary degrees from several universities (ten) in various countries (seven); Leidy Medal (1946); Wallace Darwin Medal (1958); National Medal of Science (1970); Gregor Mendel Medal (1980); Darwin Medal, Royal Society (1987). Member of Rothschild expedition to Dutch New Guinea (1928). Jesup lecturer, Columbia University (1941). Curator, American Museum of Natural History (1944–1953); Alexander Agassiz professor of zoology, Harvard University (1953–1975); director, Museum of Comparative Zoology, Harvard University (1961–1970). Author, *Systematics and the Origin of Species* (1942); *Animal Species and Evolution* (1963); *Principles of Systematic Zoology* (1969); *One Long Argument* (1991). Interviewed by Grant Rich (10/21/94). Age 90.

McCarthy, Eugene. Male. b. 3/29/16. Politician, author, writer. American. U.S. Congressman (1949–1959). U.S. Senator (1959–1970). Democratic presidential candidate (1972). Independent presidential candidate (1976). Author, *The Limits of Power* (1967); *The Ultimate Tyranny* (1980); others. See his *Up 'Til Now* (1987). Interviewed by Kevin Rathunde (11/16/90). Age 74

McNeill, William. Male. b. 10/31/17. Historian, teacher. American (b. Canada). Professor, University of Chicago (1947–1987). Fellowships from the Fulbright, Rockefeller, and Guggenheim foundations. President, Demos Foundation (1968–1980); American Historical Association (1985). Recipient, National Book Award (1964). Author, *The Rise of the West* (1963); *Plagues and Peoples* (1976); *The Pursuit of Power* (1982). Interviewed by Kevin Rathunde (8/10/90). Age 72.

Milner, Brenda. Female. b. 7/15/18. Neuropsychologist. Candian (b. England). Research contributions on temporal-lobe function and memory disorders and the effects of unilateral brain lesions on cerebral organization. Recipient, Izaak Walton Killam Prize (1983); Hermann von Helmholtz Prize (1984); Ralph Gerard Prize (1987); Grand Dame of Merit, Order of Malta (1985). Author of numerous scholarly articles. Interviewed by Carol A. Mockros (1/5/94). Age 75.

Murphy, Franklin. Male. b. 1/29/16; d. 6/16/94. Media, university, and arts administrator. American. Dean of the School of Medicine, University of Kansas (1948–1951); chancellor, University of Kansas (1951–1960); chancellor, UCLA (1960–1968); chairman of the Board and CEO, Times Mirror Company (1968–1981); chairman of the Board of Trustees, National Gallery of Art, Los Angeles County Museum of Art; Carnegie Foundaton for the Advancement of Teaching. Recipient, the First Class of the Order of the Sacred Treasure (Japan, 1982); Officer's Cross of the Order of Merit of the Federal Republic of Germany (1983); Officer of the National Order of the Legion of Honor (France, 1985); Andrew W. Mellon

Medal, National Gallery of Art (1991). Interviewed by Keith Sawyer (9/24/92). Age 76.

Neugarten, Bernice. Female. b. 2/11/16. Social scientist, professor. American. Pioneer in the field of adult development and aging. Member of various advisory boards; Fellow, American Council on Education (1939–1941); past president, Gerontological Society. Recipient, International Association of Gerontology Kleemier Award (1972); Brookdale Award (1980); Sandoz International Prize (1987); American Psychological Association Lifetime Contribution Award (1994). Coauthor or editor, *Society and Education* (1957); *Personality in Middle and Late Life* (1964); *Middle Age and Aging* (1968); *Adjustment to Retirement* (1969); *Social Status in the City* (1971); *Age or Need? Public Policies for Older People* (1982); others. Interviewed by Carol A. Mockros (1/20/93). Age 76.

Noelle-Neumann, Elisabeth. Female. b. 12/19/16. Communications researcher, businessperson, teacher. German. Recipient, Grosses Bundesverdienstkreuz (Germany, 1976); Helen S. Dinerman Award, World Association for Public Opinion Research (1990). Professor of journalism, University of Mainz. Founder and director of the first German survey research institute, Institut fur Demoskopie Allensbach (1947–). Author, *The Germans: Public Opinion Polls, 1967–1980* (1981); *Die Schweigespirale: Öffentliche Meinung—unsere soziale Haut* (1980; translated as *The Spiral of Silence: Public Opinion—Our Social Skin*, 1984); others. Interviewed by Mihaly Csikszentmihalyi (4/28/90). Age 73.

Norman, Donald A. Male. b. 12/25/35. Cognitive scientist, author. American. Recipient, Excellence in Research award, University of California (1984). Professor of Psychology, University of California, San Diego (1966–); professor and founding chair, Department of Cognitive Science, UCSD (1988–). Chairman and founding member, Cognitive Science Society. Author, *Learning and Memory* (1982); *Human Information Processing* (2d ed., 1977); *The Design of Everyday Things* (1989); *Turn Signals Are the Facial Expressions of Automobiles* (1992). Editor, Cognitive Science Series (Lawrence Erlbaum Associates, 1979–); *Cognitive Science Journal* (1981–1985). Interviewed by Keith Sawyer (9/25/92). Age 56.

Offner, Frank. Male. b. 4/8/11. Electrical engineer, inventor, businessman. American. Accomplishments include applications of transistorized measuring devices and the development of differential amplifier; medical instrumentation that made possible the measurements of the electrocardiogram, the electroencephalogram, and the electromyogram. Developed the only successful heat-homing missiles produced during World War II. Recipient, Professional Achievement Citation Award (1991). Interviewed by Mihaly Csikszentmihalyi, Carol A. Mockros, and R. Keith Sawyer (2/19/92). Age 81.

Pais, Abraham. Male. b. 5/19/18. Physicist, professor. American (b. Holland). One of the founders of particle physics. Recipient, Oppenheimer Prize (1979); Physics Prize of the Netherlands (1992); Andrew Gemant Award (1993); Science Medal of the Royal Dutch Academy of Sciences (1993). Author of numerous scholarly articles as well as scientific biographies of Bohr and Einstein. His books include *Inward Bound* (1986); *Niels Bohr's Times* (1991); *Subtle Is the Lord* (1983), and *Einstein Lived Here* (1994). Interviewed Carol A. Mockros (4/13/94). Age 75.

Pauling, Linus. Male. b. 2/28/01; d. 8/19/94. Chemist, activist, teacher. American. Recipient, Nobel Prize in chemistry (1954) for research on the nature of the chemical bond and its application to the structure of complex substances; Nobel Prize in peace (1962). Author, *The Nature of the Chemical Bond and the Structure of Molecules and Crystals* (1939); *Vitamin C and the Common Cold* (1971); *No More War!* (1958). See *Linus Pauling: A Man and His Science,* by Anthony Serafini (1989). Interviewed by Kevin Rathunde (11/20/90). Age 89.

Peterson, Oscar. Male. b. 8/15/25. Jazz pianist, composer. Canadian. Founder of Advanced School of Contemporary Music; chancellor, York University (1991–). Played in Johnny Holmes Orchestra (1944–1949); appeared with Jazz at the Philharmonic, Carnegie Hall (1949); toured United States and Europe (1950–); toured Soviet Union (1974). Recipient, several honorary degrees; award for piano, *Down Beat* magazine (thirteen times); *Metronome* magazine Award (1953–1954); Edison Award (1962); Order of Canada (1974); Diplome d'honneur Canadian Conference of the Arts (1975); Grammy Award, four times. Founded Oscar Peterson Scholarship, Berklee School of Music, Boston (1982). Composer, *Canadiana Suite, Hymn to Freedom, Fields of Endless Day, City Lights.* Author, *Jazz Exercises and Pieces: Oscar Peterson New Piano Solos.* Interviewed by Grant Rich (9/20/94). Age 69.

Prigogine, Ilya. Male. b. 1/25/17. Chemist. Belgian (b. Russia). Recipient, Nobel Prize in chemistry (1977) for his contributions to nonequilibrium thermodynamics; Rumford gold medal, Royal Society of London (1976). Decorated commander, Order of Arts and Letters (France, 1984). Author of many technical books and articles. Author (general science), *From Being to Becoming* (1980); *Order Out of Chaos* (with Isabelle Stengers, 1984). Interviewed by Jeanne Nakamura (10/29/95). Age 78.

Rabinow, Jacob. Male. b. 1/8/10. Electrical engineer, inventor. American (b. Russia). Recipient, Edward Longstreth Medal, Franklin Institute (1959); Harry Diamond Award, Institute of Electrical and Electronics Engineers (1977). Holder of more than 200 patents in diverse fields, including optical character recognition technology, mail-sorting machinery, automatic regulators, and motors. See his *Inventing for Fun and Profit* (1990). Interviewed by Jeanne Nakamura (5/16/93). Age 83.

Randone, Enrico. Male. b. 12/21/10. Lawyer, insurance executive. Italian. Worked at the Assicurazioni Generali (1937–); president and chairman of the Board since 1979. See *Il Leone di Trieste,* by C. Lindner and G. Mazzuca (1990). Interviewed by Mihaly Csikszentmihalyi (5/13/91). Age 80.

Reed, John. Male. b. 2/7/39. Banker, philanthropist. American. CEO, Citicorp. Member of the Board, Sloan-Kettering Cancer Center; MIT; Spencer Foundation; Russell Sage Foundation; Center for Advanced Study in the Behavioral Sciences. Interviewed by Keith Sawyer (4/15/92). Age 53.

Riesman, David. Male. b. 9/22/09. Social scientist, lawyer, teacher. American. Recipient, Tocqueville Prize, French Academy (1980). Professor of Sociology, Harvard University. Author, *The Lonely Crowd* (in collaboration with Reuel Denney and Nathan Glazer, 1950); *The Academic Revolution* (with Christopher Jencks, 1968); *The Perpetual Dream* (with Gerald Grant, 1978); others. See his "Becoming an Academic Man" in *Authors of Their Own Lives: Intellectual Autobiographies of Twenty American Sociologists,* edited by Bennett M. Berger (1990, pp. 22–74). Interviewed by Mihaly Csikszentmihalyi (6/20/90). Age 80.

Rubin, Vera. Female. b. 7/23/28. Observational astronomer. American. Known for her work in determining that visible matter provides only a fraction of the overall mass of the universe. Member, National Academy of Sciences; Council of American Astronomical Society (1977–1980); editorial board of *Science* magazine (1979–). Recipient, National Medal of Science (1993). Past president, Committee on Galaxies, International Astronomical Union. Associate editor, *Astronomical Journal* (1972–1977); *Astrophysical Journal of Letters* (1977–1982). Author of more than 125 scientific papers published in specialist journals and books on the dynamics of galaxies. Interviewed by Carol A. Mockros (10/10/92). Age 64.

Salk, Jonas. Male. b. 10/28/14; d. 6/23/95. Biologist, philosopher, author. American. Recipient, Congressional Gold Medal (1955); Presidential Medal of Freedom (1977). Developer of the first successful vaccine against poliomyelitis (1955). Author of many scientific papers. Author (philosophy), *The Survival of the Wisest* (1973); *Anatomy of Reality* (1983); others. See *Breakthrough: The Saga of Jonas Salk,* by Richard Carter (1965). Interviewed by Kevin Rathunde (5/1/91). Age 76.

Sarton, May. Female. b. 5/3/12; d. 7/16/95. Writer. American (b. Belgium). Recipient, Golden Rose Award for poetry (1945); Levinson Prize for Poetry (1993). Author, *Selected Poems of May Sarton* (1978); *Mrs. Stevens Hears the Mermaids Singing* (1965); others. See her *Plant Dreaming Deep* (1967); *Journal of a Solitude* (1973); others. Interviewed by Jeanne Nakamura (4/25/94 and 4/26/94). Age 81.

Schuler, Gunther. Male. b. 11/22/25. Composer, conductor, author, educator. American. Teacher at Manhattan School of Music (1950–1963);

head of composition department, Tanglewood; president, New England Conservatory of Music (1967–1977); founder of two music labels. Composer, *Quartet for Four Double Basses* (1947); *Seven Studies on Themes of Paul Klee* (1959); Spectra (1960); *The Visitation* (opera, 1966); *Horn Concerto No. 2* (1976); *On Light Wings* (piano quartet, 1984); *A Bouquet for Collage for Clarinet, Flute, Violin, Cello, Piano, and Percussion* (1988). Recipient, several honorary degrees; Creative Arts Award, Brandeis University (1960); Guggenheim grant (1962, 1963); Rodgers and Hammerstein Award (1971); Friedman Award (1988); Pulitzer Prize in music (1994). Author, *Musings: The Musical Worlds of Gunther Schuller* (1985); *The Swing Era: The Development of Jazz, 1930–1945* (1989). Interviewed by Grant Rich (11/17/94). Age 68.

Sebeok, Thomas. Male. b. 11/9/20. Linguist, teacher. American (b. Hungary). Professor of linguistics, Indiana University. Recipient, Distinguished Service Award, American Anthropological Association (1984). President, Linguistic Society of America (1975); Semiotic Society of America (1984). Author, *Perspectives in Zoosemiotics* (1972), *Structure and Texture: Selected Essays in Cheremis Verbal Art* (1974); *The Play of Musement* (1981); others. Editor, *Style in Language* (1960); others. Interviewed by Keith Sawyer (8/28/92). Age 71.

Shankar, Ravi. Male. b. 4/7/20. Sitar player, composer. Indian. Director of All-India Radio's instrumental ensemble (1949–1956); toured extensively around the world; founder of Kinnara School of Indian Music, Los Angeles. Recipient, honorary degrees from the University of California and Indira Kala Sangeet University; Indian National Academy of Music, Dance, and Drama (1962); National Academy of Recording Arts and Sciences (1966); UNICEF; the Presidential Padma Bhushan Award. Composer, two concertos for sitar and orchestra (1970, 1976). Several ballet and film scores. See *Raga* (full-length film on his life and music, 1972); *My Life, My Times* (autobiography, 1978); *The Great Shankars: Uday, Ravi* (1983). Interviewed by Grant Rich (5/2/94). Age 73.

Smith, Bradley. Male. b. 6/30/10. Photojournalist, author. American. Exhibitions of his photographs by the Museum of Modern Art, others. Freelance photographer for *Life, Paris Match, Time*, and other magazines (1942–1965). Author, *Japan: A History in Art* (1964); *Erotic Art of the Masters* (1974); *France: A History in Art* (1984); others. Interviewed by Mihaly Csikszentmihalyi (4/17/91). Age 83.

Snow, Michael. Male. b. 10/10/29. Artist, jazz musician, cinematographer. Canadian. Professor of advanced film at Yale University (1970). Recipient, Guggenheim fellowship (1972); honorary degrees from Brock University, Ontario (1976), and Nova Scotia College of Art and Design (1987); Order of Canada (1983). Paintings exhibited at the World Exposition of 1967, The Ontario Gallery of Art, The National Gallery of Art; others. Piano solos and band performance recorded on CCMC label, others. See *The Michael Snow Project* (1993); *Presence and Absence: The Films of Michael Snow, from 1956 to 1991* (1995). Interviewed by Mihaly Csikszentmihalyi (5/11/94). Age 64.

Spock, Benjamin. Male. b. 5/2/03. Pediatrician, psychiatrist, author, activist. American. Recipient, Gold medal (crew) 1924 Olympic Games; Family Life Book Award (1963); Thomas Paine Award, National Emergency Civil Liberties Committee (1968). Author, *The Common Sense Book of Baby and Child Care* (1946; 6th rev. ed., 1992 [with Michael Rothenberg]); *A Better World for Our Children* (1994); others. Presidential candidate, Peoples Party (1972). See his *Spock on Spock: A Memoir of Growing Up with the Century* (with Mary Morgan Spock, 1989). Interviewed by Kevin Rathunde (7/13/91). Age 88.

Spock, Mary Morgan. Female. b. 11/27/43. Author, activist. American. Author, *Stepparenting* (1986); *Spock on Spock: A Memoir of Growing Up with the Century* (with Benjamin Spock, 1989). Interviewed by Kevin Rathunde (7/13/91). Age 47.

Stern, Richard. Male. b. 2/25/28. Writer, teacher. American. Recipient, National Institute of Arts and Letters Fiction Award (1968); American Academy and Institute of Arts and Letters Medal of Merit for the Novel (1985). Author, *Golk* (1960); *Natural Shocks* (1978); *Noble Rot* (1990); others. See *Richard Stern,* by James Schiffer (1993). Interviewed by Nicole Brodsky, Mihaly Csikszentmihalyi, and Sean Kelley (2/11/94). Age 65.

Stigler, George. Male. b. 1/17/11; d. 12/1/91. Economist, teacher. American. Professor of economics, University of Chicago. Recipient, Nobel Prize in economics (1982) for work on the economic theory of information and theory of public regulation; National Medal of Science (1987). Author, *The Organization of Industry* (1968); *Essays in the History of Economics* (1965); others. See his *Memoirs of an Unregulated Economist* (1988). Interviewed by Mihaly Csikszentmihalyi and Kevin Rathunde (6/7/90). Age 79.

Strand, Mark. Male. b. 4/11/34. Writer. American (b. Canada). U.S. Poet Laureate, Library of Congress (1990–1991). Recipient, Edgar Allan Poe Award, Academy of American Poets (1974); Bollingen Prize in Poetry (1993). Author, *Sleeping With One Eye Open* (1964); *The Continuous Life* (1990); others. Interviewed by Kevin Rathunde (7/4/91). Age 57.

Trachinger, Robert. Male. b. 11/26/23. Broadcast executive, educator. American. Vice president, ABC-TV (1978–1985). Winner of Emmy Awards for documentaries (1966–1968). Responsible for the development and first broadcast use of slow-motion videotape, hand-held and underwater cameras. Professor of communications at UCLA for twenty-three years.

International lecturer and consultant. Fulbright scholar (1985–1986). Interviewed by Kevin Rathunde (11/20/90). Age 67.

Weisskopf, Viktor. Male. b. 9/19/08. Physicist, author, teacher. American (b. Austria). Recipient, Max Planck Medal (Germany, 1956); National Medal of Science (1980); Enrico Fermi Award (1988). Director General, European Organization for Nuclear Research (1961–1966). Author of many scientific publications. Author (general science), *Knowledge and Wonder* (1962); others. See his *The Joy of Insight* (1991). Interviewed by Kevin Rathunde (2/22/91). Age 82.

Wheeler, John A. Male. b. 7/9/11. Physicist, professor. American. Known for his work on black holes. Guggenheim Fellow (1949–1950). Recipient of the Morrison Prize (1947); Albert Einstein Prize (1965); Enrico Fermi Award (1968); Franklin Medal (1969); National Medal of Science (1971); Herzfeld Award (1975); Niels Bohr International Award (1982); Oersted Medal (1983); Oppenheimer Memorial Prize (1984). In addition to numerous professional papers, some of his publications include: *Geometrodynamics* (1962); *Spacetime Physics* (1966); *Gravitation* (with C. Misner and K. Thorne, 1972); *Frontiers of Time* (with W. Zurek, 1979); and *Quantum Theory and Measurement* (1983). Interviewed by Carol A. Mockros (11/17/92). Age 81.

Whitman, Marina. Female. b. 3/6/35. Economist, teacher. American. Vice president and chief economist (1979–1985) and vice president and group executive (1985–1992), General Motors Corporation. Recipient, Catalyst Award (1976); Columbia University Award for Excellence (1984). Member, Council of Economic Advisers (1972–1973). Author, *Government Risk-Sharing in Foreign Investment* (1965); *Reflections of Interdependence: Issues for Economic Theory and U.S. Policy* (1979); others. Interviewed by Jeanne Nakamura (5/25/94). Age 59.

Wilson, Edward O. Male. b. 6/10/29. Biologist, teacher. American. Recipient, National Medal of Science (1976); Crafoord Prize, Royal Swedish Academy of Sciences (1990); Gold Medal, World Wide Fund for Nature (1990); Pulitzer Prize for general nonfiction (1979 and 1991). Author, *Sociobiology: The New Synthesis* (1975); *On Human Nature* (1978); *The Ants* (with Bert Hölldobler, 1990); others. See his *Naturalist* (1994). Interviewed by Grant Rich (12/2/94). Age 65.

Woodward, Comer Vann. Male. b. 11/13/08. Historian, writer, professor. American. Leading historian of the American South. Recipient, the Bancroft Prize (1951); National Academy Institute Arts and Letters Literature Award (1954); Pulitzer Prize (1982); Life Work Award, The American Historical Society (1986); Gold Medal for History (1990). Member, the American Academy of Arts and Letters. Author of numerous scholarly articles and eleven books, including, *Tom Watson, Agrarian Rebel* (1938); *Origins* *of the New South* (1951); *The Strange Career of Jim Crow* (1955). Editor, *Mary Chestnut's Civil War* (1981). Also see M. O'Brien, "C. Vann Woodward and the Burden of Southern Liberalism," *The American Historical Review* (1973). Interviewed by Carol A. Mockros (3/15/93). Age 84.

Yalow, Rosalyn. Female. b. 7/19/21. Medical physicist. American. Recipient, Nobel Prize in physiology and medicine (1977, with Roger Guillemin and Andrew Schally) for contributions to the discovery and the development of the radioimmunoassay procedure; National Medal of Science (1988). President, Endocrine Society (1978–1979). Author of many scientific publications. Interviewed by Mihaly Csikszentmihalyi (3/14/92). Age 70.

Zeisel, Eva. Female. b. 11/11/06. Ceramic designer. American (b. Hungary). Recipient, NEA senior fellowship (1983); The Order of the Star Award (Hungarian People's Republic, 1987). Traveling retrospective exhibition through the United States and Canada organized by Le Château Dufresne and the Smithsonian Institution (1984). Castleton dinnerware set exhibited by MoMA (1946). Ceramic design instructor, Pratt Institute (1939–1953); artistic director, A. T. Heisey [glass factory] (1953); industrial design instructor, Rhode Island School of Design (1959–1960). See *Eva Zeisel: Designer for Industry,* by Martin P. Eidelberg (1984). Interviewed by Mihaly Csikszentmihalyi and Kevin Rathunde (1/28/91). Age 84.

SUMMARY OF INTERVIEWEES
(Several persons could have been listed under more than one heading)

Arts and Humanities
Historians
Davis, Franklin, McNeill, Woodward
Media
Anderson, Gruenenberg, Konner, Noelle-Neumann, Trachinger
Performers and Composers
Asner, Blackwood, Hart, Peterson, Schuler, Shankar
Philosophers and Critics
Adler, Booth, Sebeok
Poets
Domin, Faludy, Hecht, Strand
Visual Artists and Architects
Baskin, N. Holton, Johnson, Kurokawa, Lanyon, Smith, Snow, Zeisel
Writers
Davies, Gordimer★, L'Engle, Levertov, S. LeVine, Livi, Mahfouz★, Sarton, Stern

Sciences

Biologists and Physicians

Commoner, Gould, Klein, Lederberg★, Mayr, Salk, B. Spock, Wilson, Yalow★

Chemists

Eigen★, I. Karle, J. Karle★, Pauling★★, Prigogine★

Economists

K. Boulding, Stigler★, Whitman

Physicists and Astronomers

Bardeen★★, Bethe★, Burbidge, Butler, Chandrasekhar★, Dyson, G. Holton, Maier-Leibnitz, Pais, Rubin, Weisskopf, Wheeler

Psychologists and Social Scientists

Campbell, Coleman, R. LeVine, Loevinger, Milner, Neugarten, Norman, Riesman

Business and Politics

Activists

E. Boulding, Henderson, Honig, M. Spock

Business and Philanthropy

Galvin, Harris, Mahoney, Murphy, Randone, Reed

Inventors

MacCready, Offner, Rabinow

Politics

Gardner, McCarthy

★Denotes recipient of Nobel Prize.

APPENDIX B

INTERVIEW PROTOCOL USED IN THE STUDY

PART A: CAREER AND LIFE PRIORITIES

1. *Of the things you have done in life, of what are you most proud?*
   a. To what do you attribute your success in this endeavor? Any personal qualities?

2. *Of all the obstacles you have encountered in your life, which was the hardest to overcome?*
   a. How did you do it?
   b. Any that you did not overcome?

3. *Has there been a particular project or event that has significantly influenced the direction of your career? If so, could you talk a little about it?*
   a. How did it stimulate your interest?
   b. How did it develop over time?
   c. How important was this project/event to your creative accomplishments?
   d. Do you still have interesting, stimulating experiences like this?

4. *What advice would you give to a young person starting out in [subject's area]?*
   a. Is that how you did it? If not how is your current perspective different from the way you started?

b. Would you advise [concerning importance of field]:

few social contacts or many? Mentors, peers, colleagues?
establish your own identity early or late?
work with leading organizations?

c. Would you advise [concerning importance of domain]:

specialize early or late?
focus on leading ideas or work on periphery?

d. Would you advise [concerning importance of person]:

intrinsic versus extrinsic reasons?
tie work to personal values or separate?

5. *How would you advise a young person on why it is important to get involved in [subject's area]?*
  a. Is that why it was important to you? If not, how is your current perspective different?

6. *How did you initially become involved or interested in [subject's area]? What has kept you involved for so long?*

7. *Have there been points when what you were doing became less intensely involving—seemed less interesting or important to you? Can you describe a time that stands out?*
  a. What were the circumstances?
  b. What did you do?

Part B: Relationships
1. *If there has been a significant person (or persons) in your life who has influenced or stimulated your thinking and attitudes about your work. . .*
  a. When did you know them?
  b. How did you become interested in them (e.g., did you actively pursue them)?
  c. How did they influence your work and/or attitudes (e.g., motivation, personal or professional values)?
  d. In what ways was he/she a good and/or bad teacher?
  e. What kinds of things did you talk to this person about (e.g., personal, general career-related, specific problems)?
  f. What did you learn from them? How to choose what problems to pursue? Field politics and marketing yourself?

2. *Is it important for you to teach and work with young people?*
   a. Why?
   b. What are you interested in trying to convey to them? Why?
   c. How do you do this?

3. *When you interact or work with a young student, can you assess whether they will be likely to leave the field or become successful in the field?*
   a. Do you recognize people who are likely to be creative in their future work? How? What characteristics do they have?

4. *Do you notice differences between men and women students/young people and male and female colleagues in the field? If so,*

   in interests?
   in ability? creativity?
   in the way they approach learning?
   in the way they interact with other people/colleagues?
   in how they define success and achievement?
   in their personal goals and values?
   in their professional goals and values?

5. *What advice would you give a young person on how to balance their private life (i.e., family, other concerns not related to work) with [subject's area]?*
   a. Is that how you did it? If not, how is your current perspective different?

   importance of other kinds of life skills?
   relative importance of career in early or later life?

Peers and Colleagues
1. *At any time in your life, have your peers been particularly influential in shaping your personal and professional identity?*

2. *In what way(s) have colleagues been important for your personal and professional identity and success?*

Family
1. *In what way(s) do you think your family background was special in helping you to become the person you are?*

2. *How did you spend most of your free time as a child? What kinds of activities did you like to do? With peers? parents? siblings? alone?*

3. *In what way(s) have your spouse and children influenced your goals and career?*

PART C: WORKING HABITS/INSIGHTS

1. Where do the ideas for your work generally come from?

 a. From:

 reading?
 others?
 your own previous work?
 life experiences?

 b. What determines (how do you decide) what project or problem you turn to when one is completed?

 c. Have there been times when it's been difficult to decide what to do next? What do you do?

2. How important is rationality versus intuition in your work? Describe.

 a. Are there two different styles in your work (e.g., one more "rational" and the other more "intuitive")?

 b. Do you think it's important to "go with your hunches" or "trust your instincts"? Or are these usually wrong/misleading?

 c. Do you have better success with a methodical, rigorous approach to your work?

 d. Do you think about work during leisure time? e.g., did you ever have any important insights during this "off" time?

 e. How many hours of sleep do you usually get? Do you tend to do your best work early in the morning or late at night?

 f. Have you ever had a useful idea while lying in bed, or in a dream?

3. How do you go about developing an idea/project?

 a. Do you write rough drafts? Outlines? How often do you rewrite?

 b. Do you publish your work right away or wait awhile?

4. Can you describe your working methods?

 a. How do you decide what mail to answer, interviews to do, etc.?

 b. Do you prefer to work alone or in a team?

5. Overall, how is the way you go about your work different now from the way you worked twenty years ago?

 a. What if any changes have there been over the years in the intensity of your involvement in [subject's area]?

 b. What about changes in the way you think and feel about it?

6. Have you experienced a paradigm change in your work? Describe.

PART D: ATTENTIONAL STRUCTURES AND DYNAMICS

1. At present, what task or challenge do you see as the most important for you?
   a. Is that what takes up most of your time and energy? If not, what does?

2. What do you do about this? [probe for field/domain/reflection]

3. Do you do this primarily because of a sense of responsibility, or because you enjoy doing this? Describe.
   a. How has this changed over the years?

4. Are you planning to make any changes in how actively you work in [subject's area]?

5. If we had spoken to you thirty years ago, what different views of the world and yourself would you have had?

6. Have there been some personal goals that have been especially meaningful to you over your career? If yes, could we talk about some of the most significant?
   a. How did your interest in this goal begin?
   b. How did it develop over time? (Now?)
   c. How important was this goal to your creative accomplishments?

NOTES

Page

CHAPTER 1

Page

1   **This book** is in many respects a sequel to two previous ones: *Flow: The Psychology of Optimal Experience*, a study of the conditions that make life enjoyable and meaningful, and *The Evolving Self*, which deals with the evolutionary implications of human life and experience. The present volume describes and interprets the lives of a number of exceptional individuals who have found ways to make flow a permanent feature of their lives, and at the same time contribute to the evolution of culture.

This book also takes its place within the contemporary literature on creativity. I would like to mention here some of the work of colleagues that has influenced me, to set the context in which the present contribution belongs, and to provide a brief glimpse of the "state of the art" in the field of creativity research. I should make clear that this is not intended as a review of the by now immense literature in the field, but simply as an introduction to those active scholars and centers of scholarship that have, one way or the other, contributed to my thinking on the subject.

To make this picture as vivid as possible in my mind—and, hopefully, in the reader's—I will start with a mental map of the locations where research on creativity is currently vigorous, starting with the northwest corner of the United States and proceeding south, and then east and north, before moving to centers outside the United States.

I shall begin with Dean Keith Simonton at the University of California, Davis, who has pursued for several decades his historiometric studies of creativity. More than any single scholar, Simonton has written extensively about the quantitative trends correlated over time with creative achievements (e.g., Simonton 1984, 1990a).

Some of the earliest studies of the personality of creative people—mostly architects and artists—was done at the University of California, Berkeley, by D. W. McKinnon and his students. This line of work was continued by Frank Barron, and then by David Harrington at the University of California, Santa Cruz (MacKinnon 1962; Barron 1969; Harrington 1990).

At the Claremont Colleges in Southern California, Robert Albert and Mark Runco have been doing longitudinal studies of students presumed to be creative. Runco is also the founding editor of *The Creativity Research Journal*, one of the two journals that define the field (Albert 1983; Runco 1994).

At the University of New Mexico, Vera John-Steiner has followed the development of creative ideas by analyzing scientific notebooks, and has focused on the collaboration in groups that achieve breakthroughs (John-Steiner 1985).

Moving now to the Midwest, the University of Chicago has also had a long tradition of studying creativity in schools (Getzels and Jackson 1962), among artists (Getzels and Csikszentmihalyi 1976), and currently in a variety of different fields, as the present volume attests.

At Michigan State University, Robert Root-Bernstein and his team continue to mine the interviews with eminent scientists that Bernice Eiduson started collecting in 1958 (Eiduson 1962; Root-Bernstein 1989).

At Carnegie-Mellon University, Herbert Simon and his colleagues have experimented with computer programs that are supposed to reproduce the mental processes involved in creative discoveries (Langley, Simon, Bradshaw, and Zytkow 1987; Simon 1988).

Paul Torrance at the University of Georgia has been running a very productive laboratory for the study of creativity in children (Torrance 1962, 1988). In North Carolina, the Center for Creative Leadership has been applying knowledge to stimulate creativity in businesses and organizations.

At Columbia University in New York, Howard Gruber and his associates continue doing careful analyses of the lifelong creative work of single individuals (Gruber 1981; Gruber and Davis 1988).

Further north in Buffalo, New York, The Center for Creative Studies supports research, consults with businesses, and publishes the other journal of the field, *The Journal of Creative Behavior* (Isaksen, Dorval, and Treffinger 1994; Parnes 1967).

Robert J. Sternberg at Yale University is one of the most influential and prolific theorists and researchers on human cognition, including creativity (e.g., Sternberg 1986, 1988).

As one would expect, the Boston area is rife with scholars involved in creativity research. First and foremost is Howard Gardner at Harvard, whose long-standing presence in the field was recently crowned with a masterful study of seven outstanding geniuses of our century (Gardner 1988, 1993). David Perkins at Project Zero has studied for a long time the cognitive processes involved in creative thinking (Perkins 1981; Weber and Perkins 1992). Also at Harvard is Teresa Amabile, who has studied extensively creativity in children and has begun to study creativity in businesses and organizations (Amabile 1983, 1990). Next is David Feldman at Tufts University, who pioneered the study of prodigies and has developed the concept of domains in the study of cognitive development (Feldman 1980, 1994).

And finally, closing the circle of this imaginary map of the United States, at the University of Maine in Orono, psychologist Colin Martindale applies historiographical methods to the waxing and waning of creativity in the arts; his work is similar to Simonton's in California (Martindale 1989, 1990).

Of researchers outside the United States, I have had the good fortune of exchanging many ideas with István Magyari-Beck from Budapest, who has argued for some time that we need a new discipline of "creatology" to avoid the current, often parochial, one-dimensional approaches to the topic (Magyari-Beck 1988, 1994). The perspectives of Fausto Massimini of the University of Milan have had a deep influence on my understanding of cultural evolution (Massimini 1993; Massimini, Csikszentmihalyi, and Delle Fave 1988), as well as many other issues. In Israel, Roberta Milgram continues the psychometric tradition of creativity testing developed by Torrance (Milgram 1990).

Of course, I must repeat that these references are only the tip of the iceberg and include only those investigators who are active in the field and whose work I know firsthand.

We share 98 percent of our genetic makeup with chimpanzees. The estimates of how much of our genetic makeup we share with the chimps varies from 94 to 99 percent (Dozier 1992, Diamond 1992).

5 **Creativity in history.** Although people have been creative all through history, they seldom realized it. For instance, for many centuries Egyptian civilization continued to lead in the arts and in technology, yet their ideology stressed extreme faithfulness to tradition. In medieval Europe many saints and philosophers broke new ground in lifestyle and in ways of thinking, yet they tended to attribute their inventions to having rediscovered God's will, rather than to their own ingenuity. According to traditional Christian thought, only God was creative; men were created but could not create. Creativity was a very minor concern of psychology until very recently. In 1950, when J. P. Guilford became president of the American Psychological Association, he gave his inaugural lecture on the importance of studying creativity in addition to intelligence. Ironically, Guilford's involvement with the subject came as a result of funding from the Department of Defense. During World War II the air force decided that intelligence tests were not sufficient to select the best pilots, those who could respond innovatively to emergency situations. Thus the needs of warfare spurred Guilford's research in originality and flexibility, which in turn stimulated decades of study in creativity (Feldman 1994, pp. 4–7).

6 **Thirty years of research.** I first started studying creativity in 1962, with my doctoral dissertation on the creative process in a group of art students. Many journal articles resulted, and the book *The Creative Vision,* which introduced new concepts and methods to the study of creativity, especially the focus on "problem finding" (Getzels and Csikszentmihalyi 1976). The "systems view" of creativity is something I developed much later, in 1988, and have elaborated since with the cooperation of students and colleagues, especially David Feldman of Tufts University and Howard Gardner of Harvard (Csikszentmihalyi 1988; Feldman, Csikszentmihalyi, and Gardner 1994; Gardner 1994).

7 **Cultural evolution.** That creativity is to cultural evolution as genetic mutation is to biological evolution is an idea I first encountered in reading Donald T. Campbell's essay on the evolution of knowledge (Campbell 1960). An earlier introduction to this way of thinking came from Teilhard de Chardin's speculative but stimulating epic, *The Phenomenon of Man* (Teilhard 1965).

The concept of a **meme,** analogous at the cultural level to a gene on the biological level, was adopted from Richard Dawkins (1976). These issues are further discussed in Csikszentmihalyi (1993, 1994).

8 **Creativity with a small c.** Just at the present time, a debate rages in the field concerning the definition of creativity—see the last 1995 issue of the *Creativity Research Journal.* At question is whether an idea

or product needs social validation to be called creative, or whether it is enough for the person who has the idea to feel that it is creative. This is an old conundrum, which almost half a century ago Morris Stein (1953) tried to resolve by dividing the phenomenon into *subjective* and *objective* phases. Despite its antiquity, the question is still unresolved, and strong arguments have been advanced on both sides. My preference would be to approach creativity as a subjective phenomenon, but unfortunately I see no realistic way of doing so. No matter how much we admire the personal insight, the subjective illumination, we cannot tell whether it is a delusion or a creative thought unless we adopt some criterion—of logic, beauty, or usefulness—and the moment we do so, we introduce a social or cultural evaluation. Hence I was led to develop the systemic perspective on creativity, which relocates the creative process outside the individual mind.

I realize that to do so goes against a powerful axiom of the times. These days we take it for granted that every person has a right to be creative, and that if an idea seems surprising and fresh to you, it should be counted as creative even if nobody else thinks so. With apologies to the Zeitgeist, I will try to demonstrate why this is not a very useful assumption.

8   **Attention is a limited resource.** Many psychologists have remarked on the fact that every intentional act must be attended to, and that the capacity to pay attention is limited (e.g., Hasher and Zacks 1979; Kahneman 1973; Simon 1969; Treisman and Gelade 1980). In my opinion, this fact is one of the most fundamental constraints on human behavior, which explains a great variety of phenomena ranging from why we strive so hard to acquire labor-saving devices to why we become resentful if we feel our friends don't pay enough attention to us (Csikszentmihalyi 1978, 1990; Csikszentmihalyi and Csikszentmihalyi 1988).

10   **Creative people are often considered odd.** Studies of the traits widely ascribed to creative people include "impulsive," "nonconformist," "makes up the rules as he or she goes along," "likes to be alone," and "tends not to know own limitations." The least typical traits of creative people include "is practical," "is dependable," "is responsible," "is logical," "is sincere" (MacKinnon 1963; Sternberg 1985; Westby and Dawson 1995).

11   **Two contradictory sets of instructions.** Until about thirty-five years ago, the leading psychological theories, such as behaviorism and psychoanalytic theory, assumed that human behavior was directed exclusively by "deficit needs," such as the desire to feed, have sex, and so on. More recently, under the influence of "humanistic" psycholo-

gists like Abraham Maslow and Carl Rogers, the importance of positive drives for self-esteem and self-actualization began to be taken more seriously (e.g., Maslow 1971; Rogers 1951). It is interesting to note that this shift was greatly helped by studies of laboratory monkeys and rats, who turned out to be as motivated to do work by the chance to explore and experience novelty as they were by the opportunity of getting food. These findings suggested the existence of "exploratory drives" and a "need for competence," which changed forever the deficit-driven picture of human behavior (White 1959). See also Csikszentmihalyi (1975, 1990, 1993).

12   **Creativity in danger of being stifled.** A good short summary of this problem was given by Gerhardt Casper, president of Stanford University, in a speech given at the Industry Summit of the World Economic Forum at Stanford, on September 18, 1994: "Government and industry seem to be increasingly preoccupied with the search for technology transfer shortcuts," he said, instead of "support of original investigations of the first rank and the investment in education and training that goes with it. . . . We can readily purchase mediocrity, which will lead to nothing other than more mediocrity."

12   **Extracurricular activities.** The importance of activities outside the classroom in stimulating talented teenagers, and in keeping their motivation focused, was apparent in a recent longitudinal study (Csikszentmihalyi, Rathunde, and Whalen 1993). Inner-city youth depend even more on "nurturing settings" outside the school where they can experience a sense of responsibility together with freedom (Heath and McLaughlin 1993). Recently, Root-Bernstein, Bernstein, and Garnier (1995) have shown that creative scientists report significantly wider interests and more physical and artistic activities (painting, drawing, writing poetry, walking, surfing, sailing, etc.) than their less creative peers.

Exceptional individuals. Given the advanced age of some of the respondents, a few have passed away between the time of the interview and the present writing. John Bardeen, Kenneth Boulding, James Coleman, Robertson Davies, Linus Pauling, and Jonas Salk are no longer with us. In the text I write about all respondents as if in an extended present, which is of course appropriate considering that the impact of these individuals' lives will continue in the memory of the culture for a long time to come.

The selection process. One limitation of this study is that most of the respondents are Caucasian Americans, Canadians, or Europeans, and few members of other ethnic groups or cultures ended up in the sample. For instance, only two African-Americans and a sprinkling of Asians were represented. This would be a problem if the creative process varied fundamentally by ethnicity. My impression is that it does not, except that access to fields and domains, and the ways fields and domains operate, will vary by culture just as it varies in time and by social class within the same culture. This would be in accord with the conclusions of the Japanese psychologist Maruyama (e.g., 1980), according to which the variation in origi- nality within cultures is much greater than the variation across them. In terms of the systems model to be introduced in chapter 2, I would say that the original contribution made by the *person* is likely to be similar across cultures, while the contributions of the *field* and the *domain* will bear the distinctive stamp of the culture in which the creative process takes place. The same is true of gender differences: Within any given discipline women will use mental processes similar to those men use to reach creative results, but the differences in socialization, training, and opportunities available to men and women in a given social system may impact on the fre- quency and kind of creative contributions made by the two genders.

15 **The strategy of disproof.** The example of the single white raven and its implications for scientific epistemology obviously derive from Karl Popper's arguments, which although discredited as a *history* of sci- ence (which Popper never claimed to be doing anyway), are still unsurpassed as the *logical foundations* of it (Popper 1959).

16 **I accept stories at face value.** Acceptance of respondents' reports—qualified by the usual skepticism a scientist must bring to the object of study—is the particular prejudice I bring to the interpreta- tion of the data. Prejudice, which generally has pejorative connota- tions, is used here in the sense developed by the philosopher Jurgen Habermas, who argued that none of us could avoid being prejudiced. But by being reflective, we can to a certain extent overcome the biases that otherwise would follow from our prejudices (Habermas 1970; Robinson 1988).

During the heyday of Victorian optimism in the nineteenth century, it became a widely shared tacit assumption—or prejudice—that humankind, if not perfect, was well on its way to perfection. The great contribution of critical thinkers such as Marx or Freud has been to show that, on the contrary, human action was rife with selfishness, irrational- ity, and denial. Their insights have been expanded upon and refined by the perspectives of behaviorism, sociobiology, and countless other "isms." The pessimism implied in these theories has further gained cred- ibility as a result of the senseless evil that wars and ideologies have wrought in the last hundred years. So the new prejudice permeating our culture is now 180 degrees removed from the previous one, and holds that every human action is self-seeking, irrational, and not to be trusted.

In my opinion, neither of these extreme positions is very useful. To reconcile the opposite poles of the dialectic, we must recognize that our behavior is largely determined by ancient genetic instructions designed for self-protection and self-replication, and by more recent cultural instructions we have learned uncritically from the cultural milieu. At the same time, it makes no sense to deny that new memes, or ideas that have emerged through time—such as the concepts of humanity, of democracy, of nonviolence—can and do direct behavior toward new goals. My own prejudice is that this relative flexibility of human adaptation, or *creativity,* makes it possible to avoid the backward-looking pessimism of the currently accepted reductionistic explanations of behavior, and to entertain the hope for a genuine evolution of humankind.

17   **The tendency of the social sciences to de-bunk human motivations** (together with many other wise observations), is discussed by Hannah Arendt (1956).

CHAPTER 2

23   **Creative insight.** A recent collection of papers that pretty much covers the subject of what insight consists of—but restricted to psychological, "inside the head" approaches—is the volume edited by Robert Sternberg (1995); see, e.g., Csikszentmihalyi and Sawyer (1995) on specifically creative insights. The **systems model** (sometimes called DIFI—for Domain, Individual, Field, Interaction framework) was originally developed by Csikszentmihalyi (1988a, 1990) and further elaborated in Feldman, Csikszentmihalyi, and Gardner (1994).

25   **Brilliant.** One of the prejudices of our times is that a person who acts in unusual ways or who is involved in the arts must be creative. For instance, in advertising companies the department in charge of designing and producing the ads is usually called *creative,* and those who work in it are known as the *creatives.* While certainly there are many advertising artists who are genuinely creative, their frequency is not necessarily greater than that of creative accountants, technicians, or librarians working in the same companies. They might, however, be more *brilliant,* in the sense used here.

Personal creativity. In psychological and educational circles, what is referred to as creativity is almost always of this kind. Tests that measure fluency or flexibility of thought, or teachers' rating of the originality of children's drawings, do not measure creativity as I use the term in this book, but only the tendency to produce unusual responses, which may or may not lead to what I call here genuine creativity. Among psychologists, Howard Gruber has argued often and eloquently that we just confuse matters by applying the term "creative" to clever children and to glib test-takers (e.g., Gruber and Davis 1988).

26   **Leonardo**'s character has been often dissected (e.g., Reti 1974); for **Newton**'s see Westfall (1980) and Stayer (1988), and for Thomas **Edison**, Wachorst (1981). It is not that these and other great geniuses were tragically flawed; rather, outside their particular range of accomplishments they were just ordinary—in other words, outside of their work they failed to display that brilliance which popular opinion is so anxious to attribute to them.

27   **Genius.** Among scientists of this century, a few—for instance Richard Feynman and John von Neumann—have gained a reputation among their peers as being geniuses. This reputation seems to be based not so much on the importance of their contributions, as on the exceptional facility with which they could see and solve problems that their peers had a much more difficult time comprehending. Usually individuals who are thought to be geniuses also have unusual, sometimes photographic memory. It is probable that such persons have rare neurological talents. Nevertheless, such talents alone do not guarantee creativity. Geniuses also often cultivate personal mannerisms that set them apart from their peers, and that impress their audience as being signs of uniqueness (e.g., Feynman playing the bongo drums, or Picasso earnestly endeavoring to play out in his own life the erotic fantasies of the bourgeoisie).

29   **Fluctuations in the attribution of creativity.** Brannigan (1981) was one of the first sociologists who explored systematically the way in which a new discovery or invention had to be legitimized by recognized authorities before it could be seen as valid. He argues, for instance, that Columbus's discovery of America would have remained a relatively trivial event, and not even counted as a "discovery," except for the official recognition bestowed on it by the administrators of the Spanish king, by cartographers, by the Church, by scholars, and so on. Kosoff (1995) develops the same idea even farther—perhaps too far, inasmuch as he views creativity exclusively as a process of attribution and impression management, neglecting entirely the substantive contribution of the person.

32   **Golden years of the Renaissance.** For the list of artworks completed in Florence during the first quarter of the fifteenth century and an evaluation of their quality, see, for example, Burckhardt (1926). The discussion that follows relies heavily on the remarks on the period found in Hauser (1951) and Heydenreich (1974).

36 **Hauser.** The quote is from Hauser (1951, p. 41). A similar conclusion
is reached by Heydenreich, who writes about the same historical
period (1974, p. 13): "the patron begins to assume a very important
role: in practice, artistic production arises in large measure from his
collaboration." The same argument holds for creative production in
other domains as well.

37 **Extrasomatic instructions.** I am basing my ideas here mostly on
the work of Fausto Massimini. An example of extrasomatic instructions are the laws contained in the various political constitutions that
the two hundred or so sovereign nations of the world have adopted.
Massimini and Calegari (1979) analyzed these constitutions as if they
were chromosomes containing a great number of genetic instructions;
specific laws are nested in the constitutions as genes are in the chromosome. They also show that it is possible to trace groups of laws to
their original "ancestral strains" in the Magna Charta, and more
recent documents like the U.S. Constitution. In other words, information coded in memes rather than genes has begun to direct human
behavior (see also Massimini 1979, 1993; Csikszentmihalyi and Massimini, 1985).

39 **Creativity and age.** The relationship between age and creative
achievements in various domains was first studied by Lehman (1953)
and Dennis (1966). For more recent studies, see Over (1989) and
Simonton (1988, 1990c).

40 **Morality as a domain.** Whether it will ever be a tightly structured
domain or not, morality at long last is receiving the attention it
deserves from psychologists. Until recently, under the influence of Jean
Piaget and Lawrence Kohlberg, most scholars confined themselves to
studying moral judgments, and how children learned to make them.
The newly developing domain attempts to study actual moral behavior
(e.g., Damon 1995; Gilligan, Ward, and Taylor 1988).

42 **Scarcity of attention.** The argument suggests that, contrary to general belief, what limits creativity is not the lack of good new memes
(i.e., ideas, products, works of art), but the lack of interest in them.
The constraint is not in the supply but in the demand. This again is
one of the consequences of the aforementioned limits on attention.
Unfortunately, most attempts to enhance creativity are focused on the
supply side, which may not only not work but is likely to make life
more miserable for a greater number of neglected geniuses. We still
have very little formal knowledge about how to enhance the demand
side of creativity, although obviously entrepreneurs and philanthropists
have always had good practical knowledge in this matter.
A creative person must convince the field. Everyone who studies

creativity has remarked on this requirement (e.g., Simonton refers to it as *persuasion* [1988, p. 417]). But usually the necessity of "selling" one's ideas is seen as something that comes after the creative process ends and is separate from it. In the systems model, the acceptance of a new meme by the field is seen as an essential part of the creative process. For a recent set of papers that take seriously the importance of the social context, see Ford and Gioia (1995).

44   **The number of students in theoretical physics at the University of Rome.** These are the numbers I remember mentioned by my friend Nicola Cabibbo, who took over the chair of physics in Rome around that time. A similar fate awaited the field of sociology in the late 1960s and early 1970s, when in the aftermath of the student unrests and the Vietnam War huge numbers of students in the United States decided to major in sociology. I was then teaching in the department of sociology and anthropology at Lake Forest College, and in a few years the number of majors increased from less than ten to over a hundred. Other institutions experienced similar bursts of interest in the domain. One consequence was that in order to accommodate the tenfold increase in students, colleges hired teachers who were often not well trained and had only a faint grasp of the domain. This, in turn, resulted in a chaotic confusion that almost wrecked the field. A similar point was made by Robert LeVine concerning child development research, where in the same period the expansion of the field brought in a great number of poorly trained academics who embraced uncritically the then fashionable cognitivist theories of Piaget and Chomsky (LeVine 1991). Unassimilated novelty can be as dangerous for the survival of a domain as no novelty at all.

45   **The Romanian government.** A former student who spent several months in Transylvania collecting ethnographic material in the 1980s has described the efforts of the Romanian Ministry of Culture, whose representatives tried to retrain Hungarian, Szekler, Moldavian, and German villagers to weave, decorate, and sing songs according to Romanian patterns instead of using their traditional forms of artistic expression. Such policies are the cultural equivalent of "ethnic cleansing"; here it is not the phenotype of the genes that is being killed but only the foreign memes.

CHAPTER 3

52   **Or for artists.** A good example are two Renaissance painters who worked for the Medicis, Filippo Lippi (1406–1469) and Giovanni Angelico (1400–1455). They both started out as friars and became

famous for their exquisitely spiritual paintings of saints and Madonnas. Lippi, however, abandoned the monastery and became a riotous drunkard and libertine—he eventually eloped with a nun and had a child with her. At one point his patron, Cosimo de'Medici, decided to lock him up in his studio to make sure he finished a canvas he had been paid to do, but Filippo escaped anyway at night by knotting bed sheets together and lowering himself from the window to join a party. But no matter how dissolute his behavior, he continued all his life to paint sweet religious pictures. Giovanni, on the other hand, remained a meek monk who prayed devotedly for divine inspiration every time he took up the brush. After he died, people started referring to him as Beato Angelico, although he was never officially canonized by the Church. From the work the two men left behind one might have surmised they were identical twins, rather than persons with diametrically opposite temperaments.

Temperamental differences may be responsible, however, for why two persons exposed to the same domain will choose different aspects of it to work in, or why one will approach it in a reductionistic mode, while the other will have a more holistic approach.

53 **Pierre Bourdieu.** The influential notion of "cultural capital" was developed by this French sociologist (Bourdieu 1980).

56 **In the 1960s.** The changes in the kind of personality that art teachers thought was appropriate for art students, and the effect this had on art students, are described in Getzels and Csikszentmihalyi (1976).

The personality of artists (and creative individuals in general). Scholars (including the present one) have tried to describe the personality traits peculiar to creative people, and some of their conclusions are to a certain extent valid, at least within our particular historical context. It is likely that such traits as sensitivity, openness to experience, self-sufficiency, lack of interest in social norms and social acceptance, and—for artists—a tendency toward manic depression might be useful in increasing the likelihood that the person will try to innovate in his or her domain (e.g., Albert and Runco 1986; Andreasen 1987; Barron 1969, 1988; Cattell and Drevdahl 1955; Cross, Cattell, and Butcher 1967; Csikszentmihalyi and Getzels 1973; Getzels and Csikszentmihalyi 1968, 1976; MacKinnon 1964; Piechowski and Cunningham 1985; Roe 1946, 1952). However, I am now convinced that such unipolar traits are less accurate in describing the personality of creative individuals than the dialectical notion of complexity.

57 **Complexity.** The concept of complexity is central to many of my previous writings, especially *The Evolving Self* (Csikszentmihalyi 1993). In this context I am using the term in a similar but much more

restricted sense, without the extensive theoretical implications I usually intend it to convey. The flexible, adaptive personality style it describes shares similarities with other traits described by psychologists, but it is not identical with any of them. For example, Jack Block's concept of *ego resiliency* (Block 1971, 1981), which includes a tendency toward adaptability and resourcefulness, could be seen as very similar; however, ego-resilient people are strong on one-dimensional traits such as integrity, dominance, and self-acceptance, which might not be the best way to describe creative individuals who are also prone at times to insecurity and self-doubt.

Previous researchers have often attributed seemingly contradictory traits to creative people, such as "openness to experience" and "preference for challenge and complexity" (e.g., Russ 1993, p. 12). But these traits are seen as separate, or orthogonal to each other, instead of representing variations along a continuum. Closer to my notion of complexity is Dennett's (1991) view of consciousness, and Ornstein's (1986) concept of "multimind," or the brain's tendency to integrate separate and often conflicting neural sequences, thus often producing incongruent or contradictory thoughts and actions within the same person. Perhaps creative individuals, for whatever reason, are more prone to accept and to leverage this feature of the mind.

Carl Gustav Jung. See, for instance, Jung (1969, 1973)

59 **Longitudinal studies.** The first longitudinal study of exceptionally gifted children was conducted by Lewis M. Terman at Stanford University, who followed the vicissitudes of one thousand children with very high IQs throughout their lives, a study that is still being continued. See, for instance, Terman (1925), Oden (1968), and Sears (1980) for the outcomes of these investigations.

Later studies. Jacob W. Getzels and Philip Jackson (1962) were the first to compare children who scored high on IQ tests but not on creativity tests with children who scored high on creativity tests but not on IQ tests; they found that the two groups were quite different. For example, the high IQ children were more conventional and extrinsically motivated, while the highly creative ones were more rebellious and intrinsically motivated. As one might expect, teachers preferred the first kind. More recent work on this topic is summarized by Westby and Dawson (1995).

60 **Howard Gardner** (1993) studied seven exemplary creative geniuses of this century.

The distinction between **convergent** and **divergent** thinking was first made by J. P. Guilford, the pioneer in the modern psychological study of creativity, who claimed that divergent thinking was peculiar to creativity, and who developed the first tests to measure it, which are still being used (Guilford 1950, 1967). Paul Torrance subsequently contributed greatly to the measurement of divergent thinking (Torrance 1988); for recent reviews of the relationship between divergent thinking and creativity, see Baer (1993) and Runco (1991).

62 **Vasari.** The first set of biographical sketches of artists was written by this Florentine historian (who was also a decent artist himself). One strength of his work is that Vasari (1550) knew personally many of the Renaissance artists whose lives he chronicled.

65 **A recent issue of** *Newsweek.* For the article on John Reed, see Levinson (1994).

Extroversion and introversion. This polarity is one of the oldest in personality psychology. It was initially adopted by C. G. Jung, and is now considered one of the five basic traits along which individuals differ. The major work on this concept was done by the German-British psychologist Hans J. Eysenck (1952, 1973), and current systematic studies of extroversion and introversion have been influenced by the research of Costa and McCrae (1978, 1984).

Teens who can tolerate being alone. For a recent account of the difference it makes to the development of talented teenagers whether they can stand solitude, see Csikszentmihalyi, Rathunde, and Whalen (1993).

70 **Androgyny.** There is ample evidence that talented and creative individuals show traits usually associated with the opposite sex, and express the traits of their own sex less strongly than the average person. In my own work, such findings were reported in Getzels and Csikszentmihalyi (1976) and in Csikszentmihalyi, Rathunde, and Whalen (1993). See also Spence and Helmreich (1978). It is likely that this tendency is responsible for the currency of rumors about the homosexuality of creative individuals like Leonardo and Michelangelo. Such attributions are always difficult, because they rely heavily on interpretation and often project current meanings on behaviors that in the past had a very different significance. Although there might be a tendency toward homosexuality for creative persons in some fields under certain sociocultural conditions, the currently widespread belief that the two are linked is probably exaggerated.

73 **Psychopathology and addictions** in artists and writers. See, for instance, the recent reports by Andreasen (1987), Claridge (1992), Cropley (1990), Jamison (1989), and Rothenberg (1990). Despite the clear relationship one finds nowadays between some forms of creativity and some forms of pathology, I am convinced that this is an accidental rather than an essential connection. In other words, if creative

musicians are often addicted to drugs and playwrights tend to get clinically depressed, this is more a reflection of the historical conditions in which they have to work than of the work itself. This was, to a certain extent, also the argument of the psychoanalysts Ernst Kris (1952) and John Gedo (1990). Certainly many great artists seem to have avoided psychopathology, and even enjoyed superior mental health: for instance, writers Chekhov, Goethe, and Manzoni; composers Bach, Handel, and Verdi; and visual artists Monet, Raphael, and Rodin.

CHAPTER 4

77   **The creative process.** As the statement by Galvin illustrates, the way in which creative results come about can be approached from two main directions. The first asks the "how" question (in this case, *anticipation*), which focuses on the mental, or cognitive, steps that lead to novel outcomes through the framing of new questions. Most research on creativity takes this approach. The second direction asks the "why" question (in Galvin's terms, *commitment*), which deals with the affect and the motivation that drives a person to innovate. Outside of psychoanalytic writers, few scholars have taken this approach, even though everyone agrees about its importance. I feel uneasy about drawing too sharp a distinction between the cognitive and the motivational aspects of creativity (or any other mental process). It seems to me that the two are so closely intertwined that to separate them sharply interferes with a real understanding of what is going on. This was the crux of my disagreement with Herbert Simon, who believes that a rational, computer modeled sequence of thought adequately represents actual historical creative processes, which, in my opinion, are largely arational (e.g., Csikszentmihalyi 1988b; Simon 1988).

79   **The steps of the creative process.** The demarcation of the cognitive steps involved in the production of novelty (i.e., the "how" of creativity) was first clearly formulated by Wallas (1926). Some scholars recognize three steps (i.e., preparation, incubation, insight), while others mention as many as five (i.e., preparation, incubation, insight, evaluation, elaboration).

  Another relevant concept is that of *intuition*, or the "vague anticipatory perception that orients creative work in a promising direction" (Policastro 1995, p. 99), a process that presumably takes place between the phases of incubation and insight.

80   **1 percent inspiration.** European colleagues tell me that the quip about creativity being 99 percent perspiration was first made by the German poet Johann Wolfgang von Goethe, who died fifteen years before Thomas Edison was born. I could not substantiate this claim, but

even though Goethe has said many insightful things about creativity, it seems to me that this particular aphorism fits Edison's mentality better.

81   **Darwin.** Howard Gruber has written the classic account of the psychology of Darwin's creative process, based on a close analysis of the notebooks in which Darwin recorded his thoughts as they unfolded throughout his active life (Gruber 1981).

Dyson's role in the development of quantum electrodynamics is discussed in the recent book by Schweber (1994), who argues that Dyson should have shared the Nobel Prize awarded to Tomonaga, Schwinger, and Feynman in 1965.

84   **Notebooks.** It is not only writers who keep diaries and notebooks of daily experiences. Scientists also keep lab notes or other records that will help them to think through their findings and ideas. A perhaps extreme example is the virologist D. Carleton Gajdusek, who was awarded the 1976 Nobel Prize in physiology and medicine, whose notebooks cover about 600,000 single-spaced typewritten pages, a third of which has already been published (Gajdusek 1995).

"Do you know a novel about happiness?" There is a good short story by Italo Calvino, "The Adventure of a Poet," on the theme of how difficult it is to write about happiness (Calvino 1985). It is true that the world literature is filled with tragedies, while the opposite of tragedy—i.e., the story of a deserving person getting his or her just dues—exists only in Horatio Alger–type narratives that fail to make the grade as great literature. (There are, however, great comedies.) My impression for why this is so is that the situation might be the opposite of what Tolstoy said, that happiness is repetitive and unhappiness unique. Happiness is such a private and idiosyncratic experience that it is almost impossible to communicate it, and the writer must resort to trite clichés to describe it. On the other hand, unhappiness is so pervasive and uniform that everyone can immediately recognize it, so the writer is freed up to use style and imagination to embroider on unhappy themes, confident that the reader will be able to empathize with the subject.

86   **Studies of creative scientists.** Ann Roe (1951, 1953) was among the first psychologists to study creative scientists, mostly from a motivational perspective (the "why" question). Another classic investigator in the same vein has been Bernice Eiduson (Eiduson 1962). The French mathematician Jacques Hadamard wrote a classic account of the cognitive aspects of creativity in his domain (Hadamard 1949), and the biochemist Hans Adolf Krebs, whose research explained how living organisms produce energy, described the creative process in physiology and medicine (Krebs and Shelly 1975). Several scientists have left excellent accounts of their working methods, including some of

those who took part in this study; for example, Freeman Dyson, Gerald Holton, John Wheeler, and E. O. Wilson.

93   **Wars are notorious for affecting the direction of science.** Dean Keith Simonton is the psychologist who has done the most extensive surveys of the relationship between historical conditions, such as wars and other forms of conflict, and creativity. His historiographical methods are based on the secondary analysis and compilation of thousands of historical facts on the one hand, and frequencies of creative productivity (e.g., books, musical compositions, inventions) on the other. See, for instance, Simonton (1990b). For a slightly different approach, which looks at the relation between forms of political power and creativity, see Therivel (1995).

94   **Twentieth century.** The history of creativity in the twentieth century is well illustrated by Howard Gardner's biographical account of seven representative geniuses of our times (Gardner 1993).

95   **Presented and discovered problems.** The psychologist Jacob W. Getzels, my mentor in graduate school, became impressed by the many accounts of creative individuals that emphasized the importance of *problem finding*—as opposed to *problem solving*—in the creative process. He then developed a model of problem formulation based on the distinction between discovered and presented problems (Getzels 1964). The model was further elaborated and applied to research with creative artists (e.g., Getzels 1975, 1982; Getzels and Csikszentmihalyi 1976). This perspective has become a useful one for studying creativity; see, for instance, the recent collection of studies edited by Runco (1994).

100   **The functions of idle time.** That incubation helps to make a connection between a highly salient but repressed experience and its expression in a form acceptable by the superego was developed by Freud in his essays on Leonardo da Vinci's childhood and Michelangelo's sculpture of Moses (Freud 1947, 1955). These essays spawned a large literature (e.g., Kris 1952; Rothenberg 1979). The classic treatment of creativity by Arthur Koestler (1964) is also heavily influenced by this perspective.

In a similar vein, the more creative scientists interviewed by Eiduson and Root-Bernstein differed from the less creative ones in that they reported more often that their ideas arose while dreaming, or while working on a different but related problem (Root-Bernstein, Bernstein, and Garnier 1995). It is difficult to know, however, to what extent such reports are shaped by received notions about how the creative process "should" unfold; a difficulty that obviously applies to my study as well.

The alternative explanation of why idle time is necessary is based on a model of mental processing that stresses random associations of ideas that may take a great deal of time to result in useful combinations (e.g., Campbell 1960, 1974; Johnson-Laird 1988; Simonton 1988)—somewhat akin to the millions of monkeys typing at random needed to produce a Shakespearean masterpiece by chance—or it involves connections that while unconscious are still based on logical associations (e.g., Dreistadt 1969; Barsalou 1982).

101 **Serial and parallel processing** of information. For a basic introduction to this topic, see Rumelhart et al. (1986).

CHAPTER 5

108 **Programmed for creativity.** That people prefer to describe what they enjoy doing most with the phrase "designing or discovering something new" was a result of the first study of optimal experience I conducted (Csikszentmihalyi 1975). The dual motivational system, programmed for survival on the one hand and for evolution on the other, is discussed in Csikszentmihalyi (1985, 1993).

109 **Entropy.** Here I am using the term in its more usual meaning, as the inability of a system to do work. It is different from *psychic entropy,* which is the state of consciousness characterized by inner disorder, negative emotions, or simply the inability to engage in purposeful action. Its opposite is *psychic negentropy,* or flow, which describes an ordered state of consciousness, positive emotions, and the ability to engage in intentional action (see Csikszentmihalyi 1978, 1982).

110 **The flow experience.** The description of the common experiential state reported by people who enjoyed various activities such as rock climbing, chess, dancing, and so on was first provided in Csikszentmihalyi (1975). A wide range of subsequent studies on flow conducted by researchers in many different cultures was reported in Csikszentmihalyi and Csikszentmihalyi (1988). See also Csikszentmihalyi (1993), Csikszentmihalyi and Rathunde (1993), Massimini and Inghilleri (1986, 1993), and Inghilleri (1995). George Klein (1990) collected a number of enlightening essays from artists and scientists describing the flow they experienced in their creative work.

116 **Separating bad ideas from good ones.** Sir Peter Medawar, the British virologist who was such a keen reporter of the creative process in his field, held that the central skill involved in creativity was to grasp which were the soluble problems (Medawar 1967). Several respondents in our study mentioned the same thing, sometimes referring back to Medawar's idea, thereby demonstrating how difficult it is to separate a direct experience from a received opinion.

117 **The barrier of entropy.** Professor Frank Lambert, a chemist, has

suggested to me that the difficulty in entering flow bears an interesting resemblance to the activation energy that certain metastable physical systems require in order to mantain a higher internal energy state. For instance, iron tends to corrode into iron oxide, or rust, when exposed to air or water, thereby losing some of its internal energy. But it will maintain its higher-energy, metastable condition if external energy is added prior to its degrading; for example, if the iron is painted or turned into steel (Lambert 1995). The phenomenological parallel is that without psychic energy expended in learning to control consciousness, the mind tends to fall into random, low-energy states. While one must make an effort to focus attention to enter the flow state, as soon as one is in it, external distractions are much less likely to disrupt concentration, and even great expenditures of physical and mental energy are experienced as if they were effortless. It remains to be seen whether there is more to the similarity between these two entirely different processes than superficial appearance.

122 **Intrinsic motivation.** The importance of intrinsic rewards has been realized relatively recently by psychologists, who until the 1960s considered only the satisfaction of genetically programmed needs to be rewarding. Currently among the leading researchers in this area are Amabile (1990) and Deci and Ryan (1985). See also Csikszentmihalyi and Rathunde (1993).

123 **The more flow, the more happiness.** See, for instance, Csikszentmihalyi and Nakamura (1989), Csikszentmihalyi and Wong (1991), Wells (1988), Adlai-Gail (1994), and Moneta and Csikszentmihalyi (1995). But if a person experiences flow in activities that are destructive or lack complexity, or if one becomes addicted to a single flow activity at the expense of a balanced life, flow might have negative consequences; see Csikszentmihalyi and Larson (1978) and Csikszentmihalyi (1985b).

125 **Children grow up believing.** In a current research of young people's transition from school to work, we find that of a national cross-section of more than four thousand teenagers 15 percent would like to become professional athletes (the number one choice), 4 percent would like to become musicians, and 6 percent actors. In other words, if we consider professional athletes as being primarily entertainers, at least one out of four adolescents aspires to a career in entertainment (Bidwell, Csikszentmihalyi, Hedges, and Schneider 1995).

Chapter 6

128 **Being in the right place.** For some of the effects of the physical environment on psychological functioning, see Gallagher (1993).

134 **Bellagio.** The history of this Italian town in Lombardy and its many creative visitors is in Gilardoni (1988).

139 **The macroenvironment.** One approach to the relationship between social structural variables and creativity is the series of historiometric analyses by Simonton (e.g., 1975, 1984). Another is the qualitative analysis of the relationship between artistic creativity and sociocultural factors in contemporary America by Freeman (1993).

141 **Pekka.** Juhani Kirjonen, a colleague from the University of Jyväskylä in Finland, told me about Pekka, whom I didn't have the good fortune to meet personally.

Joe M. was one of the subjects of my earliest studies of flow conducted twenty-five years ago (see Csikszentmihalyi 1975).

142 **Symbolic ecology in the home.** The study of more than one hundred households and the objects in them that were special to their owners is described in Csikszentmihalyi and Rochberg-Halton (1981).

143 **The car as a "thinking machine."** This and other conclusions about the place of cars in our symbolic ecology are based on a study I conducted for Nissan U.S.A. in 1991.

145 **Career change every ten years.** In his analysis of the biographies of great geniuses, Howard Gardner (1993) concluded that major breakthroughs in their work occurred once every ten years. Presumably these two observations—career change and timing of new masterpieces—reflect the same cycle of creative work.

CHAPTER 7

151 **Childhood and creativity.** Of the many studies concerned with the early experiences of creative individuals—most of which, by necessity, use biographical accounts of long-dead individuals and hence are often of dubious authenticity—one might mention Freud's reconstruction of Leonardo's infancy and childhood (Freud 1947); and the summary of the biographical evidence on the childhoods of three hundred eminent persons by Goertzel and Goertzel (1962). Some of the analyses of the sibling position of creative individuals are by Zajonc (1976), Albert (1983), and Albert and Runco (1989).

154 **Giotto's childhood.** The biographical note on Giotto is from Semenzato (1964, p. 7).

158 **Interest.** As mentioned earlier, until recently psychologists were not very interested in the topic of interest. This has changed, however; see, for instance, Renninger, Hidi, and Krapp (1992) and Schiefele (1991). We still know next to nothing about individual differences in interest, that is, whether some children are more interested than others in gen-

eral, or why some become interested in one topic and others in another topic.

161 The influence of parents. See, for example, Harrington, Block, and Block (1992) and Csikszentmihalyi and Csikszentmihalyi (1993). Despite the fact that recent scholarship in family studies has abandoned the notion that the parents' effects on children are primary in favor of a systemic perspective that sees the family interaction as having the most important effect (Grotevant 1991), I still believe that parents influence children more than the other way around, and largely independently of any interaction effects.

168 Missing fathers. The essay from which the quotation is taken is in Klein (1992).

171 Jean-Paul Sartre's aphorism about a father's gift to his son is in *Les Mots* and is quoted in Klein (1992, p. 162).

172 The Mirror of Retrospection. Many of the surveys conducted by Elisabeth Noelle-Neumann (e.g., 1985) show that adults who are satisfied with their present condition report having had more idyllic childhoods. The successful young artist who revised his past was first described in the volume by Getzels and Csikszentmihalyi (1976), when his childhood was still unproblematic.

174 The influence of a teacher. The importance of single individuals— parents, teachers, peers, mentors, spouses, students—in helping along the career of creative individuals is examined in Mockros (1995) and Mockros and Csikszentmihalyi (1995).

Eugene Wigner's recollections of his high school math teacher are found in his autobiography (Wigner 1992). The influence of teachers in the Lutheran high school in Budapest has also been described by Hersh and John-Steiner (1993).

176 For the conflicts peculiar to **talented teenagers,** see Csikszentmihalyi, Rathunde, and Whalen (1993).

178 Artists uninterested in academic subjects. A thorough description of the values held by artists in general and as they relate to academics can be found in Getzels and Csikszentmihalyi (1976).

CHAPTER 8

187 Recent studies. I am referring here in particular to the survey studies of sexual behavior completed by colleagues at the University of Chicago (Laumann et al. 1994).

199 Erik Erikson. Erikson's classic description of his eight psychosocial stages of development is in Erikson (1950).

206 Wigner. The quote is from Wigner (1992, p. 254).

208 "Life theme." The concept of life themes, or the cognitive representations we develop of our goals and of the narrative of our lives, was first developed in Csikszentmihalyi and Beattie (1979). See also Csikszentmihalyi (1990, pp. 230–40).

CHAPTER 9

211 Creativity peaks in the third decade of life. The early studies of age changes in creativity were by Lehman (1953) and Dennis (1966). See also Simonton (1990c) and the brief summary in Rybash, Roodin, and Hoyer (1995).

Quantity and quality. What older individuals accomplish is not determined only—or even primarily—by the limitations of biological aging, but also by the personal attitudes and social opportunities concerning old age. There is accumulating evidence that longevity, health, physical performance, and social achievements in old age can be greatly improved by adopting appropriate values and behaviors; see, for instance, the conclusions of Dr. Walter Bortz (1991).

213 The distinction between fluid and crystallized intelligence was introduced by Horn (1970). Some of the most careful investigations of changes in mental performance with age were done by K. Warner Schaie and his associates (e.g., Schaie 1990, 1994). See also Labouvie-Vief (1985).

215 Risk taking in later life. It has been suggested that riskier problems in science are more typically addressed either by established scientists who can afford to do so, or by those not established at all who have very little to lose (Zuckerman and Lederberg 1986; Zuckerman and Cole 1994), a propensity that fits Sternberg and Lubart's (1991) economic investment theory of creativity.

224 Integrity. The quotation is from Erikson (1968, p. 140). The psychiatrist George Vaillant has recently suggested that an important developmental stage between generativity and integrity is the one he calls "Keeper of the meaning," a task that confronts a person after midlife and that involves selecting and passing on to the next generation the wisdom one has learned (Vaillant 1993). This stage is one that the people in this book seem well prepared to meet.

231 Anthropic principle. For an introduction to the concept see Barrow and Tipler (1986) and Gribbin and Rees (1989).

CHAPTER 10

238 Why are we interested in literature? I heartily agree with Umberto Eco on this matter: ". . . it is easy to see why fiction fascinates us. It offers us the opportunity to employ limitlessly our faculties for perceiving the world and reconstructing the past . . . it is through fiction that we adults train our ability to structure our past and present experience" (Eco 1994, p. 131). In their cross-cultural interviews, the team of investigators led by Fausto Massimini found that reading was one of the most often mentioned sources of flow worldwide—often the main source (e.g., Massimini, Csikszentmihalyi, and Delle Fave 1988).

241   Problem-finding process. One of the most pronounced differences between artists whose work was judged to be creative by expert judges and those whose work was deemed not creative is that the former approached an experimental drawing task not knowing what they wanted to draw, whereas the less creative artists started with a clear idea of what they wanted to do from the beginning (Getzels and Csikszentmihalyi 1976). The former "discovered" their problem in the process of drawing, in interaction with the medium and the developing image; the latter toiled on a problem that could already be visualized before the creative process started. This kind of open-ended process that leads to discovery was typical of the working methods of the group reported on in this study, and is well described here by Mark Strand. For a recent update on studies of the relationship between problem finding and creativity, see Runco (1994).

242   "I could never stay in that frame of mind. . . ." In fact, staying in the flow state for long periods is almost impossible. People who are fortunate to have a vocation or a hobby that is enjoyable and all-involving may experience flow every day, and sometimes for long periods. But even the most adept must take occasional breaks because of hunger, sleep, fatigue, or the sheer exhaustion that follows from the extreme concentration of the flow experience.

245   Struggles with the field. It is probable that the less binding the rules of a domain are, the more free the field is to exploit young people who want to be recognized and advance in it. For this reason, an actress is more likely to be expected to sleep with a producer in order to get ahead than a young scientist is, because in science the value of a person's contribution can be more clearly established with references to the rules of the domain. But of course this does not mean that even the most rigorously organized domains, such as mathematics, can keep the field entirely free of exploitation, politics, and personal vendettas.

252   Assimilating the style of predecessors. There is a story about Pablo Picasso, who in his maturity was asked by an interviewer why he had spent so much time as a young man imitating the style of the great masters of painting. "If I had not imitated them," Picasso is supposed to have answered, "I would have had to spend the rest of my life imitating myself." Certainly Picasso cannot be accused of being a traditionalist, but even he recognized that without mastering the best achievements of a domain, one is left only with one's naked talents, having to reinvent the wheel without tools.

254 **Too much encouragement.** L'Engle's contention that too much parental encouragement can be an obstacle to the development of a child's talent makes sense on two counts: In the first place, praise tends to heighten self-consciousness, which in turns interrupts the flow experience; thus it is important to reserve whatever praise one wants to give until the child's episode of involvement with the talent area is over. Second, and more important, is the fact that parental encouragement often takes the form of heightening the child's awareness of extrinsic rewards, thereby undermining the intrinsic rewards of the activity. For instance, if the parents keep stressing "You have to keep practicing that piano or you will never play in Carnegie Hall," the child learns that the reason for playing is to get future recognition and success, not the present enjoyment of music. Unfortunately, indiscriminate praise is often advised as a parental technique that will raise children's self-esteem (e.g., McKay and Fanning 1988), as if a self-esteem based on spurious praise was worth having. See also Damon (1995, chapter 4) for a similar argument. But feedback—including praise—that is directed at concrete details of the performance can be very useful; see Dweck's (1986) distinction between "learning" and "performance" goals, and Deci and Ryan's (1985) distinction between "informational" and "controlling" feedback.

255 *Owning Your Own Shadow.* L'Engle is referring here to an aspect of the recent revival of interest in Carl Gustav Jung's thought, which includes the concept of the "shadow," or the dialectical opposite of the traits a person usually acknowledges and displays (Jung 1946, 1968). This dark side of the personality can cause severe inner conflict if it remains repressed. For a contemporary interpretation of the influence of the shadow on consciousness and behavior, see the edited volume by Abrams and Zweig (1991) and O'Neil (1993).

261 **"I would say that the chief obstacle is—oneself."** The quote by Stern in which he describes vanity, pride—the rubbishy parts of oneself—as being the greatest obstacles in one's life is a good illustration of what it means to "own one's shadow," as referred to in the preceding note.

CHAPTER 11

266 **Alexander von Humboldt** (1769–1859). His five-volume description of the cosmos was translated from German into English in the last decade of the last century (von Humboldt 1891–93). Some people claim that our understanding of nature took a wrong turn when Dar-

win's vision of natural selection, which was very compatible with the competitive capitalistic ideology of late Victorian England, prevailed over Humboldt's more systemic vision. Whether this is true or not, the issue suggests how social systems might influence fields in the shaping of domains; in this case, the competitive Victorian social milieu might have recognized itself in Darwin's theory, and encouraged naturalists to adopt it as the dogma of biology.

Sociobiology. This perspective for explaining human behavior, based on an elaboration of Darwinian principles, has been perhaps the most momentous paradigm shift affecting the social sciences in the second half of this century. While Freudians explain our actions with reference to repressed sexual desire, Marxists in terms of conflicts brought about by unequal control of the means of production, Skinnerians in terms of learned responses to pleasant stimuli, sociobiologists explain them in terms of the reproductive advantages that different actions give. In other words, other things being equal, we choose to do things that give us a greater chance to leave offspring who, in their turn, will grow up to leave offspring. Simple as this assumption sounds, it can be applied to a very wide range of actions, and with a certain amount of mathematical precision. The widespread impact of this concept is in large part due to the work of E. O. Wilson (1975).

267 **Ernst Mayr.** Some of the relevant references are to Mayr (1947), James Watson (1980), Konrad Lorenz (1966), Ellsworth Huntington (1945), and William Hamilton (1964).

270 **Thomas Kuhn.** The seminal description of the sudden changes that transform domains, namely "paradigm shifts," is in Kuhn (1970). Before it was thought that science proceeded slowly by logical steps based on prior knowledge rather than radical reformulations.

279 **"I still remember . . ."** The quote is from the Kleins' autobiographical essay (Klein and Klein 1989, p. 7).

280 **"My own boss . . ."** Ibid., p. 14.

289 **Responsibility to the living world.** This attitude is well expressed in the concept of "biophilia" (Wilson 1984), as well as in the writings of the other life scientists discussed in this chapter, e.g., Salk (1983) and Klein (1992).

CHAPTER 12

295 **"Most people in the university work for the admiration of their peers."** Commoner here puts his finger on a problem that is typical of the history of fields in general. At first, they are constituted to solve a genuine problem: The priesthood exists for the sake of providing meaning to people's lives, doctors for the sake of curing disease,

the army to protect us from enemies, the universities to teach specialized knowledge . . . but as time passes, each institution unconsciously changes its priorities to that of aggrandizing and preserving itself. This is the kind of "mimetic exploitation" I discuss in Csikszentmihalyi (1993, pp. 109–14) and which we have to learn to avoid lest the culture become stagnant.

296 **Those who are not properly socialized.** One of the paradoxes of creativity is that a person must be socialized to a field, that is, learn its rules and expectations, yet at the same time remain to a certain extent aloof from it. A person who identifies too strongly with the field and its problems has no incentives to break into new territory, and is not interested in exploring knowledge that lies outside the boundaries of the domain. This is why creative persons are so often marginal, with one foot in the field and one outside of it (compare Therivel 1993).

298 **C. S. Peirce.** According to the pragmatic philosopher Charles Sanders Peirce, an act of *recognition* is one in which the object is simply assimilated to previous conceptual schemas, and nothing new happens in the mind; whereas an act of *perception* is one in which the object stimulates new thoughts or feelings that result in the expansion of consciousness (Peirce 1931). This distinction is echoed in the teachings of the Yaqui Indian sorcerer Don Juan, described by Carlos Castaneda, one of whose basic techniques involved breaking down the conventional conceptual categories of experience (Castaneda 1971).

Lack of formal education. Dean Simonton's research into the background of recent historical figures suggests that the most creative ones reached the sophomore year of a college education. More education than that seems to be as detrimental as less (Simonton 1990a). Of course, in some domains it is impossible to get started, let alone make a creative contribution, without advanced degrees.

302 **"I learned through the school of hard knocks. . . ."** The quote by Henderson gives a very good prescription of how one can build flow and intrinsic rewards into the plan of an organization, an important topic about which we unfortunately know very little as yet.

315 **Simple cultures' views of the cosmos.** See, for example, Massimini and Delle Fave (1991).

316 **Myth of Gaia.** A review of how the idea that the Earth is a self-correcting organism developed can be found in Joseph (1990). The original Gaia hypothesis, stating that the temperature of the planet and the chemistry of the gases surrounding it are produced and maintained by the sum of living organisms, was developed by James E. Lovelock (1979).

CHAPTER 13

318 The axemaker's gift. See Burke and Ornstein (1995) and also Csikszentmihalyi (1993, chapter 5).

319 The reasons for the decline of the **Maya** civilization were discussed at a recent archaeological meeting reported in the *San Francisco Chronicle* of April 12, 1995, p. A7.

322 Géza Róheim. Róheim was a psychoanalytically trained ethnographer who studied, among other native cultures, that of the Australian aborigine. He became convinced that the ideal condition of existence was that of inorganic matter, and that life forms, including human life, were transient forms of irritation or disease (Róheim 1945). In this sense, his views are diametrically opposed to those expressed by Wilson (1984). Such conflicting ways of interpreting the same phenomena are a good example of the sort of differences based on conflicting metaphysical assumptions that Popper (1959) claimed could not be resolved scientifically.

325 Standards for the arts. For an overview of one of these approaches see Smith (1989) and Dobbs (1993).

327 Supportive families. While there is no direct evidence about the relationship between family practices and creativity, the retrospective interviews collected by Benjamin Bloom (1985) with scientists and artists reveal an enormous amount of parental investment in their gifted children. See also Harrington, Block, and Block (1992). In general, a combination of parental love and discipline seems to work best in nurturing the development of talent in children (e.g., Baumrind 1989; Rathunde and Csikszentmihalyi 1993).

331 Self-esteem of Asian and African-American students. The popular belief is that disadvantaged minorities suffer from low self-esteem, and if only their self-esteem could be raised, their academic performance and success in general would improve. But the facts seem to be different. For instance, the self-esteem of African-American students tends to be higher than that of Caucasians, which in turn is higher than Asian students'—in inverse proportion to their academic achievement (Bidwell, Csikszentmihalyi, Hedges, and Schneider, still in press). The reasons for this are not so difficult to understand, if one keeps in mind William James's (1890) formula for self-esteem: the ratio of achievements over expectations. If expectations are very high, as they usually are among Asian-Americans, one would expect their self-esteem to be low, even when their achievements are relatively high.

Expectations of American and Asian parents. How Asian-American families communicate high expectations for academic achievement is described in Sue and Okazaki (1990), Schneider et al. (1992), Stevenson and Stigler (1992), and Asakawa and Csikszentmihalyi (1995).

333 **10 percent of thirteen-year-olds wanted to be architects.** This is one of the findings of a cross-national sample of American adolescents (Bidwell, Csikszentmihalyi, Hedges, and Schneider, still in press).

335 **Benvenuto Cellini (1500–1571).** One English translation of the autobiography of this exemplar of a Renaissance artist is in Cellini (1952).

340 **Getty Center for Education in the Arts.** Discipline-Based Arts Education, or DBAE for short, is the method for teaching art in schools that was developed under the auspices of the Getty Center. See, for instance, Alexander and Day (1991) and Dobbs (1993).

341 **Four-year-olds who learn calculus.** An example of a child with extremely precocious mathematical gifts is described in Feldman (1986).

Creativity and material advantages. Occasionally it is claimed that creative persons have no interest in material success; this, in my opinion, is a romanticized exaggeration of the obviously strong intrinsic motivation such people possess. In line with the complexity of their personality, one should not expect that the strong intrinsic motivation of creative individuals needs to exclude an interest in fame and fortune. Recently Sternberg and Lubart (1991) have proposed an "economic" theory of creativity, based on the maxim "buy low, sell high"; in other words, the notion that creativity involves, at least in part, interest in developing unpopular ideas that might eventually catch on.

342 **Science students see only the drudgery of the discipline.** Whereas students talented in music and the arts report a much more positive quality of experience than average when engaged in music and art, students talented in math and science report a much lower than normal quality of experience when they are doing math and science. In other words, they are less happy, less motivated, and have a less positive self-esteem when working on their talent than when doing other things (Csikszentmihalyi and Schiefele 1992; Csikszentmihalyi, Rathunde, and Whalen 1993). The reason for this is partly that art and music are more immediately enjoyable than math and science; but in large part it is due to our attitudes toward "hard" academic subjects, and the way we teach them.

CHAPTER 14

345 **Obstacles are internal.** Many of these internal obstacles are the result of the fact that our nervous system cannot handle more than a limited amount of information in consciousness at the same time, and so we cannot attend to more than a few things at a time (see Csik-

szentmihalyi 1978, 1990; Hasher and Zacks 1979; Kahneman 1973; Simon 1969; Treisman and Gelade 1980).

347 Try to be surprised. As we know, creative individuals tend to face experience with openness, bordering on what Goethe called "naïveté." This suggestion is similar to the one Don Juan gave to his apprentices, a practice he called "stopping the world" (Castaneda 1971). It consists in registering sensory stimuli without labeling them according to culturally defined conventions; for instance, looking at a tree without thinking of it as a "tree," or letting any previous knowledge about trees enter into consciousness. It turns out that this exercise is extremely difficult, if not impossible to carry out (compare Peirce's concept of *perception*). The suggestion to be surprised by what one encounters during the day is a less radical version of "stopping the world."

Try to surprise at least one person. Of course, I am not advocating that one should become obnoxious or pushy. Some individuals need so much attention to confirm their importance—or even exis tence—that they will do anything to get it: talking loudly, acting flamboyantly, defying conventions, getting involved in risky behavior. The difference between such behavior and what I am suggesting is that the latter is a tool for learning about oneself and the world, a way of broadening one's repertoire of experiences, a means for generating novelty.

348 When there is nothing specific to do, our thoughts soon return to the most predictable state. The conclusion that the natural state of the mind is chaos is based on my studies with the Experience Sampling Method, which show that when people are alone with nothing to do, their thoughts tend to become disordered and their moods negative (e.g., Csikszentmihalyi 1992; Csikszentmihalyi and Larson 1984; Kubey and Csikszentmihalyi 1990). As the neuropsychologist George Miller said, "The mind survives by ingesting information" (Miller 1983, p 111); when there is no information to keep it in an ordered state, the mind begins to lose control of attention, at least temporarily. As with most such generalizations, this rule does not apply to everyone: Individuals who have learned to control their minds even in the absence of external inputs of information—by learning a symbolic system and its operations, such as prayer, meditation, mathematics, poetry—can avoid the entropy of solitude, and even enjoy it.

350 The most mundane activities. On the average, we spend almost 40 percent of our waking life doing "maintenance" activities, such as washing, dressing, eating, and cleaning (e.g., Kubey and Csikszentmihalyi 1990). These are not productive activities that generate income or some tangible product, nor are they leisure activities we do because they are inherently enjoyable. Maintenance activities involve routines we must do repeatedly just to survive (e.g., eat) and to get along with others (e.g., wash and dress). Many people feel that this part of life is "wasted" because it is neither fun nor productive. Thus it would improve the quality of life greatly if one were able to transform even a small portion of this wasted time into enjoyable experience.

Gardening. A good example of how gardening produces flow is described in a research report of a study conducted in Germany by Dieter Reigber (1995).

352 Relinquishing control over attention. One of the most often heard comments on my concept of psychic energy is that it applies only to Western cultures, and that the highest achievements of Eastern religions and philosophies depend not on the control of attention, but on the contrary—its surrender. I think this objection is based on a misunderstanding of what the process of "surrendering" or "relinquishing" control of attention entails. In my opinion, these processes of surrender are among the most difficult acts of control that consciousness can accomplish. Given the naturally chaotic state of the mind, to achieve the affectless, unfocused consciousness of the mystic requires enormous effort and long training. Therefore, I believe that at least in this respect East and West are similar: In both cultures, the highest psychic accomplishments depend on the control of attention.

355 Microenvironments. How the personal space one creates and the objects one surrounds oneself with affect a person's self is discussed in Csikszentmihalyi and Rochberg-Halton (1981) and Rudmin (1991).

357 Learning the dynamics of one's emotions. The therapeutic use of the Experience Sampling Method to record one's activities and experiences is described in Delespaul (1995), Delle Fave and Massimini (1992), and deVries (1992). The ESM makes it possible for the psychiatrist or therapist (and, by extension, for the patient also) to assess the patient's quality of life, and to propose changes in activities and habits that might improve it.

369 The creativity of artists in this century. The argument that modern artists express the lack of trust in values and beliefs previously taken for granted is certainly not a new one; my own small addition to this argument is in Csikszentmihalyi (1992b).

370 Prodigies. The best account of exceptionally gifted children is by Feldman (1986). A recent book by Ellen Winner (1996) summarizes the truths and the myths about the development of gifted children.

References

Abrams, J., and C. Zweig, eds. 1991. *Meeting the shadow*. Los Angeles: St. Martin's Press.

Adlai-Gail, W. 1994. The flow personality and its correlates. Ph.D. diss., University of Chicago.

Albert, R. S., ed. 1983. *Genius and eminence*. New York: Oxford University Press.

Albert, R. S., and M. A. Runco. 1986. The achievement of eminence: A model based on a longitudinal study of exceptionally gifted boys and their families. In *Conceptions of giftedness*, edited by R. J. Sternberg and J. E. Davidson. New York: Cambridge University Press, pp. 332–60.

Alexander, K., and M. Day. 1991. *Discipline-based art education: A curriculum sampler*. Santa Monica, Calif.: Getty Center for Education in the Arts.

Amabile, T. M. 1983. *The social psychology of creativity*. New York: Springer Verlag.

———. 1990. Within you, without you: The social psychology of creativity, and beyond. In *Theories of creativity*, edited by M. A. Runco and R. S. Albert. Newbury Park, Calif.: Sage Publications, pp. 61–91.

Andreasen, N. C. 1987. Creativity and mental illness: Prevalence rates in writers and their first-degree relatives. *American Journal of Psychiatry* 144:1288–92.

Arendt, H. 1956. *The human condition*. Chicago: University of Chicago Press.

Asakawa, K., and M. Csikszentmihalyi. 1995. Quality of experience and internalization of values in Asian American adolescents: An exploration of educational achievements. Submitted for publication.

Baer, J. 1993. *Creativity and divergent thinking*. Hillsdale, N.J.: Lawrence Erlbaum.

Barron, F. 1969. *Creative person and creative process*. New York: Holt, Rinehart, & Winston.

———. 1988. Putting creativity to work. In *The nature of creativity*, edited by R. J. Sternberg. New York: Cambridge University Press, pp. 76–98.

Barrow, J. D., and F. J. Tipler. 1986. *The anthropic cosmological principle*. New York: Oxford University Press.

Barsalou, L. W. 1982. Context independent and context dependent information in concepts. *Memory and Cognition* 10:82–93.

Baumrind, D. 1989. Rearing competent children. In *Child development today and tomorrow*, edited by W. Damon. San Francisco: Jossey-Bass, pp. 349–78.

Bidwell, C., M. Csikszentmihalyi, L. Hedges, and B. Schneider. *Images and experiences of work in adolescence*. In press.

Block, J. 1971. *Lives through time*. Berkeley, Calif.: Bancroft Books.

———. 1981. Some enduring and consequential structures of personality. In *Further explorations in personality*, edited by A. I. Rabin, J. Aronoff, A. M. Barclay, and R. A. Zucker. New York: John Wiley and Sons, pp. 27–43.

Bloom, B. 1985. *Developing talent in young people*. New York: Ballantine Books.

Bortz, W. M. 1991. *We live too short and die too long*. New York: Bantam Books.

Bourdieu, P. 1980. La capital social. *Actes de la Recherche en Sciences Sociales* 31:2–4.

Brannigan, A. 1981. *The social basis of scientific discoveries*. New York: Cambridge University Press.

Burke, J., and R. Ornstein. 1995. *The axemaker's gift*. New York: Putnam.

Burckhardt, J. 1926. *Die Kultur der Renaissance in Italien*. Leipzig.

Calvino, I. 1985. *Difficult loves*. New York: Harcourt, Brace, Jovanovich.

Campbell, D. T. 1960. Blind variation and selective retention in creative thought as in other knowledge processes. *Psychological Review* 67:380–400.

———. 1974. Evolutionary epistemology. In *The philosophy of Karl Popper*, edited by P. A. Schlipp. La Salle, Ill.: Open Court, pp. 413–63.

———. 1995. In what sense does a selectionist model provide epistemological "justification" for scientific belief? Paper presented at The Epistemology Group meeting, The Evolution of Knowledge and Invention, Royal Society of Arts, London, United Kingdom, May 24.

Castaneda, C. 1971. *A separate reality.* New York: Simon & Schuster.

Cattell, R. B., and J. E. Drevdahl. 1955. A comparison of the personality profile (16PF) of eminent researchers with that of eminent teachers and administrators. *British Journal of Psychology* 46:248–61.

Cellini, B. 1952. *The autobiography of Benvenuto Cellini.* Translated by D. B. Guralnik. Cleveland, Ohio: Fine Editions Press.

Claridge, G. 1992. Great wits and madness. In *Genius and eminence.* 2d ed. Edited by R. S. Albert. New York: Pergamon Press, pp. 329–50.

Costa, P. T., Jr., and R. R. McCrae. 1978. Objective personality assessment. In *The clinical psychology of aging,* edited by M. Storandt, I. C. Siegler, and M. F. Elias. New York: Plenum.

———. 1984. Personality as a lifelong determinant of well-being. In *Emotion in adult development,* edited by C. Z. Malatesta and C. E. Izard. Beverly Hills, Calif.: Sage, pp. 141–58.

Cropley, A. 1990. Creativity and mental health in everyday life. *Creativity Research Journal* 3:167–78.

Cross, P. G., R. B. Cattell, and H. J. Butcher. 1967. The personality pattern of creative artists. *British Journal of Educational Psychology* 37:292–99.

Csikszentmihalyi, M. 1975. *Beyond boredom and anxiety.* San Francisco: Jossey-Bass.

———. 1978. Attention and the wholistic approach to behavior. In *The stream of consciousness,* edited by K. S. Pope and J. L. Singer. New York: Plenum, pp. 335–58.

———. 1982. Towards a psychology of optimal experience. In *Review of personality and social psychology: 3,* edited by L. Wheeler. Beverly Hills, Calif.: Sage Publications, pp. 13–36.

———. 1985a. Emergent motivation and the evolution of the self. In *Motivation in adulthood,* edited by D. Kleiber and M. H. Maehr. Greenwich, Conn.: JAI Press, pp. 93–113.

———. 1985b. Reflections on enjoyment. *Perspectives in Biology and Medicine* 28:469–97.

———. 1988a. Society, culture, person: A systems view of creativity. In *The nature of creativity,* edited by R. J. Sternberg. New York: Cambridge University Press, pp. 325–39.

———. 1988b. Motivation and creativity: Towards a synthesis of structural and energistic approaches to cognition. *New Ideas in Psychology* 6:159–76.

———. 1990. *Flow: The psychology of optimal experience.* New York: HarperCollins.

———. 1992a. Öffentliche meinung und die psychologie der einsamkeit. In *Öffentliche meinung: Theorie, methode, befunde*, edited by J. Wilke. Freiburg: Verlag Karl Alber, pp. 31–40.

————. 1992b. Imagining the self: An evolutionary excursion. *Poetics* 21:153–67.

————. 1993. *The evolving self: A psychology for the third millennium*. New York: HarperCollins.

————. 1994. Memes vs. genes: Notes from the culture wars. In *Changing the world*, edited by D. H. Feldman, M. Csikszentmihalyi, and H. Gardner. Westport, Conn.: Praeger, pp. 159–72.

Csikszentmihalyi, M., and O. Beattie. 1979. Life themes: A theoretical and empirical investigation of their origins and effects. *Journal of Humanistic Psychology* 19:45–63.

Csikszentmihalyi, M., and I. Selega Csikszentmihalyi, eds. 1988. *Optimal experience: Psychological studies of flow in consciousness*. New York: Cambridge University Press.

————. 1993. Family influences on the development of giftedness. In *The origins and development of high ability*. Chichester, UK: Wiley (Ciba Foundation Symposium 178), pp. 187–206.

Csikszentmihalyi, M., and J. W. Getzels. 1973. The personality of young artists: A theoretical and empirical exploration. *British Journal of Psychology* 64:91–104.

Csikszentmihalyi, M., and R. Larson. 1978. Intrinsic rewards in school crime. *Crime and Delinquency* 24:322–25.

————. 1984. *Being adolescent: Conflict and growth in the teenage years*. New York: Basic Books.

Csikszentmihalyi, M., and F. Massimini. 1985. On the psychological selection of bio-cultural information. *New Ideas in Psychology* 3:115–38.

Csikszentmihalyi, M., and J. Nakamura. 1989. The dynamics of intrinsic motivation. In *Handbook of motivation theory and research*. Vol. 3, *Goals and Cognitions*, edited by R. Ames and C. Ames. New York: Academic Press, pp. 45–71.

Csikszentmihalyi, M., and K. Rathunde. 1993. The measurement of flow in everyday life. *Nebraska Symposium on Motivation* 40:58–97.

Csikszentmihalyi, M., and E. Rochberg-Halton. 1981. *The meaning of things*. New York: Cambridge University Press.

Csikszentmihalyi, M., and K. Sawyer. 1995. Creative insight: The social nature of a solitary moment. In *The nature of insight*, edited by R. J. Sternberg and J. E. Stevenson. New York: Cambridge University Press, pp. 329–64.

Csikszentmihalyi, M., and U. Schiefele. 1992. Arts education, human development, and the quality of experience. In *Arts in education. The ninety-first yearbook of the Society for the Study of Education*, edited by G. Reimer and

R. A. Smith. Chicago: University of Chicago Press, pp. 169–91.

Csikszentmihalyi, M., and M. Wong. 1991. The situational and personal correlates of happiness: A cross-national comparison. In *The social psychology of subjective well-being,* edited by F. Strack, M. Argyle, and N. Schwartz. London: Pergamon Press, pp. 193–212.

Csikszentmihalyi, M., K. Rathunde, and S. Whalen. 1993. *Talented teenagers.* New York: Cambridge University Press.

Damon, W. 1995. *Greater expectations: Overcoming the culture of indulgence in America's homes and schools.* New York: The Free Press.

Dawkins, R. 1976. *The selfish gene.* New York: Oxford University Press.

Deci, E. L., and M. Ryan. 1985. *Intrinsic motivation and self-determination in human behavior.* New York: Plenum.

Delespaul, P. A., and E. G. 1995. *Assessing schizophrenia in daily life. The experience sampling method.* Maastricht: UPM.

Delle Fave, A., and F. Massimini. 1992. The ESM and the measurement of clinical change: A case of anxiety disorder. In *The experience of psychopathology: Investigating mental disorders in natural settings,* edited by M. W. de Vries. Cambridge, UK: Cambridge University Press, pp. 280–99.

Dennett, D. C. 1991. *Consciousness explained.* Boston: Little, Brown.

Dennis, W. 1966. Creative productivity between the ages of twenty and eighty years. *Journal of Gerontology* 21:1–18.

deVries, M. W., ed. 1992. *The experience of psychopathology: Investigating mental disorders in natural settings.* Cambridge, UK: Cambridge University Press.

Diamond, J. 1992. *The third chimpanzee.* New York: HarperCollins.

Dobbs, S. M. 1993. *The DBAE handbook.* Santa Monica, Calif.: The J. P. Getty Trust.

Dozier, R. W., Jr. 1992. *Codes of evolution.* New York: Crown.

Dreistadt, R. 1969. The use of analogies in incubation in obtaining insight in creative problem solving. *Journal of Psychology* 71:158–75.

Dweck, C. S. 1986. Motivational processes affecting learning. *American Psychologist* 41:1040–48.

Eco, U. 1994. *Six walks in the fictional woods.* Cambridge, Mass.: Harvard University Press.

Eiduson, B. T. 1962. *Scientists: Their psychological world.* New York: Basic Books.

Erikson, E. H. 1950. *Childhood and society.* New York: Norton.

———. 1968. *Identity: Youth and crisis.* New York: Norton.

Eysenck, H. J. 1952. *The scientific study of personality.* London: Routledge & Kegan Paul.

———. 1973. *Eysenck on extroversion.* New York: John Wiley and Sons.

Feldman, D. H. 1980. *Beyond universals in cognitive development.* Norwood, N.J.: Ablex.

————. 1986. *Nature's gambit*. New York: Basic Books.

————. 1994. Creativity: Proof that development occurs. In *Changing the world: A framework for the study of creativity*, by D. H. Feldman, M. Csikszentmihalyi, and H. Gardner. Westport, Conn.: Praeger, pp. 85–102.

Feldman, D. H., M. Csikszentmihalyi, and H. Gardner. 1994. *Changing the world: A framework for the study of creativity*. Westport, Conn.: Praeger.

Ford, C. M., and D. A. Gioia, eds. 1995. *Creative action in organizations*. Thousand Oaks, Calif.: Sage Publications.

Freeman, M. 1993. *Finding the muse: A sociopsychological inquiry into the conditions of artistic creativity*. New York: Cambridge University Press.

Freud, S. 1947. *Leonardo da Vinci: A study in psychosexuality*. New York: Random House.

————. [1908] 1955. The Moses of Michelangelo. Edited and translated by J. Strachey. *The Standard Edition* 8:211–38.

Gajdusek, D. C. 1995. Early inspiration. *Creativity Research Journal* 7:341–49.

Gallagher, W. 1993. *The power of place: How our surroundings shape our thoughts, emotions, and actions*. New York: HarperPerennial.

Gardner, H. 1988. Creativity: An interdisciplinary perspective. *Creativity Research Journal* 1:8–26.

————. 1993. *Creating minds*. New York: Basic Books.

————. 1994. More on private intuitions and public symbol systems. *Creativity Research Journal* 7:265–75.

Gedo, J. 1990. More on creativity and its vicissitudes. In *Theories of creativity*, edited by M. Runco and R. Albert. Newbury Park, Calif.: Sage Publications, pp. 35–45.

Getzels, J. W. 1964. Creative thinking, problem-solving, and instruction. In *Theories of learning and instruction* (63rd yearbook of the National Society for Education), edited by E. R. Hilgard. Chicago: University of Chicago Press, pp. 240–67.

————. 1975. Problem finding and the inventiveness of solution. *Journal of Creative Behavior* 9:12–118.

————. 1982. The problem of the problem. In *Question-forming and response consistency*, edited by R. M. Hogarth. San Francisco: Jossey-Bass, pp. 37–44.

Getzels, J. W., and M. Csikszentmihalyi. 1968. On the roles, values, and performance of future artists: A conceptual and empirical exploration. *Sociological Quarterly* 9:516–50.

————. 1976. *The creative vision*. New York: Wiley Interscience.

Getzels, J. W., and P. Jackson. 1962. *Creativity and intelligence: Explorations with gifted students*. New York: J. Wiley and Sons.

Gilardoni, L. 1988. *Bellagio: An historical perspective*. Milan: A. Pizzi Editore.

Gilligan, C., J. V. Ward, and J. M. Taylor. 1988. *Mapping the moral domain: A contribution of women's thinking to psychological theory and education.* Cambridge, Mass.: Harvard University Press.

Goertzel, V., and M. G. Goertzel. 1962. *Cradles of eminence.* Boston: Little, Brown.

Gribbin, J., and M. Rees. 1989. *Cosmic coincidences: Dark matter, mankind, and anthropic cosmology.* New York: Bantam.

Grotevant, H. D. 1991. Child development within the family context. In *Child development today and tomorrow,* edited by W. Damon. San Francisco: Josscy-Bass, pp. 34–51.

Gruber, H. 1981. *Darwin on man.* Chicago: University of Chicago Press.

Gruber, H., and S. N. Davis. 1988. Inching our way up Mount Olympus: The evolving systems approach to creative thinking. In *The nature of creativity,* edited by R. J. Sternberg. New York: Cambridge University Press, pp. 243–70.

Guilford, J. P. 1950. Creativity. *American Psychologist* 14:205–8.

———. 1967. *The nature of human intelligence.* New York: McGraw-Hill.

Habermas, J. 1970. On systematically distorted communication. *Inquiry* 13:360–75.

Hadamard, J. 1949. *The psychology of invention in the mathematical field.* Princeton, N.J.: Princeton University Press.

Hamilton, W. D. 1964. The genetic evolution of social behavior. *Journal of Theoretical Biology* 7:1–52.

Harrington, D. M. 1990. The ecology of human creativity: A psychological perspective. In *Theories of creativity,* edited by M. Runco and R. S. Albert. Newbury Park, Calif.: Sage Publications, pp. 143–69.

Harrington, D. M., J. H. Block, and J. Block. 1992. Testing aspects of Carl Rogers's theory of creative environments: Child-rearing antecedents of creative potential in young adults. In *Genius and eminence,* edited by R. S. Albert. New York: Pergamon Press, pp. 195–208.

Hasher, L., and R. T. Zacks. 1979. Automatic and effortful processes in memory. *Journal of Experimental Psychology* 108:356–88.

Hauser, A. 1951. *The social history of art.* New York: Vintage.

Heath, S. B., and M. W. McLaughlin, eds. 1993. *Identity and inner city youth: Beyond ethnicity and gender.* New York: Teacher's College.

Hersh, R. and V. John-Steiner. 1993. A visitor to Hungarian mathmatics. *Mathmatical Intelligencer* 19, no. 2:13–26.

Heydenreich, L. H. 1974. *Il Primo Rinascimento.* Milano: Rizzoli.

Hidi, S. 1990. Interest and its contribution as a mental resource for learning. *Review of Educational Research* 60:549–71.

Horn, J. L. 1970. Organization of data on life-span development of human abilities. In *Life-span developmental psychology: Research and theory,* edited by L. R. Goulet and P. B. Baltes. New York: Academic Press, pp. 423–66.

Humboldt, A. von. 1891–93. *Cosmos: A sketch of the physical description of the universe.* Translated by E. C. Otte. London: George Bell & Sons.

Huntington, E. 1945. *Mainsprings of civilization.* New York: J. Wiley and Sons.

Inghilleri, P. 1995. *Esperienza soggettiva, personalità, evoluzione culturale.* Torino: UTET.

Isaksen, S. G., K. B. Dorval, and D. J. Treffinger. 1994. *Creative approaches to problem solving.* Dubuque, Iowa: Kendall-Hunt.

James, W. 1890. *Principles of psychology.* Vol. 1. New York: Henry Holt.

Jamison, K. R. 1989. Mood disorder and patterns of creativity in British writers and artists. *Psychiatry* 52:125–34.

John-Steiner, V. 1985. *Notebooks of the mind.* Albuquerque: University of New Mexico Press.

Johnson-Laird, P. N. 1988. Freedom and constraint in creativity. In *The nature of creativity,* edited by R. Sternberg. New York: Cambridge University Press, pp. 202–19.

Joseph, L. E. 1990. *Gaia: The growth of an idea.* New York: St. Martin's Press.

Jung, C. G. 1946. The fight with the shadow. *Listener,* Nov. 7.

———. 1968. *Analytical psychology: Its theory and practice.* New York: Vintage.

———. 1969. *On the nature of the psyche.* Princeton, N.J.: Princeton University Press.

———. 1973. *Memories, dreams, reflections.* New York: Pantheon.

Kahneman, D. 1973. *Attention and effort.* Englewood Cliffs, N.J.: Prentice-Hall.

Klein, G. 1992. *Pietà.* Cambridge, Mass: MIT Press.

Klein, G., and E. Klein. 1989. How one thing has led to another. *Annual Review of Immunology* 7:1–33.

Klein, G., ed. 1990. *Om Kreativitet och Flow.* Stockholm: Brombergs.

Koestler, A. 1964. *The Act of creation.* New York: Macmillan.

Kosoff, J. 1995. Explaining creativity: The attributional perspective. *Creativity Research Journal* 8:311–66.

Krebs, H. A., and J. H. Shelly. 1975. *The creative process in science and medicine.* Amsterdam: Elsevier.

Kris, E. 1952. *Psychoanalytic explorations in art.* New York: International Universities Press.

Kubey, R., and M. Csikszentmihalyi. 1990. *Television and the quality of life.* Hillsdale, N.J.: Lawrence Erlbaum.

Kuhn, T. S. 1970. *The structure of scientific revolutions.* Chicago: University of Chicago Press.

Labouvie-Vief, G. 1985. Intelligence and cognition. In *Handbook of the psychology of aging*. 2d ed. Edited by J. E. Birren and K. W. Schaie. New York: Van Nostrand Reinhold.

Lambert, F. L. 1995. Thermodynamics can be obstructed, the Second Law be damned. Los Angeles, California.

Langley, P., H. A. Simon, G. L. Bradshaw, and J. M. Zytkow. 1987. *Scientific discovery: Computational exploration of the creative process*. Cambridge, Mass.: MIT Press.

Laumann, E. O., J. H. Gagnon, R. T. Michael, and S. Michaels. 1994. *The social organization of sexuality: Sexual practices in the United States*. Chicago: University of Chicago Press.

Lehman, H. C. 1953. *Age and achievement*. Princeton, N.J.: Princeton University Press.

LeVine, R. 1991. Cultural environments in child development. In *Child development today and tomorrow*, edited by W. Damon. San Francisco: Jossey-Bass, pp. 52–68.

Levinson, M. 1994. Citi comes in from the cold. *Newsweek*, September 26, p. 50.

Lorenz, K. 1966. *On aggression*. New York: Harcourt, Brace and World.

Lovelock, J. E. 1979. *Gaia: A new look at life on earth*. New York: Oxford University Press.

MacKinnon, D. W. 1962. The nature and nurture of creative talent. *American Psychologist* 20:484–95.

———. 1963. Creativity and images of the self. In *The study of lives*, edited by R. W. White. New York: Atherton, pp. 251–78.

———. 1964. The creativity of architects. In *Widening horizons in creativity*, edited by C. W. Taylor. New York: J. Wiley and Sons.

Magyari-Beck, I. 1988. New concepts about personal creativity. *Creativity and Innovation Yearbook*, *1*, Manchester, England: Manchester Business School, pp. 121–26.

———. 1994. *Múzsák a Piacon*. Budapest: Aula Kiadó.

Martindale, C. 1989. Personality, situation, and creativity. In *Handbook of Creativity*, edited by J. Glover, R. Ronning, and C. R. Reynolds. New York: Plenum, pp. 211–32.

———. 1990. *The clockwork muse: The predictability of artistic change*. New York: Basic Books.

Maruyama, M. 1980. Mindscapes and science theories. *Current Anthropology* 21 (5):589–600.

Maslow, A. 1971. *The farther reaches of human nature*. New York: Viking.

Massimini, F. 1979. *I Presupposti teoretici e osservativi del paradigma della selezione culturale. Primo contributo: Il doppio sistema ereditario*. Milan: Ghedini.

———. 1993. I presupposti teoretici del paradigma della selezione culturale umana. In *La selezione psicologica umana: Teoria e metodo di analisi*, edited by F. Massimini and P. Inghilleri. Milano: Cooperativa Libraria Iulm.

Massimini, F., and C. Calegari. 1979. *Il contesto normativo sociale*. Milano: F. Angeli.

Massimini, F., and A. Delle Fave. 1991. Religion and cultural evolution. *Zygon* 26:27–48.

Massimini, F., and P. Inghilleri. 1986. *L'Esperienza quotidiana: Teoria e metodo di analisi*. Milano: Angeli.

———. 1993. *La Selezione psicologica umana: Teoria e metodo di analisi*. Milano: Cooperativa Libraria Iulm.

Massimini, F., M. Csikszentmihalyi, and A. Delle Fave. 1988. Flow and biocultural evolution. In *Optimal experience: Psychological studies of flow in consciousness,* edited by M. Csikszentmihalyi and I. Selega Csikszentmihalyi. New York: Cambridge University Press, pp. 60–84.

Mayr, E. 1947. *Systematics and the origin of species.* New York: Columbia University Press.

McKay, M., and P. Fanning. 1988. *Self-esteem.* New York: St. Martin's Press.

Medawar, P. 1967. *The Art of the soluble.* Chicago: University of Chicago Press.

Milgram, R. 1990. Creativity: An idea whose time has come and gone? In *Theories of creativity,* edited by M. A. Runco and R. S. Albert. Newbury Park, Calif.: Sage, pp. 215–33.

Miller, G. A. 1983. Informavore. In *The study of information,* edited by F. Machlup and U. Mansfield. New York: J. Wiley and Sons.

Mockros, C. 1995. The social construction of extraordinary selves: A lifespan study of creative individuals. Unpublished doctoral dissertation, The University of Chicago.

Mockros, C., and M. Csikszentmihalyi. 1995 (in press). The social construction of creative lives. In *Social creativity*. Vol. I. Edited by R. Purser and A. Montuori. Creskill, N.Y.: Hampton Press.

Moneta, G., and M. Csikszentmihalyi. 1995 (in press). The effect of perceived challenges and skills on the quality of subjective experience. *The Journal of Personality.*

Noelle-Neumann, E. 1985. Identifying opinion leaders. Paper presented at 38th ESOMAR Conference, Wiesbaden, Germany.

Oden, M. H. 1968. The fulfillment of promise: 40-year follow-up of the Terman gifted group. *Genetic Psychology Monographs* 77:3–93.

O'Neil, J. 1993. *The paradox of success.* New York: G. P. Putnam's Sons.

Ornstein, R. 1986. *Multimind: A new way of looking at human behavior.* Boston: Houghton Mifflin.

Over, R. 1989. Age and scholar impact. *Psychology of Aging* 4:222–25.

Parnes, S. J. 1967. *Creative behavior guidebook.* New York: Scribners.

Peirce, C. S. 1931–35. *The collected works of Charles Sanders Peirce.* Vols. 1–6. Edited by C. Hartshorne and P. Weiss. Cambridge, Mass.: Harvard University Press.

Perkins, D. N. 1981. *The mind's best work.* Cambridge, Mass.: Harvard University Press.

Piechowski, M., and K. Cunningham. 1985. Patterns of overexcitability in a group of artists. *Journal of Creative Behavior* 19:153–74.

Policastro, E. 1995. Creative intuition: An integrative review. *Creativity Research Journal* 8:99–113.

Popper, K. 1959. *The logic of scientific discovery.* New York: Basic.

Rathunde, K. (in press). Family context and talented adolescents' optimal experience in school-related activities. *Journal of Research in Adolescence.*

Rathunde, K., and M. Csikszentmihalyi. 1993. Undivided interest and the growth of talent: A longitudinal study of adolescents. *Journal of Youth and Adolescence* 22:1–21.

Reigber, D. 1995. *Glück im Garten—Erfolg in Markt.* Offenburg: Burda.

Renninger, K. A., S. Hidi, and A. Krapp, eds. 1992. *The role of interest in learning and development.* Hillside, N.J.: Erlbaum.

Reti, L., ed. 1974. *The unknown Leonardo.* New York: McGraw-Hill.

Robinson, R. 1988. Project and prejudice: Past, present, and future in human development. *Human Development* 31:158–72.

Roe, A. 1946. The personality of artists. *Educational and Psychological Measurement* 6:401–8.

———. 1951. A psychological study of physical scientists. *Genetic Psychology Monographs* 43:123–235.

———. 1952. *The making of a scientist.* New York: Dodd, Mead.

Rogers, C. 1951. *Client-centered therapy.* Boston: Houghton Mifflin.

Róheim, G. 1945. *The eternal ones of the dream: A psychoanalytic interpretation of Australian myth and ritual.* New York: International Universities Press.

Root-Bernstein, R. S. 1989. *Discovering, inventing, and solving problems at the frontiers of science.* Cambridge, Mass.: Harvard University Press.

Root-Bernstein, R. S., M. Bernstein, and H. C. Garnier. 1995. Correlations between avocations, scientific style, work habits, and professional impact of scientists. *Creativity Research Journal* 8:115–37.

Rothenberg, A. 1979. *The emerging goddess: The creative process in art, science, and other fields.* Chicago: University of Chicago Press.

———. Creativity, mental health, and alcoholism. *Creativity Research Journal* 3:179–201.

Rudmin, F. W., ed. 1991. *To have possessions.* Corte Madera, Calif.: Select Press.

Rumelhart, D. E., J. L. McClelland, and the PDP Research Group, eds.

1986. *Parallel distributed processing: Explorations in the microstructure of cognition.* Vol. 1, *Foundations.* Cambridge, Mass.: MIT Press.

Runco, M. A. 1991. *Divergent thinking.* Norwood, N.J.: Ablex.

Runco, M. A., ed. 1994. *Problem finding, problem solving, and creativity.* Norwood, N.J.: Ablex.

Russ, S. W. 1993. *Affect & creativity.* Hillsdale, N.J.: Lawrence Erlbaum.

Ryback, J. M., P. A. Roodin, and W. J. Hoyer. 1995. *Adult development and aging.* 3d ed. Dubuque, Iowa: Brown & Benchmark.

Salk, J. 1983. *Anatomy of reality: Merging intuition and reason.* New York: Columbia University Press.

Schaie, K. W. 1990. The optimization of cognitive functioning in old age: Prediction based on cohort-sequential and longitudinal data. In *Longitudinal research and the study of successful (optimal) aging,* edited by P. B. Baltes and M. Baltes. Cambridge, UK: Cambridge University Press, pp. 94–117.

———. 1994. The course of adult intellectual development. *American Psychologist* 49:304–13.

Schiefele, U. 1991. Interest, learning, and motivation. *Educational Psychologist* 26:299–323.

Schiefele, U., and M. Csikszentmihalyi. 1994. Interest and the quality of experience in classrooms. *European Journal of Psychology of Education* 9(3):251–70.

Schneider, B., J. A. Hieschima, S. Lee, and S. Plank. 1992. East Asian academic success in the United States: Family, school, and community explanation. In *Cross-cultural roots of minority child development,* edited by P. M. Greenfield and R. R. Cocking. Hillsdale, N.J.: Lawrence Erlbaum, pp. 323–50.

Schweber, S. S. 1994. *QED and the men who made it: Dyson, Feynman, Schwinger, and Tomonaga.* Princeton, N.J.: Princeton University Press.

Sears, P., and R. R. 1980. 1,528 little geniuses and how they grew. *Psychology Today,* Feb., pp. 29–43.

Semenzato, C. 1964. *Giotto.* New York: Barnes & Noble.

Simon, H. A. 1969. *Sciences of the artificial.* Boston: MIT Press.

———. 1988. Creativity and motivation: A response to Csikszentmihalyi. *New Ideas in Psychology* 6:177–81.

Simonton, D. K. 1975. Sociocultural context and individual creativity: A transhistorical time-series analysis. *Journal of Personality and Social Psychology* 32:1119–33.

———. 1984. *Genius, creativity, and leadership: Historiometric inquiries.* Cambridge, Mass.: Harvard University Press.

———. 1988. Age and outstanding achievement: What do we know after a century of research? *Psychological Bulletin* 104:163–80.

———. 1990a. *Scientific genius: A psychology of science.* New York: Cambridge University Press.

————. 1990b. Political pathology and societal creativity. *Creativity Research Journal* 3:85–99.

————. 1990c. Creativity and wisdom in aging. In *Handbook of the psychology of aging*. 3d ed. Edited by J. E. Birren and K. W. Schaie. San Diego, Calif.: Academic Press, pp. 320–29.

Smith, R. A., ed. 1989. *Discipline-based art education*. Urbana: University of Illinois Press.

Spence, J. T., and R. L. Helmreich. 1978. *Masculinity and Femininity*. Austin: University of Texas Press.

Stayer, M. S., ed. 1988. *Newton's dream*. Kingston, Ont.: McGill-Queen's University Press.

Stein, M. 1953. Creativity and culture. *Journal of Psychology* 36:311–22.

Sternberg, R. J. 1985. Implicit theories of intelligence, creativity, and wisdom. *Journal of Personality and Social Psychology* 49:607–27.

————. 1986. A triarchic theory of intellectual giftedness. In *Conceptions of giftedness*, edited by R. J. Sternberg and J. E. Davidson. New York: Cambridge University Press, pp. 223–43.

Sternberg, R. J., ed. 1988. *The nature of creativity*. New York: Cambridge University Press.

Sternberg, R. J., and J. E. Davidson, eds. 1995. *The nature of insight*. Cambridge, Mass.: MIT Press.

Sternberg, R. J., and T. I. Lubart. 1991. An investment theory of creativity and human development. *Human Development* 34:1–31.

Stevenson, H. W., and J. W. Stigler. 1992. *The learning gap: Why our schools are failing*. New York: Touchstone.

Sue, S., and S. Okazaki. 1990. Asian-American educational achievements: A phenomenon in search of an explanation. *American Psychologist* 45:913–20.

Teilhard de Chardin, P. 1965. *The phenomenon of man*. New York: Harper & Row.

Terman, L. M. 1925. *Genetic studies of genius*. Stanford, Calif.: Stanford University Press.

Therivel, W. A. 1993. The challenged personality as a precondition for sustained creativity. *Creativity Research Journal* 6:413–24.

————. 1995. Long-term effect of power on creativity. *Creativity Research Journal* 8:173–92.

Torrance, E. P. 1962. *Guiding creative talent*. Englewood Cliffs, N.J.: Prentice-Hall.

————. 1988. Creativity as manifested in testing. In *The nature of creativity*, edited by R. J. Sternberg. New York: Cambridge University Press, pp. 43–75.

Treisman, A. M., and G. Gelade. 1980. A feature integration theory of attention. *Cognitive Psychology* 12:97–136.

Vaillant, G. E. 1993. *The wisdom of the ego.* Cambridge, Mass.: Harvard University Press.

Vasari, G. [1550] 1959. *Lives of the most eminent painters, sculptors, and architects.* New York: The Modern Library.

Wachhorst, W. 1981. *Thomas Alva Edison: An American myth.* Cambridge, Mass.: MIT Press.

Wallas, G. 1926. *The art of thought.* New York: Harcourt-Brace.

Watson, J. 1980. *The double helix: A personal account of the discovery of the structure of DNA.* New York: Norton.

Weber, R. J., and D. N. Perkins. 1992. *Inventive minds.* New York: Oxford University Press.

Wells, A. 1988. Self-esteem and optimal experience. In *Optimal experience: Psychological studies of flow in consciousness,* edited by M. Csikszentmihalyi and I. Selega Csikszentmihalyi. New York: Cambridge University Press, pp. 327–41.

Westby, E. L., and V. L. Dawson. 1995. Creativity: Asset or burden in the classroom? *Creativity Research Journal* 8:1–10.

Westfall, R. S. 1980. *Never at rest: A biography of Sir Isaac Newton.* New York: Cambridge University Press.

White, R. W. 1959. Motivation reconsidered: The concept of competence. *Psychological Review* 66:297–333.

Wigner, E. 1992. *The recollections of Eugene P. Wigner.* New York: Plenum Press.

Wilson, E. O. 1975. *Sociobiology: The new synthesis.* Boston: Belknap Press.

———. 1984. *Biophilia: The human bond with other species.* Cambridge, Mass.: Harvard University Press.

Winner, E. 1996. *Gifted children: Myths and realities.* New York: Basic Books.

Zajonc, R. B. 1976. Family configuration and intelligence. *Science* 192:227–35.

Zuckerman, H., and J. R. Cole. 1994. Research strategy in science: A preliminary inquiry. *Creativity Research Journal* 7: 391–405.

Zuckerman, H., and J. Lederberg. 1986. Forty years of genetic recombination with bacteria: A post-mature discovery. *Nature* 324:629–31.

INDEX

About the author

About the book

Read on

Insights,
Interviews
& More . . .

Meet Mihaly Csikszentmihalyi

© 1990 by Michael P. Weinstein

MIHALY CSIKSZENTMIHALYI was the C. S. and D. J. Davidson Professor of Psychology and Management at Claremont Graduate University's Drucker School of Management in Claremont, California. He was director of the Quality of Life Research Center at the Drucker School. He was a former chairman of the Department of Psychology at the University of Chicago, where he also taught psychology. For more than thirty-five years was involved in research on topics related to flow. Funding for these studies has come from the Public Health Service and various private grants, primarily the Spencer Foundation. In addition, a great deal of interest in his work has been building outside academia, spearheaded by substantial articles in *Psychology Today*, the *New York Times*, the *Washington Post*, the *Chicago Tribune*, *Omni*, *Newsweek*, and elsewhere. He is the author of *Beyond Boredom and Anxiety*, *The Evolving Self*, *Flow*, and *Creativity*, and is coauthor of *The Creative Vision*, *The Meaning of Things*, and *Being Adolescent*.

Dr. Csikszentmihalyi was a member of the American Academy of Arts and Sciences, the National Academy of Education, and the National Academy of Leisure Sciences. He has been a Senior Fulbright Fellow and has sat on several boards, including the Board of Advisers for the *Encyclopaedia Britannica*. He has appeared on a number of foreign television networks, including the BBC and RAI (Italian television), and has taken part in several hour-long segments of *Nova*. ❧

Happiness Revisited
An Excerpt from *Flow*

The following passage is excerpted from Flow: The Psychology of Optimal Experience *by Mihaly Csikszentmihalyi (HarperCollins, 1993).*

Introduction

Twenty-three hundred years ago Aristotle concluded that, more than anything else, men and women seek happiness. While happiness itself is sought for its own sake, every other goal—health, beauty, money, or power—is valued only because we expect that it will make us happy. Much has changed since Aristotle's time. Our understanding of the worlds of stars and atoms has expanded beyond belief. The gods of the Greeks were like helpless children compared to humankind today and the powers we now wield. And yet on this most important issue very little has changed in the intervening centuries. We do not understand what happiness is any better than Aristotle did. And as for learning how to attain that blessed condition, one could argue that we have made no progress at all.

Despite the fact that we are now healthier and grow to be older, despite the fact that even the least affluent among us are surrounded by material luxuries undreamed of even a few decades ago (there were few bathrooms in the palace of the Sun King, chairs were rare even in the richest medieval houses, and no Roman emperor could turn on a TV set when he was bored), and regardless of all the stupendous scientific knowledge we can summon at will, people often end up feeling that their lives have been wasted, that instead of being filled with happiness their years were spent in anxiety and boredom.

Is this because it is the destiny of mankind to remain unfulfilled, each person always wanting more than he or she can have? Or is the pervasive malaise that often sours even our most precious moments the result of our seeking happiness in the wrong places? The intent of this book is to use some of the tools of modern psychology to explore this very ancient question: When do people feel most happy? If we can begin to find an answer to it, perhaps we shall eventually be able to order life so that happiness will play a larger part in it.

Twenty-five years before I began to write these lines, I made a discovery that took all the intervening time for me to realize I had made. To call it a "discovery" is perhaps misleading, for people have been aware of it since the dawn of time. Yet the word is ▶

66 When do people feel most happy? If we can begin to find an answer, perhaps we shall eventually be able to order life so that happiness will play a larger part in it. 99

appropriate, because even though my finding itself was well known, it had not been described or theoretically explained by the relevant branch of scholarship, which in this case happens to be psychology. So I spent the next quarter-century investigating this elusive phenomenon.

What I "discovered" was that happiness is not something that happens. It is not the result of good fortune or random chance. It is not something that money can buy or power command. It does not depend on outside events, but, rather, on how we interpret them. Happiness, in fact, is a condition that must be prepared for, cultivated, and defended privately by each person. People who learn to control inner experience will be able to determine the quality of their lives, which is as close as any of us can come to being happy.

Yet we cannot reach happiness by consciously searching for it. "Ask yourself whether you are happy," said J. S. Mill, "and you cease to be so." It is by being fully involved with every detail of our lives, whether good or bad, that we find happiness, not by trying to look for it directly. Viktor Frankl, the Austrian psychologist, summarized it beautifully in the preface to his book *Man's Search for Meaning*: "Don't aim at success—the more you aim at it and make it a target, the more you are going to miss it. For success, like happiness, cannot be pursued; it must ensue . . . as the unintended side-effect of one's personal dedication to a course greater than oneself."

So how can we reach this elusive goal that cannot be attained by a direct route? My studies of the past quarter-century have convinced me that there is a way. It is a circuitous path that begins with achieving control over the contents of our consciousness.

Our perceptions about our lives are the outcome of many forces that shape experience, each having an impact on whether we feel good or bad. Most of these forces are outside our control. There is not much we can do about our looks, our temperament, or our constitution. We cannot decide—at least so far—how tall we will grow, how smart we will get. We can choose neither parents nor time of birth, and it is not in your power or mine to decide whether there will be a war or a depression. The instructions contained in our genes, the pull of gravity, the pollen in the air, the historical period into which we are born—these and innumerable other conditions determine what we see, how we feel, what we do. It is not surprising that we should believe that our fate is primarily ordained by outside agencies.

Yet we have all experienced times when, instead of being buffeted by anonymous forces, we do feel in control of our actions, masters of our own fate. On the rare occasions that it happens, we feel a sense of exhilaration, a deep sense of enjoyment that is long cherished and that becomes a landmark in memory for what life should be like.

This is what we mean by *optimal experience*. It is what the sailor holding a tight course feels when the wind whips through her hair, when the boat lunges through the waves like a colt—sails, hull, wind, and sea humming a harmony that vibrates in the sailor's veins. It is what a painter feels when the colors on the canvas begin to set up a magnetic tension with each other, and a new *thing*, a living form, takes shape in front of the astonished creator. Or it is the feeling a father has when his child for the first time responds to his smile. Such events do not occur only when the external conditions are favorable, however: people who have survived concentration camps or who have lived through near-fatal physical dangers often recall that in the midst of their ordeal they experienced extraordinarily rich epiphanies in response to such simple events as hearing the song of a bird in the forest, completing a hard task, or sharing a crust of bread with a friend.

Contrary to what we usually believe, moments like these, the best moments in our lives, are not the passive, receptive, ▶

relaxing times—although such experiences can also be enjoyable, if we have worked hard to attain them. The best moments usually occur when a person's body or mind is stretched to its limits in a voluntary effort to accomplish something difficult or worthwhile. Optimal experience is thus something that we *make* happen. For a child, it could be placing with trembling fingers the last block on a tower she has built, higher than any she has built so far; for a swimmer, it could be trying to beat his own record; for a violinist, mastering an intricate musical passage. For each person there are thousands of opportunities, challenges to expand ourselves.

Such experiences are not necessarily pleasant at the time they occur. The swimmer's muscles might have ached during his most memorable race, his lungs might have felt like exploding, and he might have been dizzy with fatigue— yet these could have been the best moments of his life. Getting control of life is never easy, and sometimes it can be definitely painful. But in the long run optimal experiences add up to a sense of mastery—or perhaps better, a sense of *participation* in determining the content of life—that comes as close to what is meant by happiness as anything else we can conceivably imagine.

In the course of my studies I tried to understand as exactly as possible how people felt when they most enjoyed themselves, and why. My first studies involved a few hundred "experts"—artists, athletes, musicians, chess masters, and surgeons—in other words, people who seemed to spend their time in precisely those activities they preferred. From their accounts of what it felt like to do what they were doing, I developed a theory of optimal experience based on the concept of *flow*—the state in which people are so involved in an activity that nothing else seems to matter; the experience itself is so enjoyable that people will do it even at great cost, for the sheer sake of doing it.

With the help of this theoretical model my research team at the University of Chicago and, afterward, colleagues round the world interviewed thousands of individuals from many different walks of life. These studies suggested that optimal experiences were described in the same way by men and women, by young people and old, regardless of cultural differences. The flow experience was not just a peculiarity of affluent, industrialized elites. It was reported in essentially the same words by old women from Korea, by adults in Thailand and India, by teenagers in Tokyo, by Navajo shepherds, by farmers in the Italian Alps, and by workers on the assembly line in Chicago.

In the beginning our data consisted of interviews and questionnaires. To achieve greater precision we developed with time a new method for measuring the quality of subjective experience. This technique, called the Experience Sampling Method, involves asking people to wear an electronic paging device for a week and to write down how they feel and what they are thinking about whenever the pager signals. The pager is activated by a radio transmitter about eight times each day, at random intervals. At the end of the week, each respondent provides what amounts to a running record, a written film clip of his or her life, made up of selections from its representative moments. By now over a hundred thousand such cross sections of experience have been collected from different parts of the world. The conclusions of this volume are based on that body of data.

The study of flow I began at the University of Chicago has now spread worldwide. Researchers in Canada, Germany, Italy, Japan, and Australia have taken up its investigation. At present the most extensive collection of data outside of Chicago is at the Institute of Psychology of the Medical School, the University of Milan, Italy. The concept of flow has been found useful by psychologists who study happiness, life satisfaction, and intrinsic motivation; by sociologists ▶

Happiness Revisited *(continued)*

who see in it the opposite of anomie and alienation; by anthropologists who are interested in the phenomena of collective effervescence and rituals. Some have extended the implications of flow to attempts to understand the evolution of mankind, others to illuminate religious experience.

But flow is not just an academic subject. Only a few years after it was first published, the theory began to be applied to a variety of practical issues. Whenever the goal is to improve the quality of life, the flow theory can point the way. It has inspired the creation of experimental school curricula, the training of business executives, the design of leisure products and services. Flow is being used to generate ideas and practices in clinical psychotherapy, the rehabilitation of juvenile delinquents, the organization of activities in old people's homes, the design of museum exhibits, and occupational therapy with the handicapped. All this has happened within a dozen years after the first articles on flow appeared in scholarly journals, and the indications are that the impact of the theory is going to be even stronger in the years to come.

Overview

Although many articles and books on flow have been written for the specialist, this is the first time that the research on optimal experience is being presented to the general reader and its implications for individual lives discussed. But what follows is not going to be a "how-to" book. There are literally thousands of such volumes in print or on the remainder shelves of book-stores, explaining how to get rich, powerful, loved, or slim. Like cookbooks, they tell you how to accomplish a specific, limited goal on which few people actually follow through. Yet even if their advice were to work, what would be the result afterward in the unlikely event that one did turn into a slim, well-loved, powerful millionaire? Usually what happens is that the person finds himself back at square one, with a new list of wishes, just as dissatisfied as before. What would really satisfy people is not getting slim or rich, but feeling good about their lives. In the quest for happiness, partial solutions don't work.

However well-intentioned, books cannot give recipes for how to be happy. Because optimal experience depends on the ability to control what happens in consciousness moment by moment, each person has to achieve it on the basis of his own individual efforts and creativity. What a book can do, however, and what this one will try to accomplish, is to present examples of how life can be made more enjoyable, ordered in the framework of a theory, for readers to reflect upon and from which they may then draw their own conclusions.

Rather than presenting a list of dos and don'ts, this book intends to be a voyage through the realms of the mind, charted with the tools of science. Like all adventures worth having it will not be an easy one. Without some intellectual effort, a commitment to reflect and think hard about your own experience, you will not gain much from what follows.

Flow will examine the process of achieving happiness through control over one's inner life. We shall begin by considering *how consciousness works, and how it is controlled* (chapter 2), because only if we understand the way subjective states are shaped can we master them. Everything we experience—joy or pain, interest or boredom—is represented in the mind as information. If we are able to control this information, we can decide what our lives will be like.

The optimal state of inner experience is one in which there is *order in consciousness*. This happens when psychic energy—or attention—is invested in realistic goals, and when skills match the opportunities for action. The pursuit of a goal brings order in awareness because a person must concentrate attention on the task at hand and momentarily ▶

forget everything else. These periods of struggling to overcome challenges are what people find to be the most enjoyable times of their lives (chapter 3). A person who has achieved control over psychic energy and has invested it in consciously chosen goals cannot help but grow into a more complex being. By stretching skills, by reaching toward higher challenges, such a person becomes an increasingly extraordinary individual.

To understand why some things we do are more enjoyable than others, we shall review the *conditions of the flow experience* (chapter 4). "Flow" is the way people describe their state of mind when consciousness is harmoniously ordered, and they want to pursue whatever they are doing for its own sake. In reviewing some of the activities that consistently produce flow—such as sports, games, art, and hobbies—it becomes easier to understand what makes people happy.

But one cannot rely solely on games and art to improve the quality of life. To achieve control over what happens in the mind, one can draw upon an almost infinite range of opportunities for enjoyment—for instance, through the use of *physical and sensory skills* ranging from athletics to music to Yoga (chapter 5), or through the development of *symbolic skills* such as poetry, philosophy, or mathematics (chapter 6).

Most people spend the largest part of their lives working and interacting with others, especially with members of their families. Therefore it is crucial that one learn to *transform jobs into flow-producing activities* (chapter 7), and to think of ways of making *relations with parents, spouses, children, and friends more enjoyable* (chapter 8).

Many lives are disrupted by tragic accidents, and even the most fortunate are subjected to stresses of various kinds. Yet such blows to not necessarily diminish happiness. It is how people respond to stress that determines whether they will profit from misfortune or be miserable. Chapter 9 describes *ways in which people manage to enjoy life despite adversity.*

And, finally, the last step will be to describe how people manage *to join all experience into a meaningful pattern* (chapter 10). When that is accomplished, and a person feels in control of life and feels that it makes sense, there is nothing left to desire. The fact that one is not slim, rich, or powerful no longer matters. The tide of rising expectations is stilled; unfulfilled needs no longer trouble the mind. Even the most humdrum experiences become enjoyable.

Thus *Flow* will explore what is involved in reaching these aims. How is consciousness controlled? How is it ordered so as to make experience enjoyable? How is complexity achieved? And last, how can meaning be created? The way to achieve these goals is relatively easy in theory, yet quite difficult in practice. The rules themselves are clear enough, and within everyone's reach. But many forces, both within ourselves and in the environment, stand in the way. It is a little like trying to lose weight: everyone knows that it takes, everyone wants to do it, yet it is next to impossible for so many. The stakes here are higher, however. It is not just a matter of losing a few extra pounds. It is a matter of losing the chance to have a life worth living.

Before describing how the optimal flow experience can be attained, it is necessary to review briefly some of the obstacles to fulfillment implicit in the human condition. In the old stories, before living happily ever after the hero had to confront fiery dragons and wicked warlocks in the course of a quest. This metaphor applies to the exploration of the psyche as well. I shall argue that the primary reason it is so difficult to achieve happiness centers on the fact that, contrary to the myths mankind has developed to reassure itself, the universe was not created to answer our needs. Frustration is deeply woven into the fabric of life. And whenever some of our needs *are* temporarily met, we ▶

66 The primary reason it is so difficult to achieve happiness centers on the fact that, contrary to the myths mankind has developed to reassure itself, the universe was not created to answer our needs. 99

immediately start wishing for more. This chronic dissatisfaction is the second obstacle that stands in the way of contentment.

To deal with these obstacles, every culture develops with time protective devices—religions, philosophies, arts, and comforts—that help shield us from chaos. They help us believe that we are in control of what is happening and give reasons for being satisfied with our lot. But these shields are effective only for a while; after a few centuries, sometimes after only a few decades, a religion or belief wears out and no longer provides the spiritual sustenance it once did.

When people try to achieve happiness on their own, without the support of a faith, they usually seek to maximize pleasures that are either biologically programmed in their genes or are out as attractive by the society in which they live. Wealth, power, and sex become the chief goals that give direction to their strivings. But the quality of life cannot be improved this way. Only direct control of experience, the ability to derive moment-by-moment enjoyment from everything we do, can overcome the obstacles to fulfillment. ～

Have You Read?
More by Mihaly Csikszentmihalyi

FLOW

Psychologist Mihaly Csikszentmihalyi's famous investigations of "optimal experience" have revealed that what makes an experience genuinely satisfying is a state of consciousness called *flow*. During flow, people typically experience deep enjoyment, creativity, and a total involvement with life. In this new edition of his groundbreaking classic work, Csikszentmihalyi demonstrates the ways this positive state can be controlled, not just left to chance. *Flow: The Psychology of Optimal Experience* teaches how, by ordering the information that enters our consciousness, we can discover true happiness and greatly improve the quality of our lives.

"Important. . . . Illuminates the way to happiness."
—*New York Times Book Review*

"Explore[s] a happy state of mind called *flow*, the feeling of complete engagement in creative or playful activity."
—*Time*

Have You Read? *(continued)*

His bestselling book, *Flow*, introduced us to a radical new theory of happiness; in this breakthrough sequel, Mihaly Csikszentmihalyi shows us how to understand and overcome our evolutionary heritage in order to re-create ourselves and the world for the twenty-first century.

"[A] wise, humane inquiry."
—*Publishers Weekly*

"Csikszentmihalyi goes beyond the psychobabble and traces human behavior from the beginning of time and shows with great clarity why we do the things we do."
—*Library Journal*

Don't miss the next book by your favorite author. Sign up now for AuthorTracker by visiting www.AuthorTracker.com.